T0294947

Improvements to Air Force Strategic Basing Decisions

Constantine Samaras, Rachel Costello, Paul DeLuca, Stephen J. Guerra,
Kenneth Kuhn, Anu Narayanan, Michael Nixon, Stacie L. Pettyjohn, Nolan Sweeney,
Joseph V. Vesely, Lane F. Burgette

RAND Project AIR FORCE

Prepared for the United States Air Force
Approved for public release; distribution unlimited

For more information on this publication, visit www.rand.org/t/RR1297

Library of Congress Cataloging-in-Publication Data is available for this publication.

ISBN: 978-0-8330-9205-2

Published by the RAND Corporation, Santa Monica, Calif.

© Copyright 2016 RAND Corporation

RAND® is a registered trademark.

www.rand.org

Preface

The U.S. Air Force manages any proposed significant changes and additions regarding the location of weapon systems and personnel through its strategic basing process, and the Air Force has conducted more than 100 strategic basing decisions since 2009. Some notable recent examples of strategic basing decisions are the evaluations of which installations will host the F-35 and KC-46A major weapon systems. Air Force basing decisions are public, frequent, and occasionally contentious. Because these decisions affect force posture, local economies, and public trust, it is imperative that the decisionmaking process be objective and reproducible. The Air Force employs a three-step enterprise-wide process to ensure consistency when it makes basing decisions. In the first step, a set of criteria is developed to evaluate an installation's suitability to support a basing decision. The basing decision criteria are developed by the applicable major command (MAJCOM), approved by the Headquarters Air Force (HAF) and the Secretary of the Air Force (SecAF), and briefed to Congress. In the second step, individual basing scores for the defined criteria are assembled by the MAJCOM, and a small set of candidate bases, usually four to six, is selected for site surveys. In the final step, a preferred base is selected. The results of each step are approved by HAF and SecAF and briefed to Congress. The credibility of this process is dependent on these data calls resulting in high-quality data that are then analyzed in a transparent and objective manner.

This report is an independent analysis of the Air Force's basing process and the quality of the underlying data. It seeks to inform decisionmakers on potential improvements to the data and assessment criteria used in making basing decisions. It seeks to address three questions:

1. Are basing decision criteria aligned with Air Force intentions?
2. Are the data used in the Air Force's strategic basing decisionmaking process authoritative, consistent, and auditable?
3. Is there potential for broader Air Force strategic or portfolio-wide inputs to strengthen the basing decisionmaking process?

In answering the first question, we examined the actual data source for each base in enterprise-wide KC-46 and F-35 basing decisions. To assess the second question, we examined the desired and actual impact of each decision criterion for the past 25 basing actions. In addressing the potential for inclusion of strategic inputs, we assessed the historical impact of strategic inputs and possible mathematical techniques to incorporate such inputs into the current basing process.

The research reported here was sponsored by the U.S. Air Force and conducted within the Resource Management Program of RAND Project AIR FORCE.

RAND Project AIR FORCE

RAND Project AIR FORCE (PAF), a division of the RAND Corporation, is the U.S. Air Force's federally funded research and development center for studies and analyses. PAF provides the Air Force with independent analyses of policy alternatives affecting the development, employment, combat readiness, and support of current and future air, space, and cyber forces. Research is conducted in four programs: Force Modernization and Employment; Manpower, Personnel, and Training; Resource Management; and Strategy and Doctrine. The research reported here was prepared under contract FA7014-06-C-0001.

Additional Information about PAF is available on our website:
http://www.rand.org/paf/

This report documents work originally shared with the U.S. Air Force in October 2014. The draft report, issued in February 2015, was reviewed by formal peer reviewers and U.S. Air Force subject matter experts.

Contents

Figures

Tables

Summary

The U.S. Air Force manages any proposed significant changes and additions regarding the location of weapon systems and personnel through its strategic basing process, and the Air Force has conducted more than 100 strategic basing decisions since 2009. Some notable recent examples of strategic basing decisions are the evaluations of which installations will host the F-35 and KC-46A major weapon systems. Air Force basing decisions are public, frequent, and occasionally contentious. Because these decisions affect force posture, local economies, and public trust, it is imperative that the decisionmaking process be objective and reproducible. Unfortunately, a small number of these decisions resulted in eroded local confidence in the process. For example, news media reports suggested that the decision to base F-35s at Burlington, Vermont, was either a result of faulty data or undue political influence (Bender, 2013). Such examples highlight the need for strong empirical processes employing the best available data.

This report focuses on the procedures and data pertinent to the Air Force's strategic basing decision process. Specifically it focuses on three questions:

- Are basing decision criteria aligned with Air Force intentions?
- Are the data used in the Air Force's basing decisionmaking process authoritative, consistent, and auditable?
- Is there potential for broader Air Force strategic or portfolio-wide inputs to strengthen the basing decisionmaking process?

Current Basing Decisionmaking Process

The Air Force employs a three-step enterprise-wide process, shown in Figure S.1, to ensure consistency when it makes strategic basing decisions, such as choosing an installation to host new weapon systems, major changes in installation personnel, or hosting tenants on installations.

- First, the applicable major command (MAJCOM) develops criteria to evaluate an installation's suitability to support the proposed basing action. Criteria include, but are not limited to, an installation's ability to execute the mission, capacity to host the unit, environmental impact of the unit, and economic factors, such as locality and construction cost factors associated with individual installations. Each criterion is further broken down into a number of attributes that define that criterion. Once Headquarters Air Force (HAF) and the Secretary of the Air Force (SecAF) approve the evaluation criteria, they are briefed to Congress.
- In the second step, individual basing scores for the defined basing criteria are assembled by the MAJCOM, and all of the bases are screened to determine the appropriateness of hosting the basing action at each location. Using these scores, a small set of candidate

bases, usually four to six, is selected for on-site surveys, which are approved by the HAF and the SecAF and then briefed to Congress.

- In the final step following the site surveys, a preferred installation is selected. As in the previous steps, this is approved by HAF and the SecAF, then briefed to Congress. The basing action proceeds through National Environmental Policy Act (NEPA) compliance and is eventually implemented at the installation.

The Air Force's process has a high level of interaction with Congress, and the Air Force provides briefing updates at several points in the process. In the Army and Navy's basing process, Congress is notified after a decision has been made.

Figure S.1
Standard U.S. Air Force Strategic Basing Process and Governance

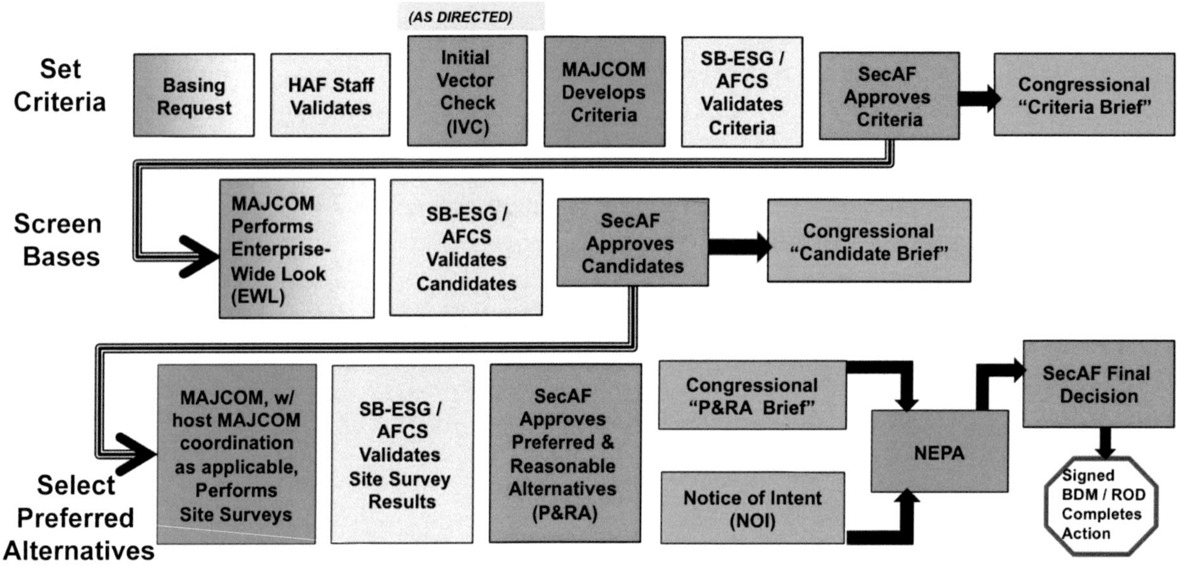

SOURCE: Pohlmeier, 2012.
NOTES: Blue boxes represent actions for MAJCOM/base-level personnel. Gray boxes represent actions for HQ/Air Staff personnel. Red boxes represent meetings/approval from SecAF/Chief of Staff of the Air Force. Green boxes represent congressional interaction/briefings. SB-ESG = Strategic Basing Executive Steering Group; AFCS = Air Force Corporate Structure; P&RA = preferred and reasonable alternatives; BDM = basing decision memo; ROD = record of decision.

Does the Current Basing Process Always Align with Air Force Objectives?

The basing criteria used to evaluate candidate bases consist of individual attributes important to the basing action under consideration. For example, is there fuel storage, runway space, and ramp space available for the units being considered? For each basing action, there are 100 total points available if an installation meets all of the criteria at the maximum desired level. This scoring system enables a comparison of installations across the enterprise. In a majority of the basing decisions analyzed, these attributes were grouped into four categories: mission, capacity, environment, and cost.

Mission criteria include attributes of an individual installation that affect the mission, such as weather, airspace, and proximity to ranges. If proximity to Army and Navy operations, training, or ranges is necessary to enable jointness between the services for a specific weapon system beddown, then the Air Force includes jointness as part of the mission criteria. Capacity criteria evaluate whether an installation's *existing* infrastructure can accommodate the proposed basing action. The environmental criteria represent the existing conditions at an installation that might inhibit accommodating the proposed mission, such as air quality, land use, and incompatible development areas due to noise or encroachment issues. The cost criteria represent the regional variation in federal employee salaries, construction capital costs, and military basic allowance for housing (BAH) rates.

To understand the basing process outcomes, we analyzed the data for 25 relevant basing actions. These data included scoring criteria, categories and weights, and points assigned from individual installation responses across the USAF enterprise. We assessed how well the weight assigned by the Air Force to each scoring category (e.g., mission, capacity) aligned with the actual influence of these categories on the choice of candidate bases using a model developed by RAND called the Generalized Boosted Model (GBM). GBM calculates the "relative influence" of each attribute on the probability of an installation either scoring in the top 10 percent of installations or becoming a site survey candidate. The model assigns relative influence to each category by iteratively attempting to replicate the actual results through a bottom-up analytical process.

The GBM results analytically validate what is an intuitive and expected outcome of the basing process as currently structured—installations that can perform the mission and that have *currently* available capacity should be competitive for a site survey. However, while our analysis shows that a good mission score is necessary to be considered for a site visit, we also found that, when it came to actually being selected for a site visit, the capacity score was usually more important. This suggests that, in practice, the mission category screens candidates, and the capacity criteria category determines site selection.

Assessing Basing Data Quality

We examined data quality along three dimensions: authoritativeness, consistency, and auditability. *Authoritativeness* measures the extent to which sources used to answer data call questions align with authoritative or credible sources. *Consistency* measures the extent to which candidate bases for a given action use the same sources to answer the same questions, and the extent to which a given base consistently answers the same question for different actions. *Auditability* measures the extent to which data sources are documented and traceable.

We found that the data used to make basing decisions are authoritative and consistent, but the auditability of the data call and response process can be improved. Therefore, a majority of "data errors" may be unrelated to data quality and result from the data collection process or human error, or potentially both.

Role of Strategy in Basing Decisions

The Air Force tends to approach basing decisions in the continental United States (CONUS) and outside of the continental United States (OCONUS) as completely separate processes. Specifically, the Air Force and the U.S. government choose and maintain overseas bases principally for their strategic and diplomatic value, while the Air Force chooses and maintains domestic bases according to other criteria, such as access to training ranges, utilizing existing capacity, and satisfying political considerations. Consequently, the CONUS basing process is influenced by the potential impact that base closures could have on their surrounding communities and the efforts of elected officials to keep bases in their districts open. As discussed in this report, the Air Force basing decision process is rooted in assessing a base's ability to support the mission of the unit being based, its capacity to support the unit, and both cost and environment issues associated with the basing action. However, the current domestic basing process does not include a portfolio-wide assessment of individual decisions or explicitly incorporate broader Air Force strategic concerns. In some cases, strategic considerations are included in the mission criteria established to evaluate installations. The lack of strategic input is a post–Cold War phenomenon. Prior to the successful employment of long-range intercontinental ballistic missiles (ICBMs), the Air Force established bases in northeastern United States to support heavy bombers. Following the introduction of ICBMs, the survivability of CONUS bases in the event of a preemptive Soviet strike shifted the Air Force toward more bases in southern portions of the United States. As the Air Force plans for new weapon systems and force structures, incorporating strategic concerns across the portfolio of basing decisions would help to maximize capabilities and minimize life-cycle costs.

Observations and Recommendations

The Air Force Should Institute an Initial Screening Process

Our analysis found that the current basing process is designed so that installations with high mission scores are advanced toward site selection, but that capacity scores largely influence the final candidate list. To reduce the cost and time of conducting the existing strategic basing process, the Air Force should consider using centralized data to answer mission-related questions that can initially screen candidate bases for suitability. With a reduced set of potential installations to consider, the detailed capacity and environmental data collection efforts could focus on those installations with competitive mission scores. An initial screening process based on mission scores would maintain equity by considering all installations in the enterprise, but collecting detailed capacity, environmental, and cost data only on installations that can best achieve the desired mission.

The Air Force could apply lessons learned from previous basing actions and military judgment to set a potential mission score threshold for screening. Setting the threshold too high could overlook bases that might otherwise be selected as candidates, while setting the threshold too low would expand the number of installations required to provide detailed capacity analyses. From our sample of 25 basing decisions, we analyzed 12 enterprise-wide look basing actions to see what mission score thresholds would have meaningfully reduced the size of the data call without omitting any bases eventually chosen as candidates for a site selection survey. Over the 12 basing actions, 1,531 total data calls were sent to CONUS installations. These data calls require personnel at MAJCOMs or headquarters to transmit, receive, and analyze the information, as well as civil engineering and other personnel at each installation to respond to each data call. For our sample, all of the installations selected for a site selection survey received at least 35 percent of the maximum possible mission score, and most received at least 50 percent. This means that, for the 12 actions in our sample, 626 to 882 data call responses would have been saved if the detailed data calls were limited to installations with high mission scores (Table S.1).

Table S.1
Reduced Data Collection Enabled by Various Mission Score Thresholds

Threshold	Data Requests to Installations	Reduction in Data Call Responses
No threshold	1,531	–
Mission score at least 50% of max possible	649	882
Mission score at least 40% of max possible	831	700
Mission score at least 35% of max possible	905	626

NOTE: Sample analyzed consisted of 12 decisions in USAF's strategic basing process that considered installations across USAF's CONUS enterprise.

Using mission criteria to screen bases prior to selecting bases for further consideration can reduce the data collection burden on the enterprise by nearly half, with no decrease in the quality of the ultimate basing decision.

Air Force Data Quality Is Sufficient to Support Decisions

The data currently used are of sufficient quality to support USAF basing decisions. Our analysis found the data to be authoritative and consistent, but the auditability of underlying data used in the basing action process is weak. It is challenging to trace specific sources used to answer questions, even when responsible parties are identified in a MAJCOM-provided data call template, as is the case for the KC-46A basing actions. There is no data call template used for other basing actions, making it even more difficult to identify underlying data sources. Improved auditability will allow for more efficient review by HAF and other parties. The combination of improved auditability and increased review should decrease the potential for human error in the basing decisionmaking process, which accounts for a majority of the publicly disputed basing decisions.

Acknowledgments

We are grateful to Acting Assistant Secretary of the Air Force for Installations, Environment and Logistics Kathleen Ferguson for her sponsorship and support throughout the execution of this research. We thank the professionals at Headquarters Air Force who provided guidance, time, and information, including Deputy Assistant Secretary of the Air Force for Installations Timothy Bridges, Col Frank Freeman III, Lt Col Dennis Burgart, Andrew Mendoza, Mark Pohlmeier, and Nugent Nguyen. We met with and corresponded with scores of individuals and organizations throughout the Air Force who gave freely of their time and perspectives. They are too numerous to list individually, but we especially thank Kevin Flood, Geno Patriarca, and Richard Rankin for repeatedly giving us so much of their time and insights.

Many RAND colleagues contributed crucial analysis, insights, critiques, and advice in the course of this research. The constant support, substantive input, and critical guidance provided by Laura H. Baldwin, former director of the Resource Management Program within RAND Project AIR FORCE, is greatly appreciated. We are grateful for the contributions, feedback, and assistance received from RAND colleagues Sean Bednarz, Grover Dunn, Katheryn Giglio, Stephanie Lonsinger, Leslie Lunger, Jason Mastbaum, Ronald McGarvey, Laura Novacic, Daniel Romano, and Anthony Rosello. We are especially thankful for the comments, suggestions, and critiques of the reviewers at RAND: Irv Blickstein, Edward Keating, and Scott Savitz. Their feedback has considerably strengthened this research.

Abbreviations

AAC	Army Air Corps
AAF	Army Air Force
AF/A1MR	USAF Manpower, Organization and Resources, Manpower Requirements Division
AF/A2R	USAF Directorate for Intelligence, Surveillance and Reconnaissance Resources
AF/A3O	USAF Director of Operations
AF/A4L	USAF Logistics, Nuclear Weapons, Munitions and Missile Maintenance Division
AF/A5X	USAF Director of Operational Planning, Policy, & Strategy
AF/A7C	USAF Civil Engineers
AF/A8P	USAF Director of Programs
AF/A8PB	USAF/Strategic Basing Division (the owner of the USAF Strategic Basing Process)
AF/A8X	USAF Director of Strategic Planning
AF/A9R	USAF Resource Analyses
AF/A10	USAF Strategic Deterrence and Nuclear Integration
AF/JA	USAF Judge Advocate General
AF/RE	USAF Reserve
AFB	Air Force Base
AFCS	Air Force Corporate Structure
AMC	Air Mobility Command
ASIP	Army Stationing and Installation Plan
BAH	basic allowance for housing
BAR	Basing Action Request
BCEG	Base Closure Executive Group
BRAC	Base Realignment and Closure
BRRP	Basing Request Review Panel
COBRA	Cost of Base Realignment Actions
CONUS	continental United States
CSAF	Chief of Staff of the Air Force
DoD	U.S. Department of Defense
GAO	U.S. Government Accountability Office
GBM	Generalized Boosted Model
GDPR	Global Defense Posture Realignment

HAF	Headquarters Air Force
ICBM	intercontinental ballistic missile
IMC	information for members of Congress
IVC	Initial Vector Check
JCS	Joint Chiefs of Staff
MAJCOM	major command
MCI	Mission Compatibility Index
MOB	main operating base
NEPA	National Environmental Policy Act
NSC-68	National Security Council Report 68
OCONUS	outside of the continental United States
OPS	operational basing
OSD	Office of the Secretary of Defense
P&RA	preferred and reasonable alternatives
RCAF	Royal Canadian Air Force
SAC	Strategic Air Command
SAF/AQX	Deputy Assistant Secretary of the Air Force for Acquisition Integration
SAF/FMB	Deputy Assistant Secretary of the Air Force for Budget
SAF/GC	Department of the Air Force General Counsel
SAF/IE	Assistant Secretary of the Air Force for Installations, Environment, and Logistics
SAF/IEB	Deputy Assistant Secretary of the Air Force for Basing and Infrastructure Analysis
SAF/IEE	Assistant Secretary of the Air Force for Installations, Environment, and Energy
SAF/IEI	Deputy Assistant Secretary of the Air Force for Installations
SAF/LL	USAF Directorate of Legislative Liaison
SAF/PA	USAF Director of Public Affairs
SAF/US(M)	USAF Office of Business Transformation
SB-ESG	Strategic Basing Executive Steering Group
SecAF	Secretary of the Air Force
SME	subject matter expert
TFI	Total Force Integration
USAF	United States Air Force
WIDGET	Web-based Installation Data Gathering and Entry Tool

1. Introduction

The U.S. Air Force (USAF) manages any proposed significant changes and additions regarding the location of weapon systems and personnel through its strategic basing process, and the Air Force has conducted more than 100 strategic basing decisions since 2009. Some notable recent examples of strategic basing decisions are the evaluations of which installations will host the F-35 and KC-46A major weapon systems. USAF basing decisions are public, frequent, and occasionally contentious. Because these decisions affect force posture, local economies, and public trust, it is imperative that the decisionmaking process be objective and justifiable. Unfortunately, a small number of these decisions resulted in eroded local confidence in the process. For example, news media reports suggested that the recent decision to base F-35s at Burlington, Vermont, was either a result of faulty data or undue political influence (Bender, 2013).

The economic importance of airbases and personnel to local stakeholders has been present from the beginning of the Air Force. Local interests influenced the location of the first airbase that the Army purchased in 1916. In its search for a suitable piece of land for an air training facility, the Army considered over a dozen locations and narrowed it down to two choices—Hampton, Virginia, or Aberdeen, Maryland—based on physical requirements. Aiming to secure the economic benefits that come with a large military base, a group of Hampton citizens tipped the balance in favor of Langley Field by offering to sell the land at a steeply discounted rate and to build a railroad line and other essential infrastructure (Brown, 1990). Installations remain desirable to local interests, and such examples highlight the need for strong empirical processes employing the best available data during the basing process. This will help maximize USAF capabilities and minimize life-cycle costs.

The Current Basing Decisionmaking Process

The Air Force employs a three-step enterprise-wide process to ensure consistency when it makes basing decisions. In the first step, a set of criteria, composed of individual criterion to be assessed for each installation, is developed to evaluate an installation's suitability to support a basing decision. The criteria are grouped by category and include, but are not limited to, mission, capacity, environment, and costs. Mission criteria include attributes of an individual installation that affect the mission, such as weather, airspace, and proximity to ranges. Capacity criteria evaluate whether an installation's *existing* infrastructure, such as hangar availability, ramp space, and classroom space, can accommodate the proposed basing action. The environmental criteria represent the existing conditions at an installation that might inhibit accommodating the proposed mission, such as air quality, land use, and incompatible development areas due to noise

1

or encroachment issues. The cost criteria represent the regional variation in federal employee salaries, construction capital costs, and military basic allowance for housing (BAH) rates, but generally do not include outlays required to support an individual basing decision. The basing decision criteria are developed by the applicable major command (MAJCOM), approved by the Headquarters Air Force (HAF) and the Secretary of the Air Force (SecAF), and briefed to Congress.

In the second step, individual basing scores for the defined criteria are assembled by the MAJCOM and compared. A small set of candidate bases, usually four to six, is selected for site surveys. This list of candidate bases is also developed by the MAJCOM, approved by HAF and SecAF, and briefed to Congress.

In the final step following the site surveys, a preferred base is selected. As in the previous steps, this is approved by HAF and SecAF and briefed to Congress. The credibility of this process is dependent on these data calls resulting in high-quality data that are then analyzed in a transparent and objective manner.

The RAND Analysis

RAND was engaged to conduct an independent analysis of the USAF's basing process. The purpose of this analysis is to inform decisionmakers on potential improvements to the quality of data and assessment criteria used in making these decisions. In considering the USAF's basing decisionmaking process, we considered three questions:

- Are basing decision criteria aligned with Air Force intentions?
- Are the data used in the Air Force basing decisionmaking process authoritative, consistent, and auditable?
- Is there potential for broader USAF strategic or portfolio-wide inputs to strengthen the basing decisionmaking process?

In answering the first question, we examined the actual data source for each base in enterprise-wide KC-46 and F-35 basing actions. We used these two basing decisions because they represent the bedding down of new major weapon systems across two lead MAJCOMs, have a broad scope, and have extensive available data. Our examination entailed contacting bases to ascertain the data source for each decision criterion. To assess the second question, we examined the desired and actual impact of each decision criterion for 25 relevant basing actions. Using statistical analysis, the actual impact of each criterion can be determined and compared with its desired impact. Finally, in addressing the potential for inclusion of strategic inputs into the basing decisionmaking process, we assessed the historical impact of strategic inputs and possible mathematical techniques to incorporate such inputs into the current basing process.

Organization of This Report

The remainder of this report is organized as follows: Chapter Two describes the current USAF strategic basing process in some detail and specifically describes where data are collected and assessed. Chapter Three assesses whether the current USAF process provides desired outcomes and explores potential process improvements. Chapter Four specifically evaluates the quality of data used in the USAF basing decisionmaking process. Chapter Five describes the challenges presented by the current basing process. Chapter Six offers a series of findings and recommendations based on our research.

A detailed analysis of the basing decision criteria evaluated in Chapter Three and a description of the 2005 Base Realignment and Closure (BRAC) process are included in Appendixes A and B, respectively. Appendix C provides additional detail on the Generalized Boosted Model (GBM) (Ridgeway, 2009) used to assess how well the weight assigned by the Air Force to each scoring category (e.g., mission, capacity) aligned with the actual influence of these categories on the choice of candidate bases. Appendix D expounds on the historical role of strategy in domestic USAF basing decisions. Appendix E describes the basing decisionmaking processes employed by the U.S. Army and U.S. Navy.

2. Current Air Force Strategic Basing Process and Governance

To standardize decisions regarding major changes to weapon systems, personnel, and tenants on installations, the Air Force developed an enterprise-wide strategic basing process in 2009 (AFI 10-503, 2010), with HAF acting as the clearinghouse for all basing actions. The Air Force Strategic Basing Division at HAF (AF/A8PB) is the designated office for managing the strategic basing process and has critical roles across the life cycle of a basing decision. The process, shown in Figure 2.1, involves numerous actors with varying incentives, information, and resources. The process can be divided into three stages: setting the criteria, screening bases, and selecting preferred alternatives.

- First, the Strategic Basing Division designates the proponent MAJCOM, who then develops criteria to evaluate an installation's suitability to support the proposed basing action and submits the criteria back to the Strategic Basing Division. Criteria include, but are not limited to, an installation's ability to execute the mission, capacity to host the unit, environmental impact of the unit, and economic factors, such as locality and construction cost factors associated with individual installations. Each criterion is further broken down into a number of attributes that define that criterion. Once HAF and the SecAF approve the evaluation criteria, they are briefed to Congress.
- In the second step, individual basing scores for the defined basing criteria are assembled by the MAJCOM, and all of the bases are screened to determine the appropriateness of hosting the basing action at each location. Using these scores, a small set of candidate bases, usually four to six, is selected for on-site surveys, which are approved by HAF and the SecAF and then briefed to Congress.
- In the final step following the site surveys, a preferred installation is selected. As in the previous steps, this is approved by HAF and the SecAF, then briefed to Congress. The basing action proceeds through National Environmental Policy Act (NEPA) compliance and is eventually implemented at the installation.

USAF's process has a high level of interaction with Congress, and the USAF provides briefing updates at several points in the process. In the Army and Navy's basing process, Congress is notified after a decision has been made, as discussed in Appendix E.

Figure 2.1
Standard U.S. Air Force Strategic Basing Process and Governance

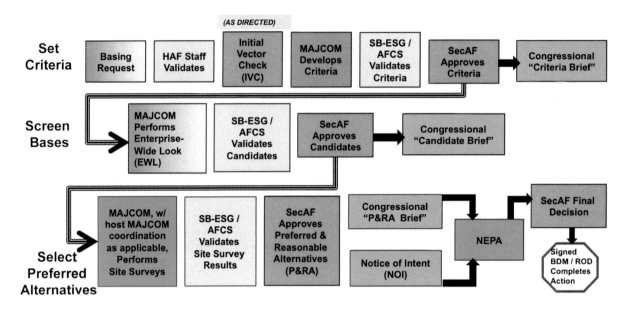

SOURCE Pohlmeier, 2012.
NOTES: Blue boxes represent actions for MAJCOM/base-level personnel. Gray boxes represent actions for HQ/Air Staff personnel. Red boxes represent meetings/approval from SecAF/Chief of Staff of the Air Force (CSAF). Green boxes represent congressional interaction/briefings. SB-ESG = Strategic Basing Executive Steering Group; AFCS = Air Force Corporate Structure.

Roles and Responsibilities

HAF actions are the responsibility of the Strategic Basing Executive Steering Group (SB-ESG). The SB-ESG reviews and evaluates USAF concepts of operations, basing objectives, criteria, policies, programming, and planning and makes recommendations to the SecAF and the Chief of Staff of the Air Force (CSAF). The SB-ESG is composed of one- and two-star general officers and civilian equivalents and chaired by the Secretary of the Air Force for Installations, Environment and Logistics (SAF/IE).[1] The Basing Requests Review Panel (BRRP) supports the SB-ESG as the first level of the Air Force Strategic Basing Structure for basing review and decisions. The BRRP is chaired by AF/A8PB and includes O-6 and civilian equivalents.[2]

The lead MAJCOM for the action submits a Basing Action Request (BAR) to AF/A8PB. For example, this action could be the relocation of a major training center, or the bedding down of a new weapon system. The requester provides the action title, requester contact information, the

---

[1] Membership of the SB-ESG includes representatives from the following: AF/A1MR, AF/A2R, AF/A3O, AF/A4L, AF/A5X, AF/A7C, AF/A8P, AF/A8X, AF/A9R, AF/A10, AF/RE, AF/JA; SAF/IEI, SAF/FMB, SAF/GC, SAF/IEE, SAF/AQX, SAF/LL, SAF/PA, SAF/US(M), and NGB/CF.

[2] Membership of the BRRP includes representatives from the following areas: AF/A1MZ, AF/A3O-B, AF/A5X, AF/A7CAI, AF/A7CIB, AF/A8PB, AF/A8X, AF/A9R, AFLOA/JACE, SAF/GCN, SAF/LLP, SAF/PAX, and SAF/IEI.

mission design series affected, the type of action, the anticipated Environmental Impact Analysis Process, the type of unit involved, the property owner, the lead MAJCOM, the lead MAJCOM action officer contact information, and the installations affected.

Some actions must be coordinated with non-USAF entities. When USAF entities propose moves onto non-USAF installations, including joint bases owned by other services, the basing proponent's lead MAJCOM submits a BAR to AF/A8PB and receives HAF approval prior to pursuing a basing option with a non-USAF entity. Once approved by HAF, AF/A8PB will coordinate the request on behalf of the proponent MAJCOM. When non-USAF entities request to move onto USAF installations, including USAF-owned joint bases, the entity's top-level headquarters element for approving a basing action submits the BAR to AF/A8PB. The proposed host MAJCOM basing office and AF/A8PB assist with the development of criteria for the enterprise-wide look. In most cases, the host MAJCOM is designated as the lead MAJCOM and the standard USAF strategic basing process is followed (Air Force Instruction 10-503, 2010, paragraph 1.1).

Basing Decision Thresholds

The first step in the strategic basing process is for HAF to validate the BAR and determine whether the request should go through the strategic basing process. AF/A8PB may enter any special interest action, regardless of size or scope, into the strategic basing process. However, typically, to be considered a basing action, the request must meet all of the following threshold criteria:

- a manpower increase or decrease greater than 50 positions on the Unit Manning Document
- a weapon system change
  - including additions, subtractions, or mission baseline design replacement
  - excluding tail number swaps and block/spiral upgrades
- an inter-MAJCOM action involving 25 or more manpower positions
- a unit which is activated or inactivated at the squadron level or higher
- the addition, subtraction, or relocation on or off base of a general officer/Senior Executive Service billet, except National Guard Bureau general officers moving under Title 32
- a non-USAF entity requesting to move onto or add 25 or more manpower positions onto USAF property.

There are exceptions that will exclude an action from the strategic basing process even if it meets the above thresholds. Exceptions include

- BRAC actions
- contingency operations
- Total Force Integration actions
- responsibilities for Total Force Integration

- real estate enhanced use leases
- *some* non–estate enhanced use real estate actions
- depot source repair actions
- banks and credit unions
- post offices
- grazing and croplands
- utility and road easements
- utility and roads granted by lease and license
- military exchange retail, warehouse, and support operations
- commissaries
- Morale, Welfare, and Recreation support activities (archery clubs, riding stables, etc.)
- base support activities (American Red Cross, labor union offices, etc.).

AF/A8PB may choose to expedite actions that are considered too urgent or too simple for the full strategic basing process. Expedited actions will skip directly to the site survey. Expedited actions must be simple, specialized, or time-sensitive. Simple actions have minimal mission and environmental impact on any installation selected to support the basing requirement and, therefore, do not need SecAF/CSAF approval. Specialized actions support unique mission requirements, which limit the enterprise of potential installations to specific locations. Time-sensitive actions already have SecAF/CSAF support for an accelerated decision timeline. Expedited processes often include actions that support special operations or are for other service or government tenants moving onto Air Force Bases. Once HAF has decided whether the request will follow the standard basing process, follow the expedited process, or falls outside of the basing process, there is an Initial Vector Check (IVC) with the SecAF.

Setting the Basing Decision Criteria

The proponent MAJCOM develops criteria for the enterprise-wide look. The process begins when the MAJCOM submits the BAR Part 1, which describes the purpose of action, the rationale, the type of facilities, the size of facilities and/or acres required, the present and projected force structure, the urgency of the proposed action, the projected environmental action, the impact if not approved, the benefit to the Air Force, and the base selection criteria. The MAJCOMs tailor criteria for each basing action as shown in Figure 2.2. SB-ESG provides broad guidance on the development of criteria and validates them before submitting them to HAF and SecAF for approval. Once approved, the criteria for this basing decision are briefed to Congress. As a result of this particular criteria system, comparisons between bases can be counterintuitive. For example, in adding up the total scores, one "Mission" point has the same impact as one "Environmental" point. As a result, a base with a relatively high "Mission" score may receive a lower grade than a base with better "Capacity" and "Environmental" scores. Appendix A discusses the details on individual basing decisions that reflect these challenges.

Figure 2.2
An Example of Basing Criteria

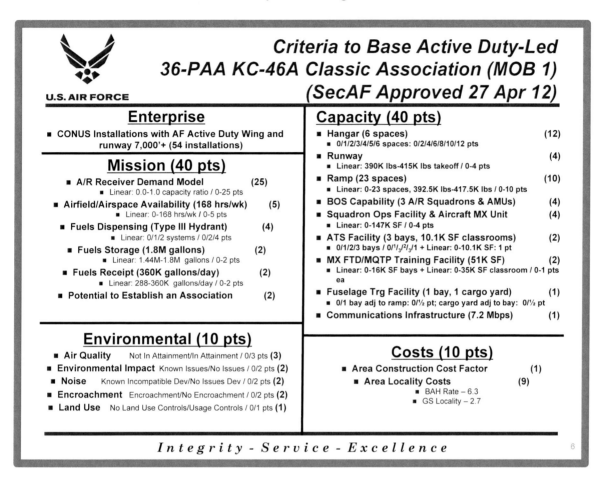

SOURCE: Pohlmeier, 2012.

Selecting Preferred Alternatives

Once the basing criteria are approved, the proponent MAJCOM performs the enterprise-wide look. A data call is submitted to all installations considered in the action. The MAJCOM collects responses and scores all bases according to agreed criteria. The scored and ranked enterprise-wide list of bases is submitted to SB-ESG/Air Force Corporate Structure (AFCS) in the BAR Part 2. The SB-ESG/AFCS then validates the candidates.

The members of the BRRP examine all of the scores and then apply military judgment to determine the number of installations to site survey and create a ranked candidate list (see Figure 2.3 for an anonymized example). BRRP members look for natural breaks and clusters in the scores, as well as for strategically important locations to create a candidate list of installations. Weapon system stakeholders and subject matter experts (SMEs) are included in this process by invitation only. The BRRP may host a conference to develop the candidate list of installations to receive site surveys. Once the SecAF approves the candidate list, the BRRP submits the site

9

survey list in alphabetical order along with the site survey control number to the proponent MAJCOM. SecAF/CSAF approves candidates prior to assigning a site survey control number. The underlying scores are data-driven and objective, but enabling BRRP members to use military judgment introduces some subjectivity into the selection of a group of candidate bases for site surveys. However, allowing the BRRP members and USAF leadership to bring their experience and tacit knowledge into the selection of a broader group of candidate installations likely enables inclusion of potentially successful locations that would otherwise have been omitted.

Figure 2.3
Ranked Candidate Bases List for KC-46A Decision

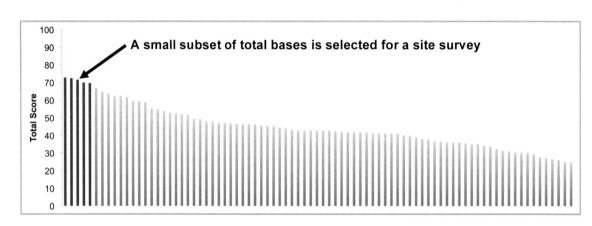

The lead proponent MAJCOM conducts site surveys of the candidate bases in coordination with the host MAJCOM and installations. After the survey is complete, the entity conducting the site survey is responsible for ensuring that the space remains available until a final basing decision is made. Follow-on surveys do not require additional approval, and they use the same site survey control number. The SB-ESG/AFCS validates site survey results.

Within 60 days of completing the last site survey, the proponent MAJCOM submits the BAR Part 3. The final BAR includes the cost, force protection requirements, the Environmental Impact Analysis Process, the evaluation of sites, the impact to existing missions, the recommendation, and alternatives. Once the preferred basing alternative is selected, SecAF makes the final decision and signs the alternative list for decision/approval, and the Base Decision Memorandum or Record of Decision. The SecAF also approves the Preferred and Reasonable Alternatives (P&RA) brief. There is a final congressional P&RA briefing and a Notice of Intent.

In the final phase of the process, USAF completes NEPA approval. NEPA approval includes the Environmental Impact Analysis Process and the Environmental Baseline Survey. The proponent MAJCOM must comply with NEPA requirements and should integrate the Environmental Impact Analysis Process during the initial planning stages to avoid delays. The final P&RA briefing is subject to the results of NEPA approval.

When SecAF signs the Basing Decision Memorandum, the action is complete and the MAJCOM is cleared to execute. Congressional engagement occurs throughout the process and includes a criteria briefing, a candidate briefing, and a P&RA briefing.

Data enter the strategic basing process at five points: when the basing criteria are proposed, when the basing criteria are finalized, when the questions are framed for the data call, when the MAJCOMs and installations respond to the data call, and, finally, when data are obtained on site visits. In Chapter Three, we examine the basing criteria and the installation scores resulting from the data call. In Chapter Four, we take a closer look at how questions are framed for the data call and the data sources used in response to the data call.

3. Is the Air Force's Basing Process Aligned with Its Objectives?

The USAF strategic basing process was developed so that the basing decisions for weapon systems and personnel could consider options across the USAF enterprise and use transparent criteria for evaluation. The intended outcome of all basing actions is the identification of a preferred alternative that is superior to other choices across criteria established for each basing action. As discussed in Chapter Two, these criteria are proposed by the lead MAJCOM, validated by the SB-ESG, approved by HAF and the SecAF, and briefed to Congress. These criteria determine the type of data collected from the candidate installations across the USAF enterprise, the level of effort required for data gathering, and, ultimately, which bases are selected for site visits as candidates for the proposed basing action. The criteria scores and weighting represent USAF priorities for the specific basing action, and, ideally, the results of basing actions would reflect the importance of these criteria. In this chapter, we evaluate the relative influence of initial scoring criteria in observed results of basing decisions, and we propose methods to reduce the level of effort required in the basing process while ensuring the alignment of results to USAF objectives.

Understanding the Basing Criteria

The basing criteria used to evaluate candidate bases across the USAF enterprise consist of individual attributes important to the basing action under consideration. Some basing actions require access to available airspace and measures of runway capabilities. Other basing actions require quantification of existing facilities on the installation that can be repurposed. These are a sample of the sometimes 20 or more criteria that are tailored for each basing action and then assigned a scoring rubric, which is used to evaluate responses received from installations. A list of criteria for a specific basing action was provided in Figure 2.2, and a list of general criteria evaluated across basing actions is listed in Table 3.1. For each basing action, there are 100 total points available if an installation meets all of the criteria at the maximum desired level. This scoring system enables a comparison of installations across the enterprise for each basing action.

Table 3.1

General Criteria Evaluated in the U.S. Air Force Strategic Basing Process

Mission	Capacity	Environment	Cost
• Airspace • Expertise and synergies • Proximity to demand, training, and requirements • Weather	• Administrative space • Base size • Child care • Classified/SAP work space • Communications • Dining • Facility power/HVAC • Fitness center • Fuel capability • Hangars • Lodging • Medical • Operations support • Ramp capability • Runway capability • Training facilities • Storage space	• Air quality non-attainment • Encroachment • Land-use controls • Noise impacts • Other environmental impacts	• Local area construction cost factors • Local basic allowance for housing rates • Government service locality pay factors

NOTES: HVAC = heating, ventilation, and air conditioning; SAP = special access programs. In some basing actions, criteria usually listed in capacity such as fuel capabilities, were listed as mission requirements. Our analysis assigned these general criteria to gain an understanding of the types of information gathered during the basing process.

The Air Force groups the individual scoring criteria into several categories. Across the basing actions we examined, these categories were primarily mission, capacity, cost, and environment. Each individual criterion has a maximum point score possible. Mission criteria include attributes of an individual installation that affect the mission, such as weather, airspace, and proximity to ranges. There are a few instances where mission criteria include infrastructure, such as the capability to receive, store, and dispense sufficient volumes of fuel, but the majority of mission criteria are largely a function of geography and outside the Air Force's control. If proximity to Army and Navy operations, training, or ranges is necessary to enable jointness between the services for a specific weapon system beddown, then the USAF includes jointness as part of the mission criteria. Capacity criteria evaluate whether an installation's *existing* infrastructure can accommodate the proposed basing action. These include criteria on available ramp space; hangars; capacity for such functions as administration, medical, and dining; and other infrastructure that could be added to an installation through construction. For some basing actions, the timing of when existing installation facilities will be ready to accommodate the proposed mission is characterized as a separate category than capacity, whereas in most basing actions, this is captured as part of the capacity category. The cost criteria used in the basing process represent the regional variation in federal employee salaries, construction capital costs, and military basic allowance for housing (BAH) rates. These criteria represent regional economic factors, which are one input to life-cycle costs of basing actions, but the potentially larger operating costs of performing a specific mission from an individual installation are not captured in these criteria. Finally, environmental criteria represent the existing conditions at an installation that might inhibit accommodating the proposed mission, such as air quality, land use, and incompatible development areas due to noise or encroachment issues.

To understand the basing process outcomes, we obtained the data for 25 basing actions that used the strategic basing process. These data included scoring criteria, categories and weights, and points assigned from individual installation responses across the USAF enterprise. Across these basing actions, there were 373 individual criteria uniquely tailored to each action. Many of these criteria sought the same general information, but questions differed based on the specific requirements of each basing action. From all of the criteria used across the 25 actions examined, we characterized 29 general areas of information gathered, as shown in Table 3.1.

Of the 100 points possible for each basing action, mission and capacity criteria were assigned more weight across the larger recent basing actions we examined, although there is still variation among the weightings. The proponent MAJCOM develops the criteria and the weights, with approval from AF/A8PB and SecAF, as discussed in Chapter Two. For the F-35 basing decision, Air Combat Command assigned mission criteria 60 points and capacity 25 points; for the KC-46A basing decision, Air Mobility Command (AMC) assigned mission and capacity criteria 40 points each. Figure 3.1 illustrates the criteria weights for each action, separated by the proponent MAJCOM.

Analyzing the Relative Influence of Basing Criteria

Using the criteria, weighting, and observed outcomes of the basing process, we characterized the relative influence of the criteria on the Air Force's final decisions. We narrowed our data sample to the 20 basing actions since 2009 that involved data calls for at least 10 installations. Of these, we focused on the 12 actions for which candidate bases had been selected by the Air Force for site surveys and hence were under final consideration for the basing action. For each of these 12 actions, the data we used included the list of installations in the data call, the scoring criteria, the scores assigned to each base for each question, and the list of selected candidates. These 12 actions include basing decisions across aircraft categories, including fighter, tanker and transport aircraft, and are representative of likely future basing decisions.

Across our data sample, we observed that the bases scoring in the top 10 percent of the total available points were often those selected for a site survey as candidate bases for the basing action. As an example, Figure 3.2 shows the anonymized total scores for each installation evaluated for the KC-46A Second Main Operating Base (MOB2) basing action; the five highest-scoring bases were selected for a site survey. This is consistent with the expected results of the USAF basing process—the highest total scores would be identified through the enterprise-wide look, and HAF and the SecAF would select groups of high-scoring candidates for site surveys.

Figure 3.1
Criteria Weights Assigned in Selected U.S. Air Force Basing Actions

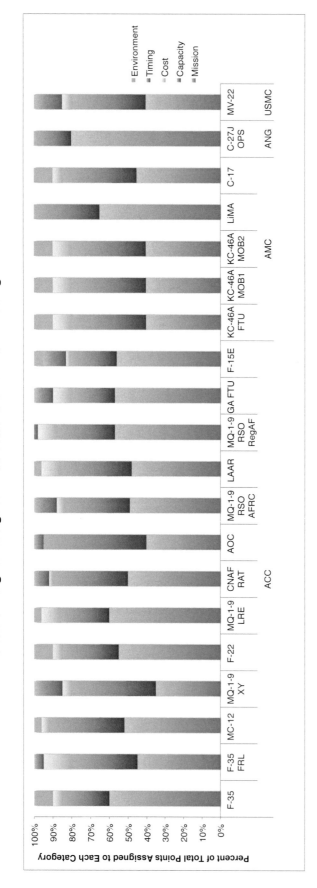

SOURCE: U.S. Air Force.
NOTES: AOC = Air and Space Operations Center; LAAR = Light Attack/Armed Reconnaissance; CNAF = Component Numbered Air Force; RAT = Rapid Augmentation Team; RSO = Remote Split Operations; LRE = Launch and Recovery Element; LiMA = Light Mobility Aircraft; FTU = Formal Training Unit; FRL = F-35 Reprogramming Laboratory.

16

Figure 3.2
Ranked Candidate Bases List for KC-46A Decision

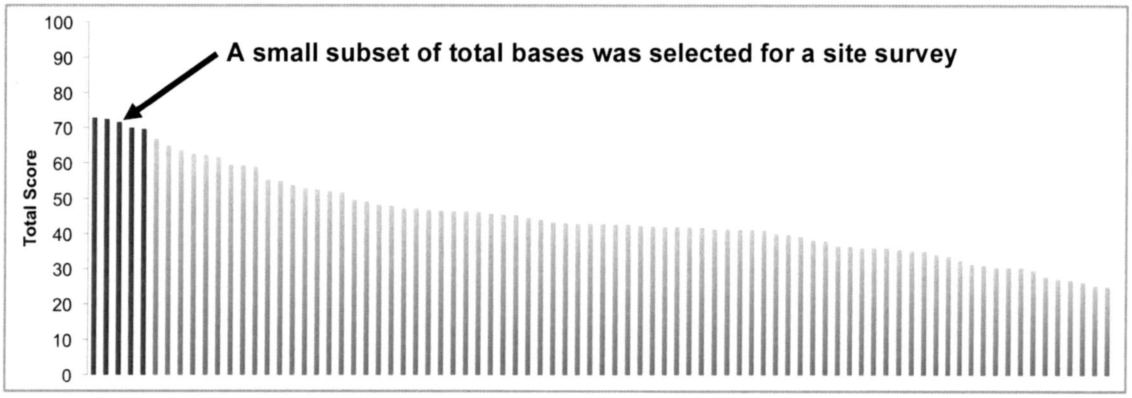

NOTE: The same figure appeared in Chapter Two as Figure 2.3.

Each installation has different weather, airspace, available facilities, and other attributes, and we observed these differences through a large variation in individual criterion scores across installations. Given the differences between installations for data elements, some criterion may be overvalued or undervalued when total scores are examined. For the 12 basing actions in our sample, we assessed the alignment between the Air Force's desired attributes, expressed as relative weight assigned by the Air Force to each scoring category (e.g., mission, capacity, cost, environment) and the actual influence of these categories for the bases chosen to advance in the basing process. To determine the effect of each category on the result, we used a model developed by RAND called the Generalized Boosted Model (GBM) (Ridgeway, 2009). A description of GBM is included in Appendix C.

Using the criteria weights, installation scores, and actual candidates selected for site surveys for each basing action, GBM calculates the "relative influence" of each attribute on the probability of an installation either scoring in the top 10 percent of installations or becoming a site survey candidate. The model assigns relative influence to each category by iteratively attempting to replicate the actual results through a bottom-up analytical process. For example, if there are 100 bases in an enterprise-wide look from which six candidates were selected, GBM initially gives each installation a 6 percent chance of becoming a candidate. On the next iteration, GBM may split the bases into installations scoring less than half of the available points in the mission category and installations scoring more than half of available points in the mission category, which alters their probabilities of becoming a candidate. Installations scoring less than half in the mission category would receive a particular (lower) probability of becoming a candidate, while the others would receive a higher probability. At the next iteration, the model splits the bases again, maybe giving an extra probability of being a candidate to installations that have a capacity score greater than half of the available points for that criterion. For each iteration, GBM automatically attempts multiple splits and adjustments. It then chooses the change that best improves the model fit to actual results, both in terms of the split and the adjustment to the probabilities on either side of the split.

During this process, GBM tracks how much the model fit was improved by each split and keeps track of which variables were split in order to achieve that gain. The improvements in model fit are normalized and termed the "relative influence" of the variable. If the mission score were a bigger driver of model fit than the capacity score, then mission score would have a higher relative influence than capacity score.

GBM avoids overfitting in two ways. First, GBM makes small changes between iterations. For each iteration, the adjustments to the probabilities are smaller than those that would optimize the fit, so changes to the probabilities are spread over the important variables. Second, GBM uses cross-validation by randomly splitting the dataset into ten groups. The model is fit on nine of the groups and predicts the outcome in the remaining group. The number of iterations that achieves the best out-of-sample prediction is considered to be the correct level of complexity, and is used to fit the entire dataset.

Applying GBM to the 12 basing actions in our dataset, we found that, for most, a good mission score was necessary to be potentially considered for a site visit, but when it came to actually being selected for a site visit, the capacity score was usually more important. We present detailed results below for the two largest basing actions under consideration by the USAF, for the F-35 Joint Strike Fighter and the KC-46A tanker. Table 3.2 shows results for the F-35 operational base actions, and Table 3.3 shows the results for the KC-46A operational base actions. For both the F-35 and KC-46A actions, the mission criteria were important to achieving a high score, but either capacity or cost criteria determined site visits. The USAF assigned a weight of 60 percent for the mission criteria for the F-35 operational base actions, and our model found that mission criteria had a relative influence of 44 percent for the F-35 first Operational Basing (OPS1) action and 68 percent for the F-35 third Operational (OPS3)[1] action to score in the top 10 percent of installations. The Air Force assigned a weight of 40 percent for the KC-46A actions, and the calculated relative influence for a top score was higher. The KC46-A first Main Operating Base (MOB1) mission criteria had a 46 percent relative influence, and the MOB2 mission criteria had a 61 percent relative influence. This means for both the F-35 and the KC-46A basing actions, scoring well on mission criteria was a relatively important factor for the bases with the top scores. Since the bases with the top scores were more likely to be further considered for a site survey, the mission criteria effectively screened out many of the candidates lacking high mission scores from being considered for a site survey.

---

[1] The F-35A OPS1 is located in the continental United States (CONUS), the OPS2 is located outside the continental United States (OCONUS), and the OPS3 is located in CONUS. OPS1 and OPS3 are examined in this report.

Table 3.2
Relative Influence of F-35 Criteria Categories on Basing Process Outcomes

	Relative Influence to Score in Top 10%				Relative Influence to Receive a Site Visit			
	Mission	Capacity	Cost	Environment	Mission	Capacity	Cost	Environment
Assigned weights	60	25	10	5	60	25	10	5
Relative influence for F-35 OPS1	44	30	18	9	11	74	8	7
Relative influence for F-35 OPS3	68	17	8	7	19	19	57	5

NOTE: F-35 OPS1 is the active duty beddown; F-35 OPS3 is the ANG beddown.

Table 3.3
Relative Influence of KC46A Criteria Categories on Basing Process Outcomes

	Relative Influence to Score in Top 10%				Relative Influence to Receive a Site Visit			
	Mission	Capacity	Cost	Environment	Mission	Capacity	Cost	Environment
Assigned weights	40	40	10	10	40	40	10	10
Relative influence for KC46A MOB1	46	28	16	9	18	66	14	3
Relative influence for KC46A MOB2	61	21	12	6	24	73	3	1

NOTE: KC46A MOB1 is the first main operating base; KC46A MOB2 is the second main operating base.

When we used GBM to assess the relative influence of criteria on which installations received site surveys, capacity criteria had far greater relative influence than the weight assigned by the Air Force for all but the F-35 OPS3 action, for which cost criteria had the greatest relative influence. This means for both the F-35 and the KC-46A basing actions, scoring well on mission criteria was not the most important factor for the selection of site surveys. Because most of the installations selected for a site survey had relatively high mission scores, capacity and cost scores became an important distinguishing variable that helped determine site selection candidates. The GBM results analytically validate what is an intuitive and expected outcome of the basing process as currently structured—installations that can perform the mission and that have *currently* available capacity should be competitive for a site survey.

Our analytical verification that the mission criteria category screens candidates and the capacity criteria category determines site selection enables a potential streamlining of the basing decisionmaking procedure. The current process relies on detailed data calls to as many as 205 installations for each basing action. As discussed previously and shown in Figure 2.2, these data calls can have 20 or more individual attributes about mission, capacity, cost, and environment, for which each installation is responsible to provide data. Since high mission scores are essential for consideration as a potential basing action candidate, the initial enterprise-wide data analysis could be limited to only bases able to fulfill the mission requirements. Collecting mission data first and screening installations by mission score would eliminate the need for labor- and time-intensive capacity data collection efforts at installations unable to support the proposed mission. Once a set of viable installations is created through this initial screening, data calls and a more detailed assessment of capacity and environmental attributes could be initiated. Cost data for regional economic factors are readily available and could be added into the analysis at any point in the analysis.

The strategy of initially screening installations for mission capabilities maintains an enterprise-wide scope, but limits the initial assessment to mission data, much of which could be collected at the headquarters or MAJCOM level. As an example, Figure 3.3 presents the mission scores for the KC-46A MOB2 basing action. Bases selected as site survey candidates for this action are highlighted in dark blue. If bases that received less than 50 percent of available mission points had been screened from this action, 37 bases would have been removed from the data call, without affecting the list of candidate bases.

Figure 3.3
Mission Scores of Bases Evaluated for the KC-46A Basing Decision

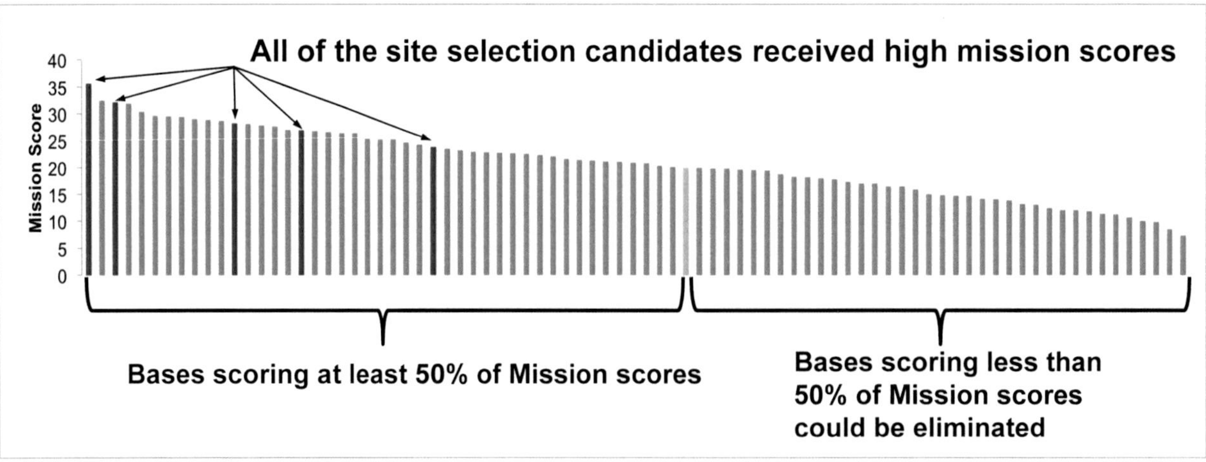

The Air Force could apply lessons learned from previous basing actions and military judgment to set a potential mission score threshold for screening. Setting the threshold too high could overlook bases that might otherwise be selected as candidates, while setting the threshold too low would expand the number of installations required to provide detailed capacity analyses. We analyzed the 12 enterprise-wide look basing actions in our sample to see what mission score thresholds would have meaningfully reduced the size of the data call without omitting any bases eventually chosen as candidates for a site selection survey. Over the 12 basing actions, 1,531 total data calls were sent to CONUS installations. These data calls require personnel at MAJCOMs or headquarters to transmit, receive, and analyze the information, as well as civil engineering and other personnel at each installation to respond to each data call. For our sample, all of the installations selected for a site selection survey received at least 35 percent of the maximum possible mission score, and most received at least 50 percent. This means that for the 12 actions in our sample, 626 to 882 data call responses would have been saved if the detailed data calls were limited to installations with high mission scores (Table 3.4).

Table 3.4
Reduced Data Collection Enabled by Various Mission Score Thresholds

Threshold	Data Requests to Installations	Reduction in Data Call Responses
No threshold	1,531	–
Mission score at least 50% of max possible	649	882
Mission score at least 40% of max possible	831	700
Mission score at least 35% of max possible	905	626

NOTE: Sample analyzed consisted of 12 decisions in USAF's strategic basing process that considered installations across USAF's CONUS enterprise.

Choosing a mission score threshold of 50 percent of the total points possible would have captured 30 of the 33 bases selected as site survey candidates for the 12 actions in our sample. There were three exceptions, demonstrating that mission score thresholds should not be set arbitrarily. Grand Forks Air Force Base (AFB) in the KC-46A MOB1 action earned 38 percent of the available mission points, Boise Air Terminal AGS earned 40 percent of the available mission points for the C-27J action and 49 percent of the available mission points for the F-35 training action. Choosing a mission score of 35 percent would have captured all 33 bases selected as candidates in our sample, and still eliminated the need for 626 data calls.

Our analysis found that the current basing process is designed so that installations with high mission scores are advanced toward site selection, but capacity scores largely influence the final candidate list. To reduce the cost and time of conducting the existing strategic basing process, the Air Force should consider using mission criteria with centralized data to initially screen candidate bases for suitability. Efforts to update and integrate USAF databases are ongoing, but priority should be given to automation and centralization of mission data to have the greatest impact on the strategic basing process. With a reduced set of potential installations to examine, the detailed capacity and environmental data collection efforts could focus only on those installations with competitive mission scores. Our analysis found that initially screening installations based on mission scores would reduce data needs and level of effort, while maintaining an enterprise-wide look and desired basing outcomes. An initial screening process based on mission scores would maintain equity by considering all installations in the enterprise but collecting detailed capacity, environmental, and cost data only on installations that can best achieve the desired mission. What we also observed but did not analyze here is that there is an underlying trade-off between mission and capacity in the existing basing process, and explicitly monetizing these trade-offs is a way to improve basing actions going forward.

4. Assessing Basing Data Quality

The Air Force has conducted more than 100 basing actions since 2009. In the previous chapter, we discussed the methodology the Air Force uses to assess and compare the relative merits of individual bases. In this chapter, we assess the quality of the underlying data used to make basing decisions, since poor data quality could lead to poor decisions, even if the employed processes are sound.

Methodology for Assessing Basing Data Quality

As part of the process for evaluating installations for a given basing action, the Air Force scores candidate installations on several *attributes* that are relevant to the particular action. (See Table 3.1 for a list of the types of attributes generally considered across basing actions.) Data calls are sent to all candidate bases requesting responses to questions pertaining to the attributes of interest. Responses to these data calls are then scored based on a predefined rubric that specifies both the relative weight of each attribute (i.e., its contribution to the overall score) and a scoring system that maps possible responses to a number of points that each response would receive.

The attributes generally fall under the four key basing *criteria*—mission, capacity, cost, and environment.[1] The weights assigned to each of the criteria and to their constituent attributes, as well as the mapping of responses to scores, vary by action. For all basing actions, the sum of the assigned weights equals 100 points.

In our assessment of the quality of basing data, we primarily focused on the data *sources* used by candidate installations across basing actions to answer data call questions. Specifically, we aimed to assess the authoritativeness, consistency, and auditability of the data sources used by candidate installations when they respond to data calls.[2] It is important to note that our assessment does *not* entail directly verifying the *accuracy of the responses provided*. In assessing data consistency, we did analyze responses to a few data call questions, but even in this aspect of our assessment, differences in tailored data call question structure made analyzing responses to all questions infeasible.

We used the KC-46A and the F-35 basing actions as case studies to gather information on the sources used to respond to the data calls. For these five basing actions, we focused on the 16 candidate installations that passed the initial screening and were chosen for site visits.[3]

---

[1] Occasionally, additional criteria, such as timeliness, are considered, but for a majority of actions, the assessed criteria are limited to the four key ones mentioned here.

[2] These metrics are defined and discussed in this chapter.

[3] Altus AFB was chosen for a site visit for both the MOB1 and Formal Training Unit actions.

Data pertaining to the KC-46A MOB1, MOB2, and Formal Training Unit tanker actions are available in a template specified by AMC and used for this specific basing action as further discussed below. In addition to requiring responses to data call questions, this template lists HAF-specified sources to use for a subset of the questions and requires that responsible entities at each installation sign off on all responses. For data call questions with no corresponding HAF-specified source, such as the KC-46A tanker actions, we contacted responsible parties listed in the data call template at the different installations to identify the data sources used. For the F-35 actions, we began at the MAJCOM level and followed a trail of contacts at the installations to identify data sources used.

Data Quality Metrics

We focused our data quality assessment around three questions:

1. Are the data sources authoritative?
2. Are the sources consistently used across installations and across basing actions?
3. Is the process through which data calls are answered auditable?

Authoritativeness measures the extent to which sources used to answer data call questions align with authoritative or credible sources. We consider a source to be authoritative if it has any of the following four *owner types*: USAF, non-USAF U.S. Department of Defense (DoD), non-DoD government, or SME knowledge. We also categorize data sources by *access level*, where the level of access could be either "limited access" or "public." Access level, while not directly related to the metrics we use for assessing data quality, could have implications for accountability. For instance, if a publicly available source is either not used or used improperly, this would have implications for the credibility of the basing process in the event that errors are discovered post-decisionmaking.

Consistency measures the extent to which candidate bases for a given action use the same sources to answer the same questions, and the extent to which a given base consistently answers the same question for different actions. To assess the latter, we analyzed responses provided to data call questions (as opposed to data sources used) when feasible.

Auditability measures the extent to which data sources are documented and traceable. Data that are auditable enable others to check specified sources and locate the data used to underpin the basing process. For the KC-46A tanker actions, approximately half of the data call questions have HAF-specified data sources associated with them. These sources are inherently auditable because they are documented in the AMC-instituted template. For the remaining KC-46A attributes and for attributes pertaining to the F-35 actions, we considered the data sources to be auditable if respondents either provided a direct source or a point of contact who was able to identify the data source.

Sources Identified for Data Call Responses

Tables 4.1–4.4 categorize data call attributes by whether pertinent sources of data are HAF-specified, the types of sources used, and the actual sources used by the majority of queried installations in our sample set.

Table 4.1
Sources Used to Answer Mission Basing Criterion Questions

Attribute	HAF-Specified/ Not Specified	Source Type	Source
Fuel storage	Specified	Limited access USAF	Base Support Plan
Fuel receipt	Specified	Limited access USAF	Base Support Plan
Temporary student housing	Specified	Limited access USAF	*Unaccompanied Housing Utilization Report from ACES*
Weapon system training facility	Specified	Limited access USAF	*Real Property Inventory Detail Report SAF-ILE(a) 7115 from ACES*
Fuselage training	Specified	Limited access USAF	*Real Property Inventory Detail Report SAF-ILE(a) 7115 from ACES*
Airfield/airspace	Not specified	Limited access DoD	*DoD Flight Information Publication (FLIP)*

Table 4.2
Sources Used to Answer Capacity Basing Criterion Questions

Attribute	HAF-Specified/ Not Specified	Source Type	Source
Housing (Military Housing Privatization Initiative)	Specified	Limited access USAF	Unaccompanied Housing Utilization Report from ACES
Permanent party housing	Specified	Limited access USAF	*Unaccompanied Housing Utilization Report from ACES*
Gym/fitness center	Specified	Limited access USAF	*Real Property Inventory Detail Report SAF-ILE(a) 7115 from ACES*
Squadron Operations/Aircraft Maintenance Unit area	Specified	Limited access USAF	*Real Property Inventory Detail Report SAF-ILE(a) 7115 from ACES*
Aircrew training system facility	Specified	Limited access USAF	*Real Property Inventory Detail Report SAF-ILE(a) 7115 from ACES*
Maintenance training facility	Specified	Limited access USAF	*Real Property Inventory Detail Report SAF-ILE(a) 7115 from ACES*
Runway dimensions	Not specified	Limited access DoD	*DoD FLIP*
Hangar spaces	Not specified	Tacit knowledge	As-built drawings, measurements
Ramp capacity	Not specified	Limited access USAF; Public DoD	Geobase maps; PCI report

25

Attribute	HAF-Specified/ Not Specified	Source Type	Source
Ramp hydrant dispensers	Not specified	Tacit knowledge	Base engineer
Medical/dental facilities	Not Specified	Tacit knowledge	Base engineer
Communications infrastructure	Not Specified	Tacit knowledge	Base engineer

Table 4.3
Sources Used to Answer Environment Basing Criterion Questions

Attribute	HAF-Specified/ Not Specified	Source Type	Source
Air quality	Specified	Public government	*EPA National Ambient Air Quality Standards*
Environmental impacts	Specified	Public government	*NPS National Register of Historic Places*
Noise	Specified	Public DoD	*Air Installation Compatible Use Zone*
Encroachment	Specified	Public government	*HUD Clear Zone and Accident Potential Zones*
Land use controls	Not specified	Public government	State-level land use policies aggregated by base engineer

Table 4.4
Sources Used to Answer Cost Basing Criterion Questions

Attribute	HAF-Specified/ Not Specified	Source Type	Source
Local area cost factor	Not specified	Public DoD	*UFC 3-701-01 DoD Facilities Pricing Guide*
Area BAH rate	Not specified	Public DoD	Defense Travel Management Office
GS locality factor	Not specified	Public DoD	Office of Personnel Management

Data Quality Assessment Findings

We found that the data used to make basing decisions are authoritative and consistent, but the auditability of the data call and response process is weak. Therefore, a majority of "data errors" may be unrelated to data quality and result from the data collection process or human error, or possibly both.

Data used to make basing decisions are derived from authoritative sources. Specifically, the majority of data call questions pertaining to the mission basing criterion are answered using standard USAF or DoD sources, and base civil engineers, who are SMEs, answer the rest. Capacity questions are answered using limited access USAF sources or SME tacit knowledge. Environment questions are answered using standard, publicly available DoD and other

26

government sources. Cost questions are answered using standard, publicly available DoD sources.

Sources used to answer data call questions are also consistent. A detailed look at responses to calls for data about weather days, runway length, and local area cost factors revealed that a given installation provided the same answers across basing actions. For the KC-46A tanker actions, HAF-specified sources are consistently used across candidate bases. When data sources are unspecified, there is little variation in the types of sources used to answer data call questions. Other attributes had notable differences in the data call question structure and were therefore not suitable for assessing consistency.

The underlying data used in the basing action process are weak in terms of auditability. It is challenging to trace specific sources used to answer questions, even when responsible parties are identified in a MAJCOM-provided data call template, as is the case for the tanker actions. There is no data call template used for other actions, making it even more difficult to identify underlying data sources.

In summary, we conclude that USAF resources would be best spent in trying to specify and standardize sources and data call protocols when possible, and to focus attention on limiting the number of candidates to which queries are sent by using a more targeted selection process, such as the one described in Chapter Three.

Recommendations to Improve the Auditability of Data Quality

Since auditability is weak, our recommendations focus on improving this aspect of the data quality. However, the benefits of implementing these recommendations may also extend to the authoritativeness and consistency of data quality.

The Air Force can increase auditability by specifying sources whenever possible, by requiring that data call responses include sources used and responsible parties, and by limiting the number of queried bases per action.

First, we recommend that the sources listed in Table 4.5 be HAF-specified across actions. Currently, there is no suggestion or requirement that installations use these sources when responding to data calls, but we found that several installations already use these sources in answering data call questions. In addition to improving auditability, specifying these sources would enhance consistency by further reducing variability in the data sources used by different installations across basing actions.

Table 4.5
Recommended Data Sources to Specify Across Basing Actions

Attribute	Potential Specified Source
Airfield/airspace	*DoD Flight Information Publication (FLIP)*
Runway dimensions	*DoD Flight Information Publication (FLIP)*
Ramp capacity	GeoBase maps; PCI report
Local area cost factor	*UFC 3-701-01 DoD Facilities Pricing Guide*
BAH rate	Defense Travel Management Office BAH Calculator
GS locality factor	Office of Personnel Management GS Locality Pay Tables

Second, we recommend that HAF standardize the data sources listed in Table 4.6. Currently, these sources are specified in the data call template for AMC actions, but they are not specified for actions originating from other MAJCOMs.

Table 4.6
Recommended Data Sources to Standardize Across Basing Actions

Attribute	Potential Standardized Source
Fuel storage	Base Support Plan
Fuel receipt	Base Support Plan
Temporary student housing	*Unaccompanied Housing Utilization Report from ACES*
Weapon system training facility	*Real Property Inventory Detail Report SAF-ILE(a) 7115 from ACES*
Fuselage training facility	*Real Property Inventory Detail Report SAF-ILE(a) 7115 from ACES*
Housing (MPHI)	*Unaccompanied Housing Utilization Report from ACES*
Permanent party housing	*Unaccompanied Housing Utilization Report from ACES*
Gym / fitness center	*Real Property Inventory Detail Report SAF-ILE(a) 7115 from ACES*
Sq ops / AMU facilities area	*Real Property Inventory Detail Report SAF-ILE(a) 7115 from ACES*
Aircrew training system facility	*Real Property Inventory Detail Report SAF-ILE(a) 7115 from ACES*
Maintenance training facility	*Real Property Inventory Detail Report SAF-ILE(a) 7115 from ACES*
Air quality	*EPA National Ambient Air Quality Standards*
Environmental impacts	*NPS National Register of Historic Places*
Noise	*Air Installation Compatible Use Zone*
Encroachment	*HUD Clear Zone and Accident Potential Zones*

Third, we recommend that a template such as the one currently used by AMC, which requires respondents to include data sources and responsible parties with their data call responses, be standardized across MAJCOMs. In addition to improving auditability, such documentation of responsible parties increases accountability by providing a means to trace responses to specific entities in charge, which could in turn reduce human error.

Of course, even if auditability were enhanced, this alone would not eliminate all data errors. For instance, outdated versions of specified sources might be accessed, or inappropriate data might be pulled from current specified sources. While there is no way to ensure an error-free process every time a basing decision is made, a couple of steps can be taken to lessen the potential for error:

- Use the mission-threshold approach described in Chapter Three to reduce the number of queried bases and the processing burden. Currently, not all queried installations have equal incentive to respond in-depth to all data calls, so some may be answered in a minimal way. Targeting only those installations that score sufficiently highly on a smaller set of mission-critical requirements could result in increased diligence in using the right data sources and providing more thorough responses to data calls.
- While steps should be taken to improve all aspects of the basing decision process to achieve the highest data quality possible, it is important to keep in mind that attributes for which related data are publicly available are more readily open to scrutiny. For instance, sources listed (either currently used or recommended) for environment-related attributes in Table 4.3 and cost-related attributes in Table 4.4 are publicly available. It is important to not just specify sources for these attributes but also to take measures to ensure that such sources are properly accessed and used.

5. Implementation Challenges and Potential Improvements

Introduction

The planning, execution, and results of the Air Force strategic basing process involve Air Force stakeholders from across the department. Data are required from each stakeholder during the basing process, and each stakeholder has incentives and preferences for specific basing outcomes. Like any decision with multiple stakeholders, there are likely potential misalignments of these incentives with obtaining optimal objectives. Similarly, stakeholders often experience implementation challenges, and have ideas for potential improvements. To gather insights on the strategic basing process across the major Air Force stakeholders, we conducted in-depth semistructured interviews with more than 40 stakeholders. These included military officers, DoD civilian managers, and Senior Executive Service SMEs. Many of these individuals have been working in basing issues for ten years or more, and deep experience with weapon systems beddowns, previous BRAC rounds, and installation planning was represented. The team conducted interviews with stakeholders at nine MAJCOM headquarters, several civil engineering offices at installations, and military construction agencies, as well as with stakeholders at HAF. The major theme of the interviews was the interviewee's experience with the strategic basing process. The interviews included discussions about the data used to support the process, challenges experienced, and recommendations for improvement. We conducted the interviews both on-site and via telephone on a not-for-attribution basis, to ensure a frank discussion of challenges and recommendations. In this chapter, we summarize major themes that emerged from the interviews and recommend potential improvements to the basing process.

Challenges and Recommendations for the Collection of Basing Data

Data Relevancy and Timeliness

It is evident from the interviews that responding to multiple data calls per year on actions where an installation is not competitive for the proposed mission is seen as superfluous by parties responsible for responding to data calls. Gathering requested basing data on installations is often time-intensive, requiring meetings and correspondence with SMEs from multiple disciplines across the local installation. Because questions are generally tailored to each action, respondents must initiate new searches for each data call. Respondents can get frustrated with continuously devoting time to data collection for basing actions that their installations are not suited for. Interviewees acknowledged that HAF's desire to look across the portfolio of bases for each action is important, but felt that it should be balanced with feasibility of supporting the proposed

31

mission. Our recommendation to initially screen the enterprise of bases for mission suitability discussed earlier would alleviate some of these concerns.

The data categories collected across the mission, capacity, cost, and environment categories in the basing process have various levels of data perishability. That is, data collected for some categories will be valid for the next few fiscal years, while the validity of other data collected will erode more rapidly. We were told that sometimes facilities noted as available by respondents during data calls often were designated for other uses by the time action from basing decision were initiated. Interviews noted that HAF generally sees all data as perishable, while MAJCOMs and installations would prefer to utilize previous data collection efforts and update these as necessary. The correct level of data perishability is a question of balance and judgment. Data collection efforts are time-intensive, but invalid data could result in undesired basing outcomes. Yet, our examination of the individual data requests revealed that most requests were not identical to previous requests in the same category, because each data call is tailored to the specific requirements of each basing action. This specialization prohibits reusing previously collected data for many categories.

Interviewees often recommended undertaking initial site surveys earlier in the basing process on an expanded list of candidates. These expanded site surveys might be able to reveal inaccuracies and potential environmental issues before a final slate of candidates is prepared for Congress. Expanded resources and personnel would be required to expand the site survey process, and the cost of undertaking earlier site surveys could be weighed against improving options for a basing action, as well as validating on-site data for use in future actions. Finally, the time delay between making basing decisions and executing a basing decision can run over fiscal year boundaries. This can interfere with funding mechanisms and force artificial restarts of decision process elements, interviewees noted.

Auditability

Interviewees agreed with our findings that enhancing the data auditability is important. Respondents may be biased for particular actions, and this influences the way they answer the questions. Interviewees stressed that, for many respondents to data calls, there are limited consequences for inaccuracy and limited ways that HAF and MAJCOMs can balance the concerns and incentives of wing commanders when installations respond to data calls. When asked about capacity availability, interviewees told us that respondents may wish to reserve capacity for other uses or may be unaware of pending uses. Inaccurate responses, whether intentional or unintentional, are often validated only during a site visit. Hence, without a site visit or enhanced auditability, the incentives and opportunity for inaccuracy exist. Interviewees also noted that, because of the timeline for basing actions, when incorrect information is identified through the site visit process, there is not an opportunity to revisit the list of other candidates and conduct site surveys on other competitive installations.

For previous basing actions, data collection was largely informal, conducted via email and phone conversations with SMEs. This is challenging for auditability, and for reconstructing sources of information, should internal or external questions arise later. A new approach that standardized data collection was pioneered by AMC for the KC-46A decision. All respondents used an online system to enter data, and respondents were required to provide contact information for themselves and the person locally verifying the data. This system greatly advances auditability and should be replicated across basing actions as appropriate. Yet, as discussed previously, listing the actual data source used is also very important for auditability and should be required.

Data Management

As the strategic basing process has matured over the past few years, an opportunity now exists to standardize data inputs and manage the collected data in a knowledge management framework. Interviewees expressed a desire for a standardized list of criteria and data sources to be established by HAF for the basing process. The group indicated that more consistent data requests on the questionnaires would hopefully lead to generating better repositories for standardized base characteristics, and minimize confusion and questions from respondents on answering data calls. Interviewees noted ambiguity about responsibility and ownership for aspects of data collection, verification, and management. The focus on HAF-driven standardization enforces the common theme of limited personnel and time availability at the MAJCOMs and installations expressed by interviewees. The group also suggested that HAF provide a single point of contact for each basing action for rapid response to questions from MAJCOMs or installations, which would take a lot of the assumptions out of the process.

Challenges and Recommendations for Criteria and Process

Developing Criteria

As described previously, assessment criteria are developed by the proponent MAJCOMs and validated by the SB-ESG. There were a wide variety of opinions and recommendations expressed by interviewees regarding improving assessment criteria, as well as scoring weights and scales. Interviewees acknowledged that MAJCOMs have incentives to tailor criteria to make installations of interest more favorable for particular basing actions. The distribution of installation final scores for basing actions is the interviewees' primary evidence of this occurring. While the incentives exist, it is important to note that optimal basing actions from the Air Force's point of view and a MAJCOM's point of view can, and often will be, similar.

There was a strong desire for consistency in criteria from interviewees, or as one interviewee put it, "We need unity of effort, a common rubric, and a common set of criteria." One interviewee suggested that the scoring criteria map to the Air Force's 12 core functions, to increase consistency between basing decisions and discouraging "gaming" of scores for individual decisions. Countering the desire for consistency, some interviewees wanted more flexibility on the criteria and scoring scales, to allow notation of issues that arise during discussions with bases. Interviewees also said that decisions are often made on the basis of available capacity, and that available capacity is too heavily weighted in many basing actions. AMC's approach of using the receiver-demand model (that simulated where tankers would have to fly to refuel fighter aircraft) in the KC-46A decision helped prioritize mission during the tanker basing process, and could serve as an example for future actions to prioritize mission impacts. It was noted there are separate Air Force and Pentagon standards for floor plans and space. While the Pentagon standards require less space per person, the Air Force standards govern in the basing process, and one interviewee felt that these are "outdated."

Interviewees reported that the cost criteria used in basing decisions are more a proxy for regional economics than basing action costs. These regional economic indicators are important, but including infrastructure costs that improve capacity scores, as well as life-cycle costs of operating at a specific location, would inform the cost portions of basing decisions.

One group of interviewees reported that environmental criteria were neither emphasized enough nor fully developed, relative to the risks that environmental aspects might introduce to successful basing actions. Environmental criteria do not evaluate a basing action within the context of the total capacity for growth at an installation, but rather whether there are issues with adding just this basing action, they stressed. The installation's available capacity for growth in water use, storm water handling, waste management, utility needs, energy production, local air pollutants, and other indicators are not evaluated. Many of these aspects are covered in installation development plans but not explicitly incorporated into the strategic basing process. In addition, the current static criteria do not evaluate the potential for a basing action to improve environmental conditions, such as replacing a more polluting weapon system with a newer, less polluting one. Finally, interviewees argued that NEPA procedures need to be initiated earlier in the basing process, so that the environmental impacts for the range of reasonable alternatives can be evaluated before the decision is made. NEPA is started when candidates are selected, but to fully evaluate alternatives, identify potential challenges, and comply with the timelines of the basing process, the NEPA process should start earlier. An interviewee stressed that the range of alternatives evaluated in NEPA documentation can be narrowed, but the narrowing needs to be "applied on a case-by-case basis and include consideration of the proponent's defined need."

Process Improvements

Interviewees pointed out that there are ways that proponents can "short-circuit" the traditional basing decision process. Most notably, with the inclusion of new assets, it's often easier to put new asset locations directly into the Program Objective Memorandum. It was indicated that this is the easiest route to align moves with a higher-level basing strategy. Alternatively, basing planners can take advantage of the Total Force Initiative process to reposition assets. However, planners indicated that sometimes friction can arise between Reserve and Guard assets and moves. Additionally, many people we talked to pointed to use of contractors rather than Air Force personnel to eliminate the 35-person trigger for the strategic basing process. Another interviewee suggested standardizing estimation of basic infrastructure and personnel requirements for different types of weapon systems, to inform standardized checklists for site surveys.

Challenges and Recommendations for Including Strategic Considerations

Strategic Posture

An overarching theme in interviewees' responses was a view that there is a lack of strategic thinking in defining a desired CONUS posture. They felt that strategic posture could be incorporated into the basing process, to inform basing decisions. Interviewees acknowledged the ability to support OCONUS operations from any CONUS location, but noted that there are strategic, performance, and life-cycle cost differences associated with each CONUS location. Strategic thinking is lacking, interviewees said, because the basing process is reactive and driven up from the MAJCOMs instead of coming down from HAF. Common points made include, "Where does the Asia rebalance show up in CONUS basing decisions?" and "What is the role of the Core Function Lead Integrators (CFLIs)?" Military judgment is infused by the CSAF, the SecAF, and the BRRP members, but not according to a codified doctrine or strategic vision. The strategy that is included is not formally articulated or quantified, which would assist in basing process transparency. Defining a CONUS posture would also eliminate uncompetitive candidates from the basing enterprise-wide looks and reduce the time and resources required in the basing process. Finally, the Air Force portfolio of CONUS installations collectively consists of a total capacity, which changes as new weapon systems are bedded down, older systems are retired, and force structure shifts. An individual basing decision to beddown a weapon system at a specific location utilizes capacity at that installation for the life of that weapon system, which may be several decades. Without a defined strategic posture that is flexible in adjusting to new threats or force structure changes, an optimal basing decision for an individual weapon system may be suboptimal across the portfolio of expected and unexpected decisions in the mid- and long-terms. Hence, the Air Force risks bedding down aircraft and weapon systems at locations that it may regret in the future.

Unique Mission Considerations

Many interviewees were SMEs and noted how the generic strategic basing process was designed for major weapon systems and required adjustment regarding other missions. One concern was with cyber basing actions and the local and regional assets needed to fulfill these missions. Interviewees suggested that units such as the 24th Air Force need to have locations in Silicon

Valley or Northern Virginia to draw on the existing cyber industrial bases. Similarly, an issue that is not included during existing basing actions is the impact on an existing industrial base by moving units elsewhere.

Satellite command-and-control (C2) basing actions also do not fit well in the traditional basing process enterprise-wide look framework. These actions "need centralization, frequency deconfliction, access to many satellite dishes, accreditation and security staff, and this can only be done at certain Air Force Bases," an interviewee told us. "If you need to build a C2 center or a space operations center, there is limited value in studying lots of bases across the enterprise," this interviewee noted. These special mission needs could be in the strategic basing process, by updating Air Force Instruction (AFI) 10-503. The current instruction says "any special interest item" requires going through the full strategic basing process. In practice, "this allows headquarters to decide what they want to study and is too vague," the interviewee told us. A potential improvement is revising the AFI 10-503 to defer to the existing MAJCOM supplement in certain special cases.

6. Conclusions and Recommendations

Under its current domestic strategic basing decision process, the Air Force considers its entire enterprise of installations for many basing actions. Each installation collects data on 30 to 40 basing criteria to support each decision. This results in a data-intensive process that requires a significant expenditure of resources across the enterprise and is directly dependent on the quality of the underlying data.

RAND evaluated the USAF basing process by considering three questions:

- Are basing decision criteria aligned with Air Force intentions?
- Are the data used in basing decisions authoritative, consistent, and auditable?
- Is there potential for broader USAF strategic or portfolio-wide inputs to strengthen the basing decisionmaking process?

In answering the first question, we identified and examined the actual data sources used by each base to respond to data calls as part of enterprise-wide KC-46 and F-35 basing decisions. To assess the second question, we examined the desired and actual impact of each basing decision criterion for the past 25 basing actions. Using statistical analysis enabled us to determine and compare the actual impact of each basing criterion relative to its desired impact. Finally, in addressing the potential for inclusion of strategic inputs into the basing decisionmaking process, we assessed the historical impact of strategic inputs.

Air Force Data Quality Is Sufficient to Support Decisionmaking

The data currently used are of sufficient quality to support USAF basing decisions. We examined data quality along three dimensions: authoritativeness, consistency, and auditability. Our analysis found that the data are authoritative and consistent but require improved auditability. Data used to make basing decisions are derived from authoritative sources. Specifically, the majority of data call questions pertaining to the mission criterion are answered using standard USAF or DoD sources, with base civil engineers, who are SMEs, answering the rest. Capacity questions are answered using limited access USAF sources or SME tacit knowledge. Environment questions are answered using standard, publicly available DoD and other government sources. Cost questions are answered using standard, publicly available DoD sources.

Sources used to answer data call questions are consistent. A detailed look at responses to calls for data about weather days, runway length, and local area cost factors revealed that a given installation provided the same answers across actions. For the KC-46A tanker actions, HAF-specified sources are consistently used across candidate bases. When data sources are unspecified, there is little variation in the types of sources used to answer data call questions.

Other attributes had notable differences in the data call question structure and were therefore not suitable for assessing consistency.

The auditability of underlying data used in the basing action process is weak. It is challenging to trace specific sources used to answer questions, even when responsible parties are identified in a MAJCOM-provided data call template, as is the case for the tanker actions. There is no data call template used for other basing actions, making it even more difficult to identify underlying data sources. Improved auditability will allow for more efficient HAF review. The combination of improved auditability and increased review should decrease the potential for human error in the basing decisionmaking process, which accounts for a majority of the publicly disputed basing decisions.

The Air Force Should Institute an Initial Screening Process

Our analysis found that the current basing process is designed so that installations with high mission scores are advanced toward site selection, but that capacity scores largely influence the final candidate list. To reduce the cost and time of conducting the existing strategic basing process, the Air Force should consider using centralized data to answer mission-related questions that can initially screen candidate bases for suitability. Efforts to update and integrate USAF databases are ongoing, but priority should be given to automation and centralization of mission data to have the greatest impact on the strategic basing process. With a reduced set of potential installations to consider, the detailed capacity and environmental data collection efforts could focus only on those installations with competitive mission scores. Our analysis found that initially screening installations based on mission scores would reduce data needs and level of effort, while maintaining an enterprise-wide look and desired basing outcomes. Using mission criteria to screen bases prior to selecting bases for further consideration can reduce the data collection burden on the enterprise by nearly half, with no decrease in the quality of the ultimate basing decision.

Appendix A. Detailed Analysis of Basing Decision Criteria

In this appendix, we examine in detail 12 basing actions with a minimum of 10 installations in the enterprise-wide look, and for which we know the list of candidate bases. For each basing action, we show the scores and ranks for the site selection candidates. We rounded scores to the nearest whole number, but considered fractional differences when calculating rank. We also show the relative influence of the score categories for each action as determined by the GBM discussed in Chapter Three.

The first basing action examined is the enterprise-wide look for the C-27J basing action, which considered 94 CONUS and OCONUS installations. The two candidate bases, Boise and Great Falls, received the top two mission scores in CONUS. Mission was weighted so highly that the Great Falls location was able to earn more total points than any other CONUS installation despite receiving zero points for capacity (Table A.1). Our GBM-based model reflects the importance of the mission score in the USAF's process by assigning it a relative influence of 92 percent for this basing action (Table A.2).

Table A.1
C-27J Candidates

Candidates	Total		Mission (Mission + Manpower + Increase Regional Lift)		Capacity (Infrastructure)	
	Score	Rank	Score	Rank	Score	Rank
Boise	52	2nd	32	3rd	20	1st (6-way tie)
Great Falls	51	3rd	51	2nd	0	3rd (66-way tie)

Table A.2
C-27J Relative Influence

	Mission (Mission + Manpower + Increase FEMA Regional Lift)	Capacity (Infrastructure)
AF weight	80	20
Relative influence site visit received	92	8

The enterprise-wide look for the F-35 one-squadron basing action considered 205 installations. Military judgment was the deciding factor in selection as a candidate base (Table A.3). The preference for bases with high capacity scores is seen in the relative influence calculated by GBM. The planned weight of the capacity category was only 25 points, but GBM assigned it a 74 percent relative influence on selection as a candidate base (Table A.4).

Table A.3
F-35 One-Squadron Candidates

Candidates	Total		Mission (Airspace + Weather)		Capacity (Facilities)		Cost		Environment	
	Score	Rank	Score	Rank	Score	Rank	Score	Rank	Score	Rank
Shaw AFB	90	7th	58	15th	23	4th (6-way tie)	4	5th (5-way tie)	4	5th (20-way tie)
Hill AFB	84	29th	52	62nd	23	4th (6-way tie)	2	26th (5-way tie)	7	3rd (29-way tie)
Mountain Home AFB	83	32nd	52	61st	18	14th (14-way tie)	3	20th (5-way tie)	10	1st (29-way tie)

Table A.4
F-35 One-Squadron Relative Influence

	Mission (Airspace + Weather)	Capacity (Facilities)	Cost	Environment
AF weight	60	25	5	10
Relative influence Top 10% of scores	44	30	18	9
Relative influence site visit received	11	74	7	8

The enterprise-wide look for the F-35 three-squadron basing action considered 205 installations. As with the F-35 one-squadron action, military judgment played an important role in candidate selection (Table A.5). However, applying the GBM technique, we see that, despite the importance of the mission category on receiving a high overall score, cost was a driving factor in the selection of candidate bases (Table A.6).

Table A.5
F-35 Three-Squadron Candidates

Candidates	Total		Mission (Airspace + Weather)		Capacity (Facilities)		Cost		Environment	
	Score	Rank	Score	Rank	Score	Rank	Score	Rank	Score	Rank
Jacksonville IAP AGS	84	11th	58	15th	15	11th (22-way tie)	4	2nd (5-way tie)	7	3rd (29-way tie)
McEntire AGB	81	22nd	57	22nd	15	11th (22-way tie)	4	2nd (5-way tie)	4	5th (20-way tie)
Burlington IAP AGS	80	27th	51	66th	15	11th (22-way tie)	4	5th (5-way tie)	10	1st (29-way tie)

Table A.6
F-35 Three-Squadron Relative Influence

	Mission (Airspace + Weather)	Capacity (Facilities)	Cost	Environment
AF weight	60	25	5	10
Relative influence top 10% of scores	68	17	8	7
Relative influence site visit received	19	19	57	5

The enterprise-wide look for the F-35 training basing action (Tables A.7 and A.8) considered 205 installations. Twenty-two bases not selected as candidates scored higher than at least one candidate base, but only two bases not selected had both higher mission and capacity scores than at least one candidate.

Table A.7
F-35 Training Candidates

Candidates	Total		Mission		Capacity		Cost		Environment	
	Score	Rank	Score	Rank	Score	Rank	Score	Rank	Score	Rank
Luke AFB	89	1st	57	1st (2-way tie)	24	1st	4	14th (3-way tie)	4	5th (20-way tie)
Holloman AFB	74	6th	56	4th	5	25th (4-way tie)	4	12th (4-way tie)	9	2nd (25-way tie)
Boise Air Terminal AGS	59	25th	29	29th	18	4th	4	6th (3-way tie)	7	3rd (29-way tie)

Table A.8
F-35 Training Relative Influence

	Mission	Capacity	Cost	Environment
AF weight	60	25	5	10
Relative influence site visit received	79	10	10	1

The enterprise-wide look for the KC-46A MOB1 basing action (Tables A.9 and A.10) considered 54 installations. The top four highest-scoring installations were selected as candidates. The capacity score had the greatest impact on the selection of bases for the KC-46A action. The top five highest-scoring bases for capacity had scores within five points of each other. However, the sixth-placed base in capacity was five points behind the fifth-placed base. None of the other score categories had a clear dividing line between installations that were or were not selected as candidates. Grand Forks' high capacity and environment scores kept it in the top four bases despite scoring below 43 other installations in its mission score. All but one of the nonselected bases scoring above Grand Forks in mission had much lower capacity scores. The exception was MacDill AFB, which beat Grand Forks in both mission and capacity, but was edged out by Grand Forks' perfect environment score.

Table A.9
KC-46A MOB1 Candidates

Candidates	Total		Mission		Capacity		Cost		Environment	
	Score	Rank	Score	Rank	Score	Rank	Score	Rank	Score	Rank
McConnell AFB	71	1st	32	5th	26	2nd	8	20th	4	7th (9-way tie)
Altus AFB	66	2nd	29	8th	23	3rd	9	2nd	6	5th (8-way tie)
Fairchild AFB	63	3rd	23	20th	26	1st	7	39th	6	5th (8-way tie)
Grand Forks AFB	55	4th	15	43rd	21	5th	9	12th	10	1st (3-way tie)

Table A.10
KC-46A MOB1 Relative Influence

	Mission	Capacity	Cost	Environment
AF weight	40	40	10	10
Relative influence top 10% of scores	46	28	16	9
Relative influence site visit received	18	66	14	3

The enterprise-wide look for the KC-46A MOB2 basing action (Tables A.11 and A.12) considered 83 installations. The top five highest-scoring installations were selected as candidates. Even though the mission and capacity scores were both worth 40 points, the capacity scores had a much wider variance across installations than the mission scores. The larger variance gave capacity scores more influence on candidate selection.

Table A.11
KC-46A MOB2 Candidates

Candidates	Total		Mission		Capacity		Cost		Environment	
	Score	Rank	Score	Rank	Score	Rank	Score	Rank	Score	Rank
Pease Int'l Trade Port	73	1st	32	3rd	30	2nd	5	76th	6	5th (13-way tie)
Forbes Field	72	2nd	27	17th	29	3rd	9	19th	8	3rd (10-way tie)
McGuire	72	3rd	36	1st	26	7th	5	70th	5	6th (14-way tie)
Rickenbacker	70	4th	24	27th	34	1st	8	32nd	4	7th (4-way tie)
Pittsburgh	70	5th	28	12th	28	4th	7	61st	7	4th (11-way tie)

Table A.12
KC-46A MOB2 Relative Influence

	Mission	Capacity	Cost	Environment
AF Weight	40	40	10	10
Relative influence top 10% of scores	61	21	12	6
Relative influence site visit received	24	73	3	1

The enterprise-wide look for the KC-46A training squadron basing action (Tables A.13 and A.14) considered 54 installations. The two candidates chosen received both the top two total

43

scores and the top two mission scores. McConnell AFB was part of a two-way tie for fourth place in the capacity category, making its high mission score the deciding factor. Both candidates received mediocre environment scores and good cost scores.

Table A.13
KC-46A Training Base Candidates

Candidates	Total		Mission		Capacity		Cost		Environment	
	Score	Rank	Score	Rank	Score	Rank	Score	Rank	Score	Rank
Altus AFB	88	1st	34	1st	39	1st	9	2nd	6	5th (9-way tie)
McConnell AFB	71	2nd	26	2nd	33	4th (2-way tie)	8	18th	4	7th (9-way tie)

Table A.14
KC-46A Training Base Relative Influence

	Mission	Capacity	Cost	Environment
AF weight	40	40	10	10
Relative influence top 10% of scores	23	44	18	15
Relative influence site visit received	33	35	27	5

The enterprise-wide look for the LiMA basing action (Tables A.15 and A.16) considered 172 installations. The top two scoring installations were selected as candidates. The two candidate bases tied for first place in the mission category. Twenty-five bases received a capacity score higher than at least one of the selected candidate bases; however, none of them could overcome the high mission scores of the two candidates.

Table A.15
LiMA Candidates

Candidates	Total		Mission (Mission + Training)		Capacity (Facilities/Infrastructure + Support Capacity)	
	Score	Rank	Score	Rank	Score	Rank
Travis	99	1st	65	1st (2-way tie)	34	2nd (2-way tie)
McGuire	95	2nd	65	1st (2-way tie)	30	4th (12-way tie)

Table A.16
LiMA Relative Influence

	Mission (Mission + Training)	Capacity (Facilities/Infrastructure + Support Capacity)
AF weight	65	35
Relative influence site visit received	79	21

The enterprise-wide look for the MC-12 basing action (Tables A.17 and A.18) considered 175 installations. The top five highest-scoring bases were selected as candidates. All five candidates had high capacity scores, but their performance in the three other categories was mixed.

Table A.17
MC-12 Candidates

	Total		Mission		Capacity		Cost		Environment	
Candidates	Score	Rank	Score	Rank	Score	Rank	Score	Rank	Score	Rank
Beale	87	1st	47	4th	36	4th (2-way tie)	1	36th	4	1st (19-way tie)
Altus	86	2nd	42	21st	38	2nd	3	16th (2-way tie)	4	1st (19-way tie)
Langley	84	3rd	44	6th	35	5th	3	14th (5-way tie)	1	4th (22-way tie)
Robins	83	4th	44	8th	32	11th	4	3rd (4-way tie)	3	2nd (30-way tie)
Key Field	83	5th	43	16th	36	4th (2-way tie)	3	17th (4-way tie)	2	3rd (56-way tie)
Whiteman	82	6th	39	40th	37	3rd	2	21st (3-way tie)	4	1st (19-way tie)

Table A.18
MC-12 Relative Influence

	Mission	Capacity	Cost	Environment
AF Weight	52	40	4	4
Relative Influence Site Visit Received	17	81	0	2

The enterprise-wide look for the MQ-1/9 Remote Split Operations regular Air Force basing action (Tables A.19 and A.20) considered 161 installations. The top three scoring bases were chosen as candidates. All three candidates scored very well in mission, capacity, and environment. Hickam placed third in total score despite coming in last in the cost category.

Table A.19
MQ-1/9 RSO Regular Air Force Candidates

Candidates	Total		Mission		Capacity (Comm Infrastructure / Facilities)		Cost		Timing	
	Score	Rank	Score	Rank	Score	Rank	Score	Rank	Score	Rank
Shaw	91	1st	48	2nd (2-way tie)	29	3rd (3-way tie)	9	7th	5	1st (13-way tie)
Davis Monthan	88	2nd	48	2nd (2-way tie)	28	4th (2-way tie)	9	18th	3	2nd (8-way tie)
Hickam	87	3rd	52	1st	29	3rd (3-way tie)	1	Last Place	5	1st (13-way tie)

Table A.20
MQ-1/9 RSO Regular Air Force Relative Influence

	Mission	Capacity (Comm Infrastructure/ Facilities)	Cost	Timing
AF weight	54	31	10	5
Relative influence site visit received	41	35	5	19

The enterprise-wide look for the MQ-1/9 RSO Air Force Reserve Command basing action (Tables A.21 and A.22) and considered 66 installations. Hurlburt Field was the only candidate chosen, even though five installations earned higher total scores. One installation, Eglin AFB, tied or beat Hurlburt Field in every category.

Table A.21
MQ-1/9 RSO AFRC Candidates

Candidates	Total		Mission		Capacity (Comm Infrastructure / Facilities)		Cost		Environment	
	Score	Rank	Score	Score	Score	Rank	Score	Rank	Score	Rank
Hurlburt Field	65	6th	42	4th	3	15th (3-way tie)	3	9th (2-way tie)	0	Last place (56-way tie)

Table A.22
MQ-1/9 RSO AFRC Relative Influence

	Mission	Capacity (Comm Infrastructure / Facilities)	Cost	Timing
AF weight	49	35	4	12
Relative influence site visit received	32	51	17	0

The enterprise-wide look for the Tactical Air Control Party (TAC-P) basing action (Tables A.23 and A.24) considered 157 installations. The top two Air Education and Training Command installations were chosen as candidates. The selected candidates had higher mission scores than any other installations in the enterprise-wide look. Neither scored any points in the timing or cost categories. For timing, the P&RA briefing states that there were "no significant issues." Although environment considerations were not part of the enterprise-wide look, the same briefing states that Lackland AFB had "low" environmental risk and Keesler AFB had "medium" risk.

Table A.23
TAC-P Candidates

Candidates	Total Score	Total Rank	Mission Score	Mission Rank	Capacity Score	Capacity Rank	Cost Score	Cost Rank	Timing Score	Timing Rank
Lackland AFB	82	1st	57	1st	25	1st	0	Last place (32-way tie)	0	Last place (136-way tie)
Keesler AFB	64	3rd	44	2nd	20	2nd (17-way tie)	0	Last place (32-way tie)	0	Last place (136-way tie)

Table A.24
TAC-P Relative Influence

	Mission	Capacity	Cost	Timing
AF weight	60	30	5	5
Relative influence top 10% of scores	59	37	3	1
Relative influence site visit received	58	28	13	0

Appendix B. Base Realignment and Closure 2005

The main report addresses the USAF's strategic basing process. Decisions regarding basing, especially closures, are frequently contentious. This appendix provides background on the BRAC process.

Introduction to BRAC 2005

DoD faced difficulties reorganizing its portfolio of military installations in the decades immediately prior to 1988. The National Environmental Policy Act (NEPA) of 1969 required that DoD use procedures that take between 10 and 18 months when undertaking any "major" installation closure or realignment and release relevant DoD plans and studies to the public (Schlossberg, 2012). Publicity generated by proposed basing actions invited litigation and congressional interference that regularly delayed or scuttled DoD plans (Schlossberg, 2012). The Defense Authorization Amendments and Base Closure and Realignment Act of 1988 (P. L. 100-526) authorized DoD to close and realign U.S. military installations without going through the process required by NEPA. The result was the first Base BRAC round in 1988. The reunification of Germany in 1990 and the imminent dissolution of the Warsaw Pact and end of the Cold War convinced Secretary of Defense Dick Cheney that further basing actions were necessary. Cheney attempted to reorganize military installations in 1990 following NEPA guidelines and without gaining congressional approval for a BRAC round but found the process too onerous. Congress subsequently passed the Defense Base Realignment and Closure Act of 1990 (P.L. 101-510), which allowed for further BRAC rounds. The BRAC Act of 1990 served as the guiding document for BRAC rounds in 1991, 1993, 1995, and 2005. Amendments were made to the act throughout this time period so that the various BRAC rounds were subtly different. The BRAC Act of 1990 has expired, leaving the NEPA process the only option currently available for making basing actions (Schlossberg, 2012). DoD has requested a new BRAC round, which would likely be similar to BRAC 2005. This section of the report details BRAC 2005, explaining the process that the 1990 Act requires.

On December 28, 2001, Congress authorized the BRAC round of 2005. The Secretary of Defense had three goals, which were announced on November 15, 2002: saving military expenses by reducing excess infrastructure, transforming the military, and fostering jointness. Prior BRACs in 1988, 1991, 1993, and 1995 focused exclusively on the first of these goals. DoD next proposed criteria to use to evaluate existing installations and potential changes to the organization of these installations. The criteria are listed in Table B.1 and were placed in the *Federal Register*, with opportunity for public comment, on December 23, 2003 (68 Fed. Reg. 74221 [2003]).

Table B.1

Department of Defense Criteria for the 2005 BRAC Round

Category	Criteria
Military value	The current and future mission capabilities and the impact on operations readiness of the total force of DoD, including the impact on joint warfighting, training, and readiness.
	The availability and condition of land, facilities, and associated airspace (including training areas suitable for maneuver by ground, naval, or air forces throughout a diversity of climate and terrain areas and staging areas for use of the Armed Forces in homeland defense missions) at both existing and potential receiving locations.
	The ability to accommodate contingency, mobilization, surge and future total force requirements at both existing and potential receiving locations to support operations and training.
	The cost of operations and the manpower implications.
Other	The extent and timing of potential costs and savings, including the number of years, beginning with the date of completion of the closure or realignment, for the savings to exceed costs.
	The economic impact on existing communities in the vicinity of military installations.
	The ability of the infrastructure of both the existing and potential receiving communities to support forces, missions, and personnel.
	The environmental impact, including the impact of costs related to potential environmental restoration, waste management, and environmental compliance activities.

SOURCE: GAO, 2013a.

DoD was required to furnish Congress with a plan for force structure, defining the size and composition of units needed to address threats to national security 20 years into the future. This requirement was new for the 2005 BRAC round. Analyses based on the force forecast and an inventory of existing installations led the Secretary of Defense to certify that there was a need for the closure and/or realignment of military bases on March 23, 2004. The secretary, as required, also certified that needed closures and realignments would yield annual net savings for each department of the military beginning no later than 2011.

Analytical teams from the individual departments of the military, together with an Infrastructure Steering Group representing seven "joint cross-service groups" investigating common such functions as medical services, and military intelligence, came up with options for the realignment and closure of bases (GAO, 2012). The Infrastructure Executive Council, a high-level body within DoD, next evaluated the collected set of options. The council approved a reduced set of recommendations and forwarded this list to the Secretary of Defense. The secretary analyzed these recommendations and submitted a list of over 200 recommendations to the BRAC Commission on May 13, 2005 (GAO, 2012).

The BRAC Commission was an independent government agency responsible for analyzing DoD recommendations related to the 2005 BRAC round. Nine commissioners managed a staff of over 100 with backgrounds in the military, various government agencies, law, and academia (GAO, 2013a). Congress has amended the original BRAC Act of 1990 at various times to limit the number of current and former DoD staff working for the commission, and to allow the commission to recommend closure of bases that DoD had not recommended be closed (Schlossberg, 2012). The commission collected comments from interested parties and expert witnesses before voting on accepting, modifying, rejecting, or adding BRAC recommendations. The BRAC Commission rejected 13 DoD recommendations outright, added five new recommendations, modified 58, and approved 119 recommendations without modification to report a final list of 182 recommendations to the President on September 8, 2005 (GAO, 2013a). The President next had the power to approve or disapprove of the recommendations, as a whole. The President approved of the recommendations on September 15, 2005. Congress next had the opportunity to cancel the BRAC round by passing a resolution disapproving of the BRAC Commission recommendations before 45 days had elapsed or before Congress adjourned. Congress did not pass the required resolution, and the recommendations became binding. The process is, and has always been, an "all-or-nothing" process in which neither Congress nor the President can modify individual BRAC recommendations. The goal is to prevent politics from adversely affecting the military reorganization process.

Next, DoD began implementation. DoD was required to provide Congress annual reports detailing recommendation-specific cost and savings estimate updates (GAO, 2013a). The implementation of BRAC 2005 recommendations was legally required to terminate on September 15, 2011, six years after the President approved of the recommendations. On this date, the latest BRAC Act expired.

Air Force Decisionmaking During BRAC 2005

The SecAF identified four goals specifically for the Air Force during the BRAC round of 2005 (Department of the Air Force, 2005):

- Transform by maximizing the warfighting capability of each squadron.
- Transform by realigning USAF infrastructure with future defense strategy.
- Maximize operational capability by eliminating excess physical capacity.
- Capitalize on opportunities for join activity.

In support of these goals, there was a general desire to consolidate resources and have larger-sized squadrons. The Air Force noted that optimal- and acceptable-sized squadrons for fighter aircraft would include 24 and 18 aircraft, respectively (Department of the Air Force, 2005). For bombers and large transport aircraft, a squadron should have 12 aircraft, while for tactical transport and tanker aircraft, 16 aircraft would be optimal and 12 acceptable (Department of the Air Force, 2005). The Air Force recommendations for BRAC 2005 involved reorganizing the force to move toward the desired sized squadrons. For example, the fraction of the C-130 fleet in squadrons of 12 or 16 aircraft was to go from 5 percent in 2006 to 83 percent in 2011 (Department of the Air Force, 2005).

The Air Force has identified 11 "Air Force basing principles" and five "Air Force basing imperatives" that were also important during the development of BRAC 2005 recommendations. These principles and imperatives are presented in Table B.2.

Table B.2
Air Force Basing Principles and Imperatives

Air Force Basing Principles

Maintain squadrons within operationally efficient proximity to DoD-controlled airspace, ranges, military operations areas, and low-level routes

Optimize the size of our squadrons—in terms of aircraft model, aircraft assigned, and crew ratios applied (e.g., same mission design series)

Retain enough capacity to base worldwide Air Force forces entirely within the United States and its territories

Retain aerial refueling bases in optimal proximity to their missions

Better meet the needs of the Air Force by maintaining/placing Air Reserve Component (ARC) units in locations that best meet the demographic and mission requirements unique to the ARC

Ensure joint basing realignment actions (when compared with the status quo) increase the military value of a function, or decrease the cost for the same military value of that function

Ensure long-range strike bases provide flexible strategic response and strategic force protection

Support the American Expeditionary Forces construct by keeping two geographically separate munitions sites

Retain enough surge capacity to support deployments, evacuations, and base repairs

Consolidate and/or co-locate older fleets

Ensure global mobility by retaining two air mobility bases and one additional wide-body capable base on each coast

Air Force Basing Imperatives

Ensure unimpeded access to polar and equatorial earth orbits

Preserve land-based strategic deterrent infrastructure as outlined by the Strategic Arms Reduction Treaty (START)

Ensure continuity of operations by maintaining airfield capabilities within the NCR to support the President of the United States, Special Airlift Missions, and foreign dignitary visits

Provide air sovereignty basing to meet the site protection and response time criteria stipulated by U.S. Northern Command and U.S. Pacific Command

Support global response by U.S. forces by keeping sufficient sovereign U.S. mobility bases along deployment routes to potential crisis areas

SOURCE: Department of the Air Force, 2005.

SAF/IE was responsible for developing recommended USAF closure and realignment actions. SAF/IE set up the Base Closure Executive Group (BCEG) to manage this process. The principal charged with carrying out necessary analyses was the Deputy Assistant Secretary for Basing and Infrastructure Analysis (SAF/IEB). A Joint Cross-Service Division was set up within SAF/IEB, responsible for coordination and communication with DoD joint cross-service groups while the Base Realignment and Closure Division was responsible for USAF-specific recommendations.

To develop its recommendations, the Air Force utilized three analytical tools developed specifically for the Air Force: the Web-based Installation Data Gathering and Entry Tool (WIDGET), the BRAC Analysis Tool, and the Air Force Cueing Tool. The first of these tools was a web-based system that stored relevant data. A senior Commission analyst leading the BRAC 2005 USAF analysis team has written an article highly critical of USAF decisionmaking during BRAC 2005. The article claims that much of the data loaded into WIDGET by the installations. in WIDGET was "erroneous or outdated," noting, for example, that an installation in Reno, Nevada, capable of supporting 12 C-130 aircraft was listed as capable of supporting only 10 (Flinn, 2006). There were, however, clear benefits from using WIDGET: Storing relevant data in one accessible location ensured that analyses were consistent and reduced the effort required to maintain and update data.

The BRAC analysis tool assigned several scores, known as mission compatability indices (MCIs), to each individual base based on how well the base was able to host missions of various types (fighter, bomber, tanker, etc.). The MCIs have been criticized as favoring active duty installations over reserve installations and for being too similar across installations, raising the possibility of "data discrepancies" unduly influencing rankings (Flinn, 2006). Here, again, there are benefits associated with the use of the Air Force's software tool: The BRAC Analysis tool ensured a consistent methodology, and data sources were used to translate data describing USAF installations into indices more directly useful for selecting bases to close and bases to realign.

The Air Force Cueing Tool used "goal programming" to select an optimal set of bases based on the MCIs from the BRAC Analysis Tool and other inputs (Department of the Air Force, 2005). This tool was critical for ensuring that the portfolio of installations the Air Force manages, taken as a whole, met certain strategic requirements. The Cueing Tool appears to be based on an algorithm solving a binary programming problem in which installations' contribution to the objective function is based on the MCI scores, and various constraints ensure that portfolio-wide strategic requirements are met.

The Air Force also utilized the Installation Visualization Tool and the Cost of Base Realignment Actions (COBRA) model, both of which are service-common software products. The Installation Visualization Tool gives the Air Force a consistent way to access imagery and geospatial data associated with installations. The COBRA model is used to estimate the costs and benefits of base realignment and closure actions. The COBRA model was developed in 1988 by the Air Force Cost Center and produced all the cost estimates used by the various branches of the military for the initial 1988 BRAC round. The COBRA model has since been used in each BRAC round and has been improved several times through the work of a COBRA Joint Process Action Team within DoD (GAO, 2005).

The Air Force also undertook a capacity analysis of its installations. Factors that would prevent bases from adding force structure were identified, along with excess space on bases.

The optimal portfolio of bases identified by the Air Force Cueing Tool and the results of the capacity analysis were used as a starting point for BCEG deliberation. The BCEG analyzed various scenarios proposed by the Base Realignment and Closure Division and the Joint Cross-Service Division in the context of the Cueing Tool results, the capacity analysis results, base-mission scores, and the USAF basing principles and imperatives.

The last three criteria for the 2005 BRAC round identified by the Secretary of Defense and listed in Table B.1, covering the economic, community, and environmental impacts of BRAC actions, were treated somewhat separately from the other criteria. Three- and four-page narratives were written for each installation, covering environmental concerns and community infrastructure (Department of the Air Force, 2005). One- and two-page datasheets were also created for each impacted installation for each recommended action (Department of the Air Force, 2005). Finally, one- to two-page summary sheets were devised for each installation to cover the cumulative environmental impacts of all BRAC recommendations (Department of the Air Force, 2005).

USAF recommendations for BRAC 2005 affected 115 installations, 76 percent of all USAF installations (Department of the Air Force, 2005). The Air Force recommended closing ten Air Force installations, including seven reserve bases (Department of the Air Force, 2005). The agency also proposed a 20 percent reduction in the number of installations with operational flying missions. By its own calculations, the Air Force was to reduce unnecessary infrastructure by 79 percent, with remaining excess kept for surge and unforeseen future missions (Department of the Air Force, 2005). The Air Force improved opportunities to host or support joint activities with other branches of the military. For example, the Air Force established a joint initial training location for the Joint Strike Fighter at Eglin AFB, near available training airspace in the Gulf of Mexico.

U.S. Government Accountability Office Criticisms of BRAC 2005

GAO was requested by Congress and subsequently mandated by law to objectively observe and report on BRAC decisionmaking processes. As recommendations from BRAC commissions are approved, GAO also must review implementation actions. More recently, the House Armed Services Committee asked GAO to investigate BRAC 2005 and develop recommendations for improving future BRAC rounds (GAO, 2013a).

GAO has repeatedly noted that DoD used a reasonable, logical process for arriving at BRAC 2005 recommendations. GAO has also stressed that DoD uses data that are "sufficiently reliable" for cost and benefit estimation, based on its independent analysis of DoD data, DoD documents, and meetings with Office of the Secretary of Defense Basing Directorate personnel. GAO has endorsed the COBRA model as a "reasonable estimator" for developing cost and benefit figures (GAO, 2013a).

The organization has, however, noted that having three goals for the 2005 BRAC, instead of the single cost-minimizing goal of prior BRAC rounds, made it more difficult for the BRAC Commission and DoD to select and implement recommendations. GAO also has drawn attention to the fact that the scale of the BRAC round of 2005 was relatively large. There were 813 total basing actions in BRAC 2005, compared with 43, 75, 163, and 106 for each of the previous BRAC rounds (GAO, 2012). Sixty percent of the actions were related to the Army National Guard and Army Reserve, a major change from prior BRAC (GAO, 2010c). According to GAO, estimating the impact of the BRAC round on bases that were to simultaneously lose and gain missions was challenging (GAO, 2010c). Some recommendations were contingent on the outcomes of external processes. For instance, Canon AFB was to be closed if the Secretary of Defense could not find a new mission for the base by the end of 2009 (GAO, 2013a). GAO believes that contingency clauses added to the complexity of overseeing BRAC recommendations.

The transformative and/or interdependent of certain recommendations also added to the complexity. For example, the Defense Logistics Agency was created and tasked with managing logistics activities throughout the military. This recommendation required high-level coordination across different branches of the military. In at least two instances, implementation of BRAC recommendations was delayed because personnel were scheduled to transfer to new facilities that were being occupied by other personnel delayed in their own transfer (GAO, 2012). Some recommendations involved many "bundled" base closures and realignments. GAO claimed these "bundled" recommendations lacked action-specific cost and benefit estimates, leading to their objection (GAO, 2013a). In a response to GAO comments, DoD noted its belief that "bundling" several actions when the actions have the same mission and purpose is warranted without being problematic (DoD comments included in GAO, 2013a). DoD noted that it provided the BRAC Commission with COBRA model outputs including action-specific cost and benefit estimates.

Particular problems were caused by construction projects and by information technology requirements. It was difficult to place personnel movements and military construction on a complementary schedule. In 2010, GAO noted that DoD was planning to complete 57 different construction projects required by 30 BRAC recommendations within three months of the legal deadline for BRAC-related projects to finish (GAO, 2010c). That schedule left little room for slippage in project schedules due to unforeseen delays, and also made the movement of personnel before the legal deadline challenging. The Army used "swing space facilities" to temporarily house personnel, but GAO noted that there was, at times, a failure to report these as costs related to BRAC actions (GAO, 2010c). One incorrect assumption that a space being built for the National Geospatial-Intelligence Agency would be a "general administrative" space rather than a more secure type of space resulted in a $726 million increase in construction costs (GAO, 2013a). The information technology costs associated with the creation of the Defense Logistics Agency increased from $30.9 million to $190 million (GAO, 2013a).

Another challenge GAO highlighted is the loss of human capital that results when civilian employees working on a mission choose not to relocate with their mission. The Army recommended moving the headquarters of the Army Material Command, the U.S. Army Forces Command, the Training and Doctrine Command, the U.S. Army Reserve Command, and the First Army, causing some officials to worry about potential loss of Headquarters staff (GAO, 2010c). Convincing technical staff to move to, and hiring technical staff at, Aberdeen Proving Ground, Naval Air Station China Lake, and Fort Sam Houston proved difficult (GAO, 2010c).

GAO also noted that BRAC recommendations calling for installation growth led to problems in certain communities, particularly with regard to transportation systems. Table B.3 lists installations scheduled to grow substantially as a result of BRAC actions.

Table B.3
Estimated Growth from All DoD Sources at and Near BRAC-Affected Military Bases
FY 2006 Through 2012, as of March 2008

Base	Total Change in Military and Civilian DoD Population	Total Change in Population of Military and Civilian DoD Dependents	Total Population Increase	Current Total Regional Population
Aberdeen Proving Ground, MD	3,400	2,200	5,600	2,512,000
Bethesda National Naval Medical Center, MD	2,500	Not available	2,500	4,331,000
Camp Lejeune, Cherry Point, and New River, NC	13,400	18,700	32,100	108,000
Eglin Air Force Base, FL	3,600	5,900	9,500	190,000
Ford Belvoir, VA	24,100	12,700	36,800	4,331,000
Fort Benning, GA	12,700	6,100	18,800	247,000
Fort Bliss, TX	28,000	41,700	69,700	722,000
Fort Bragg, NC	18,900	17,100	36,000	301,000
Fort Carson, CO	10,400	14,400	24,800	514,000
Fort Knox, KY	(2,900)	4,500	1,600	117,000
Fort Lee, VA	10,200	4,600	14,800	138,000
Fort Lewis, WA	13,500	17,400	30,900	3,422,000
Fort Meade, MD	7,000	4,200	11,200	2,512,000
Fort Sam Houston, TX	10,900	6,100	17,000	1,416,000
Fort Sill, OK	3,700	(400)	3,300	81,000
Fort Riley, KS	10,900	15,000	25,900	109,000
Marine Corps Base Quantico, VA	3,600	1,000	4,600	202,000
Redstone Arsenal, AL	7,900	2,000	9,900	291,000
Total	**181,800**	**173,200**	**355,000**	

SOURCE: GAO, 2009a.

A DoD study of the National Naval Medical Center in Bethesda, Maryland, found that 15 out of 27 nearby intersections would experience deteriorating service levels during peak periods as a result of BRAC actions (GAO, 2009a). Many of these intersections would experience a "failing" service level using Transportation Research Board criteria. The Office of Economic Adjustment bears the primary responsibility for helping communities impacted by BRAC actions. GAO noted that the Office of Economic Adjustment "is not at an appropriate organizational level within DoD" to bear this responsibility, since there is a need to gather assistance from a diverse set of federal and other government agencies. The Defense Access Roads program is authorized to fund highway improvements related to BRAC actions. Unfortunately, this program has narrow eligibility criteria and, as of 2009, had only funded work near three installations suffering from BRAC-induced traffic growth (GAO, 2009a). The situation was especially problematic in Maryland, where Aberdeen Proving Grounds, Fort Meade, and the Bethesda National Naval Medical Center were scheduled to add over 12,000 personnel in total. The state of Maryland set aside almost $100 million for intersection improvements around the three installations, but noted that as much as $470 million more may be needed to complete the projects (GAO, 2009a).

GAO has pointed out that costs to implement BRAC Commission recommendations grew over time, while estimates of cost savings shrunk. The original estimated cost of recommended changes was $21 billion (in 2005 dollars), but six years later the estimated cost had increased 53 percent, to $32.3 billion (in 2005 dollars) (GAO, 2012). It is worth noting that the Air Force was the one armed service that saw a decline in BRAC implementation costs between FY 2009 and FY 2010 (GAO, 2009b). Cost increases in this period were driven by construction costs linked to ten specific recommendations shown in Table B.4 (from GAO, 2009b). Military construction costs were responsible for most of the cost increases for seven out of the ten recommendations listed in Table B.4. Indeed, military construction cost estimates increased $1.864 billion from FY 2009 to FY 2010, while the overall costs of the BRAC round increased $2.488 billion (GAO, 2009b).

Table B.4
BRAC Recommendations with the Largest Increases in One-Time Estimated Costs
from FY 2009 to FY 2010 ($ millions)

	FY 2009 Cost Estimate	FY 2010 Cost Estimate	Net Cost Increase
Recommendation			
Realign Walter Reed Medical Center to Bethesda National Naval Medical Center, MD, and to Fort Belvoir, VA	$1,640	$2,418	$779
Realign Army Maneuver Training to Fort Benning, GA	1,509	1,763	254
Co-locate miscellaneous Office of the Secretary of Defense, defense agency, and field activity leased locations in the National Capital Region	1,194	1,440	245
Close Fort Monmouth, NJ	1,595	1,751	156
Establish San Antonio Regional Medical Center and realign enlisted medical training to Fort Sam Houston, TX	1,724	1,876	152
Realign to establish Combat Service Support Center at Fort Lee, VA	1,270	1,418	148
Relocate medical command headquarters in the National Capital Region	43	161	118
Close National Geospatial-Intelligence Agency leased locations and realign others at Fort Belvoir, VA	2,441	2,554	113
Close Fort Gillem, GA	101	160	59
Relocate Army headquarters and field operating activities in the National Capital Region	444	490	47
Total one-time estimated costs from the BRAC recommendations listed above	**$11,961**	**$14,031**	**$2,071**
Total one-time estimated costs for all recommendations	**$32,433**	**$34,922**	**$2,488**
Percentage of increase in one-time costs from recommendations listed above of all recommendations			**83%**

SOURCE: GAO, 2009b.

While expected costs increased, expected future savings decreased for the BRAC round of 2005. Estimated savings in operating expenses, for one year and over a 20-year period, went from $4.2 billion and $36 billion, respectively, to revised estimates of $3.8 billion and $9.9 billion (GAO, 2012). The end result, according to GAO, is that the costs of the changes required by the 2005 BRAC round will not be recouped until 2018. Recall that to kick off the BRAC process, the Secretary of Defense had been required to certify that the process would yield annual net savings for each department of the military beginning no later than 2011. The largest decrease in estimated savings came from the recommendation to centralize medical command headquarters in the National Capital Region. DoD's FY 2010 budget indicated that this recommendation would result in no savings but rather a net cost of $1 million per year (GAO, 2009b). Indeed, using the FY 2010 budget, GAO estimates that 76 BRAC recommendations will yield no net savings in the first 20 years after implementation (GAO, 2009b).

DoD claims that cost growth during BRAC 2005 implementation was due to "implementation investment decisions" and to "congressional direction" rather than failures in DoD processes (DoD comments included in GAO, 2013a). DoD will update the "standard factor" used to estimate information technology costs within the COBRA model and will improve COBRA user manuals to "emphasize the importance of footnoting the source documentation for personnel changes" (DoD comments included in GAO, 2013a).

In addition to the above points, estimates of the recurring savings to DoD are often based on the costs of military personnel whose positions are being eliminated even when end-strength numbers are not expected to change (GAO, 2009b). Approximately half of the initially projected recurring savings came from eliminating positions (GAO, 2007). GAO has noted its objection to this practice. GAO also recommends that DoD develop and use metrics for performance tracking during the next BRAC round, and that Congress require the Secretary of Defense to fix "targets" that DoD will achieve (GAO, 2013a). DoD objects to this recommendation, noting "The premise that we should be required to close a particular number of bases or eliminate a particular number of civilian jobs is arbitrary, counterproductive, and would undermine military capability" (Conger, 2013).

Appendix C. Generalized Boosted Model

This appendix provides additional detail on the Generalized Boosted Model (GBM) (Ridgeway, 2009) used in the main text to assess how well the weight assigned by the Air Force to each scoring category (e.g., mission, capacity) aligned with the actual influence of these categories on the choice of candidate bases. The decision to use GBM for this analysis was influenced by its successful application in Keller et al. (2014), which examined how the Weighted Airman Promotion System was functioning by using GBM to model promotion outcomes by AFSC.

To estimate the relative influence of covariates, it is necessary to first fit a GBM model of the outcome of interest as a function of the available covariates. In this context, the outcome of interest was whether each site was selected versus not, and the covariates are capacity, cost, environmental, and mission scores. The model-fitting process starts very simply, estimating equal probabilities of being selected for each observation in the analytic file. GBM then proceeds in an iterative fashion, gradually adding complexity to the model. For each iteration, the model adds or subtracts a small, constant value to the estimated probability of being selected for those bases whose scores fall above or below particular thresholds on one or two scores. Over hundreds or thousands of iterations of making such small changes to the predictions, GBM is able to recreate complex associations between the covariates and the probabilities associated with the outcomes of interest.

In this analysis, we used GBM to model the probability of a base being selected as a candidate for a basing decision (versus not being selected). GBM was used to calculate the probability through piecewise constant combinations of the explanatory variables (e.g., mission score and capacity score). For example, the first iteration may estimate that bases scoring below 90 for mission and below 15 in capacity have a slightly lower chance of being selected as a candidate for the basing action than all other bases. In this case, the model is said to "split" on mission and capacity. At the next iteration, the model may increase the probability of bases scoring 10 in environment becoming candidates. The model continues making thousands of such adjustments, each time adding or subtracting a small, constant value from the estimated probability of being selected for bases above or below the estimated thresholds.

If the GBM algorithm ran long enough, it would be able to fit the observed data perfectly— resulting in poor predictive performance. GBM uses two techniques to avoid such overfitting. First, to ensure changes to the probabilities are spread over the important variables, each iteration's adjustments to the probabilities are smaller than those that would optimize the fit. Second, GBM chooses an early stopping iteration in the algorithm using a method called ten-fold cross valuation. Cross valuation results in a simpler model than if GBM were allowed to run until it achieved a perfect fit. In ten-fold cross valuation, the dataset is randomly split into ten mutually exclusive subsets. The model is fit on nine-tenths of the data, and that model is used to predict the outcomes in the held-out tenth of the data; this process is repeated to produce out-of-sample predictions for all observations. The number of iterations that results in the best out-of-sample prediction is then used for the final GBM model that is fit to the entire dataset.

Key for this application, an additional output of GBM software is the "relative influence" of each explanatory variable on the outcome. Out of all of the variation explained by the model, the relative influence is the percentage that is attributable to each explanatory variable. This measure is insensitive to rescaling the explanatory variables, which makes it particularly appealing for this application.

To calculate the relative influence of each covariate, the software counts the percentage of splits that involve each covariate and weights this sum by the squared improvement at each step. The relative influence is standardized so that each covariate's relative influence is between zero and 100 percent and the relative influences sum to 100 percent (Friedman, 2001; Elith, Leathwick, and Hastie., 2008)

In Chapter Three, we discuss the results of our analysis for a sample of basing actions. Here we will walk through an example using the basing action to place a single squadron of F-35s. The Air Force considered 205 bases as potential locations for a squadron of F-35s. Each base was graded on its ability to support the mission (60 points), its existing capacity (25 points), cost factors (5 points), and the potential impact on the environment (10 points). We discuss these covariates in Chapter Two.

We wanted to know the relative influence of each scoring category on a base being selected for a site visit. We gave the GBM software a list of the 205 bases; their mission, capacity, cost, and environmental scores; and a field indicating if the base was selected for a site visit or not. GBM took the list of bases and randomly divided it into ten groups, each with 20 or 21 bases. GBM built a separate prediction model for each group so the model built using 184 or 185 bases' information can be assessed on the observed outcomes for the remaining 20 or 21 bases. From this exercise, GBM estimates an appropriate number of iterations or, equivalently, the appropriate model complexity. At this point, the final GBM model can be fit to produce the corresponding relative influence estimates.

For the basing action placing a single squadron of F-35s, the final GBM model estimates relative influences for capacity, mission, cost, and environment were 74 percent, 11 percent, 8 percent, and 7 percent, respectively. From these numbers, we know that the GBM model either splits most often on the capacity score, or, when it does, the model achieves relatively large improvements in model fit (or likely both). We conclude that capacity scores are substantially more important in predicting which bases were selected, even though mission accounts for more than twice as many points in the original measurement scale.

Readers interested in a technical description of GBM are encouraged to refer to Ridgeway (2009) and Section 8.1 of Friedman (2001).

Appendix D: The Role of Strategy in Basing Decisions at Home and Abroad

DoD as a whole and the USAF in particular tend to approach basing decisions in CONUS and OCONUS as completely separate processes. In part, this may be due to the widespread assumption that one is more strategically driven than the other. Specifically, the conventional assumption is that overseas bases are chosen and maintained principally for their strategic and diplomatic value, while domestic bases are chosen and maintained because of mission, capacity, cost, environmental, and political considerations. Consequently, the CONUS basing process is usually coupled with the impact that base closures could have on their surrounding communities and the efforts of local and state officials to keep bases in their districts open.

For example, the USAF's proposal to consolidate the three Alaska-based fighter squadrons at Joint Base Elmendorf-Richardson in Anchorage was met with fierce resistance by local business leaders and politicians. The logic behind USAF's proposal was that moving the F-16 Aggressor squadron currently stationed at Eielson AFB near Fairbanks to Joint Base Elmendorf-Richardson would realize operational efficiencies and thereby decrease operating costs without reducing USAF's capability in the region (USAF, 2013, p. ES-2). The Fairbanks community, however, strongly objected to the plan because Eielson would lose its permanent aircraft and be downgraded to a "warm" base—one where the base remains open but with reduced staff, systems, and capabilities. This is in contrast to a "cold" base, which is closed temporarily or permanently through the BRAC process. The community claimed downgrading to a warm base would damage the local economy by raising the level of unemployment and significantly reducing the value of local homes. Opponents also argued that the Air Force had not considered all of the costs associated with the proposal, the congested airspace in Anchorage, or the fact that basing all fighter aircraft in one location would increase USAF vulnerability ("Murkowski Gets Transparency Pledge from Next USAF Chief of Staff," 2012). The Alaska delegation succeeded in delaying the move for a year, and, in October 2013, the Air Force announced that it was abandoning the plan because of lower-than-projected savings and the impact that the move would have on the local communities near Eielson (USAF, 2013a, 2013b).

By contrast, the debate over the U.S. overseas military presence centers on how DoD needs to alter its posture to better meet emerging threats in a period of fiscal austerity. For instance, Michael O'Hanlon and Bruce Riedel argue that the United States should permanently station 150 land-based fighter aircraft in the Persian Gulf to reduce the nation's dependence on expensive aircraft carriers (O'Hanlon and Riedel, 2013). Others have challenged this proposal on the grounds that it restricts USAF flexibility by tying down a large number of its combat aircraft and increases the probability that DoD will encounter access problems with its partners in the Gulf (Pettyjohn and Montgomery, 2013). According to this view, a mixed presence of ground- and sea-based combat aircraft is the appropriate posture. Similarly, China's rapidly expanding military capabilities and its more assertive behavior prompted DoD to announce that it was pivoting (or rebalancing) toward the Asia-Pacific region (Clinton, 2011; DoD, 2012). As a part of this initiative, DoD is seeking additional access in Southeast Asia and Oceania to improve the resiliency of its posture. Debates surrounding this policy have centered on whether the components of the pivot represent a significant change to the current posture; the sustainability of this shift; its effect on other critical regions, like the Middle East; and the risks of provoking an insecure China (Manyin et al., 2012; Ross, 2012; Saab, 2013; Sutter et al., 2013).

These examples clearly reflect the conventional view that domestic political concerns figure prominently in CONUS basing decisions, while strategic factors are paramount in OCONUS basing decisions. Existing studies tend to focus on either domestic or international bases, or on only one type of basing decision, especially closure (Sturm, 1969; Benson, 1982; Brown, 1990; Sorenson, 1998; Goren, 2003; Shaw, 2004; Sorenson, 2007).

Types of Strategic Influences on Air Force Basing Decisions

According to Edward Mead Earle, "Strategy is the art of controlling and utilizing the resources of a nation—or a coalition of nations—including its armed forces, to the end that its vital interests shall be effectively promoted and secured against enemies, actual, potential or merely presumed" (quoted in Kennedy, 1992, p. 2). If basing decisions were driven solely by national strategy, therefore, the Air Force would simply select the course of action—whether opening a new base, realigning bases, or closing a base—that is expected to best achieve its objective of enhancing U.S. security. In other words, basing decisions would be made through a rational process that aligned ends, means, and costs.

Strategic influences, however, may have different time horizons. On the one hand, there are short- or immediate-term strategic factors that are focused on preparing for present security challenges. Considerations such as a base's relevance for countering a current threat, concerns about the survivability of a facility, and an administration's preferred defense strategy fall under this rubric. On the other hand, there are long-term strategic factors that are focused on preparing for a distant and unknown future. Farsighted strategic considerations try to hedge against future uncertainty by ensuring that USAF base structure remains capable of dealing with unspecified future security challenges. Factors that fall within this category include ensuring that the Air Force maintains a flexible posture, protects unique facilities and assets that are difficult to recreate, and prioritizes bases that are best capable of supporting a particular mission.

Short-Term Strategic Influences

Relevance of Base for Countering Current Security Challenges

One of the most important strategic influences on USAF basing decisions is the relevance of a particular facility for countering current security challenges. An airbase's utility for dealing with a particular threat is primarily a function of its geographic location. Because an airbase is the center of a plane's radius of flight, a base's location and the range of its aircraft determine what is within reach (Brown, 1990, p. 2; Warnock, 2004, p. 6). In short, aircraft need to be stationed at bases that are within the range of where they are expected to operate, or they will have to deploy from their home stations to forward operating locations. Accordingly, if current threats recede and new threats appear in different regions, some bases may become obsolete and new requirements may arise. Airbases that are not well situated to deal with current and emerging security challenges are often considered for closure, or they may be repurposed.

Geography—in particular, proximity to the anticipated operating area—has been especially important when deciding where to establish a new base. During World War I, for instance, the U.S. Army first developed a large number of CONUS airfields to facilitate the deployment of U.S. forces to Europe. Consequently, the vast majority of the airbases constructed during the World War I were situated East of the Mississippi River or along the Gulf of Mexico. U.S. troops would deploy from these inland locations to two clusters of facilities, on Long Island and near Washington, D.C., which were used as embarkation points to Europe (Brown, 1990, p. 48).

World War I–era Air Service bases were positioned to fight an expeditionary war against an adversary that posed no threat to the homeland. Consequently, throughout most of the interwar period there were few CONUS bases that were capable of supporting an air defense mission (Futrell, 1951, pp. 1–2). This began to change in the lead up to World War II, as U.S. officials became increasingly concerned about the threat posed by Nazi Germany and Japan. In the late 1930s, growing concerns about the destructive potential of airpower led President Franklin Delano Roosevelt to conclude that the relative geographic isolation of the United States could no longer provide sufficient protection (Reynolds, 2001, pp. 93–94). Consequently, Roosevelt asked Congress to fund measures that would enable the War Department to defend the nation and the entire Western Hemisphere from aggressors (Conn and Fairchild, 1989, p. 3; Conn, Engelman, and Fairchild, 1989).

As a part of this initiative, in 1940 the Army Air Corps (AAC) developed two air bridges—one running through the West Indies to South America and the other through Texas and Central America—composed of land and seaplane bases stretching from CONUS to Brazil. In total, the AAC built 48 airbases along these routes because of the relatively short range of the aircraft in its inventory (Conn and Fairchild, 1989, pp. 249–251; Pettyjohn, 2012, pp. 46–47).[1] Because

---

[1] The Airport Development Plan was pursued secretly because of anticipated opposition from Latin American nations. For this reason, the Air Force relied on Pan American Airways to improve the airfields under the guise of

U.S. military planners intended to use these bases to quickly deploy personnel and equipment to defend Brazil against a possible Axis attack, they selected sites located near the coasts to ease the task of supporting forces.

In an effort to bolster U.S. continental defenses, the AAC established a number of air defense bases situated near key industrial centers, such as New England. In general, therefore, CONUS air defense bases required peripheral locations near the likely avenues of an enemy's attack (USAF, 1963, p. 11). For many years this was primarily along the coastlines, but with the advent of long-range bombers that could navigate polar air routes during the early years of the Cold War, the focus shifted to the United States' northern borders (Haulman, 2004, p. 71).

Even though improvements in military technologies have extended the reach of many of USAF aircraft, AMC still uses the 3,500 nautical mile point-to-point range of its primary airlifter—the C-17—to determine its Atlantic en route basing needs today.[2] AMC draws a 3,500 nautical mile range arc representing the distance that a C-17 could fly from its east coast CONUS bases, and then draws a second 3,500 nautical mile range arc displaying the distance that a C-17 could fly from a destination in Southwest Asia. The area where these two circles overlap, which can be reached from CONUS and Southwest Asian bases, is the "sweet spot" or "lens" where AMC requires en route bases (AMC, 2010, pp. 5–6).

Vulnerability and Survivability of Base

In addition to a base's distance from anticipated operating locations, USAF officials consider whether a base is vulnerable to attack by an adversary. Combatants first recognized that airbases were inviting targets during World War I and began to experiment with a host of defensive measures to reduce their exposure.[3] Active and passive defenses increase an airbase's ability to resist an attack and the likelihood that forces stationed at the facility will survive. Yet deciding where an airbase is established—especially whether it is within or beyond the range of an adversary's weapons—is one of the most important ways of reducing vulnerability. Moreover, even in cases where an adversary's weapons can target an airbase, its location still influences warning time and therefore survivability. With early notice of an incoming attack, the aircraft at a base can disperse, increasing the probability that they escape unscathed.

During the Cold War, the Air Force in general, and the Strategic Air Command (SAC) in particular, worried about whether its forces could survive a preemptive strike by the Soviet Union. For example, after the Soviet Union successfully detonated an atomic bomb in August

---

expanding civil aviation needs on its behalf (Weathers, 1943, pp. 124–151).

[2] AMC uses different planning factors for different regions to account for variation in the quality of infrastructure and the geography. For example, in Africa there is a dearth of airfields capable of handling USAF strategic airlifters and a lack of quality aviation fuel. Because of these constraints, AMC uses the unrefueled range of a C-17 as its Africa planning factor, meaning that a C-17 can only travel 2,000 nautical miles, offload its cargo with its engine running, and then return to its original point of embarkation without refueling (AMC, 2010, p. 4).

[3] For more see Kreis, 1988, pp. 3–22.

1949, SAC decided to relocate its forces stationed in the United Kingdom (UK) from East Anglia (where they had operated from since 1948) to the British midlands northwest of London (where a string of air defense bases could protect the bombers).[4] Similarly, in 1952, U.S. Air Forces in Europe (USAFE) reached an agreement with France to relocate its tactical fighter bases from the U.S. occupation zone in West Germany to bases in the French occupation zone, which were west of the Rhine River. Given the proximity of the U.S. sector to the central front, the first USAFE fighter airbases (Furstenfeldbruck, Neubiberg, and Giebelstadt) were likely to be quickly overrun in the event of a Soviet invasion. By contrast, the French airbases (Bitburg, Landstuhl, Spangdahlem, Hahn, and Sembach) offered more protection against an attack but were still close enough to the front lines to allow USAFE's short-range tactical aircraft to support ground operations on the central front (Sturm, 1969, p. 16; Benson, 1981, p. 34).[5]

Overseas bases, however, were not the only ones vulnerable to Soviet attack. By the late 1950s, the Air Force was extremely concerned about the survivability of its CONUS bomber bases, which were overcrowded because the pace of aircraft construction had exceeded that of base construction. A 1956 USAF study on base vulnerability argued that "the reduction of SAC's ground vulnerability will remain for several years the most important single military step that can be taken to deter the sudden outbreak of general war, i.e. a surprise air attack on the United States" (Leighton, 2001, p. 447). This statement resonated with broader audiences who recognized that, at that time, the United States' nuclear deterrent relied entirely on SAC's bombers, and therefore these audiences supported efforts to improve bomber bases' survivability, including expanding radar coverage, increased active defenses, and scattering SAC's bombers across many bases.[6] The initial logic behind dispersal was straightforward: The USAF would proliferate the number of targets in the CONUS so that they exceeded the number of Soviet warheads (Haulman, 2004, p. 62). In 1955, therefore, the Air Force planned to expand from 34 to 55 bases (Leighton, 2001, pp. 446–447).

It quickly became apparent that there was no simple way of increasing the survivability of the strategic bomber force and that dispersal alone was insufficient. As SAC tried to reduce its vulnerability through a combination of active and passive defenses, the Soviets were simultaneously developing faster and longer-range bombers, which would reduce the warning time for two-thirds of SAC's bases to less an hour. Consequently, the Air Force began to move SAC toward the interior and southern part of CONUS.

Once the Soviets were on the verge of developing an intercontinental ballistic missile (ICBM), the severity of the threat to SAC's bases increased considerably. Because of the speed of missiles, SAC expected to have only 15 minutes of warning before a Soviet attack would

---

[4] The original four bases in the UK were RAF Scampton, RAF Marham, RAF Waddington, RAF Lakenheath. New bases are Fairford, Greenham Common, Brize Norton, and Upper Heyford Charles Hildreth (USAFE, 1967). See also Sturm, 1969, pp. 8–9; Benson, 1982, pp. 13–14.

[5] French bases had one significant drawback: the weather. They were plagued by frequent fog.

[6] For instance, the Killian panel in February 1955 urged dispersal program (Leighton, 2001, p. 445).

strike its CONUS targets. In the mid-1950s, SAC's response was to create an alert program that enabled some bombers to be airborne before an ICBM could reach their base because they had their weapons loaded and crews on standby for immediate takeoff (Hopkins, 1986; Leighton, 2001, p. 635). In a 1956 white paper, General Curtis LeMay proposed a number of measures to ensure the survivability of the bomber force, including increasing the number of SAC bomber bases to 101 (Leighton, 2001, pp. 635–636). The cost of establishing and maintaining so many bases was exorbitant, however, and prevented SAC from ever realizing its extensive dispersal plans. SAC's CONUS base network reached its pinnacle in 1960, with 46 bases scattered across the United States (Hopkins and Goldberg, 1986, p. 89; Haulman, 2004, p. 65).

The Kennedy and Johnson administrations adopted a different approach to dealing with SAC's growing vulnerability. According to Secretary of Defense Robert McNamara, "The introduction of ballistic missiles is already exerting a major impact on the size, composition, and deployment of the manned bomber force, and this impact will become greater in the years ahead." Therefore, "as the number of . . . ballistic missiles increases, requirements for strategic aircraft will be gradually reduced." Because "the growing enemy missile capability will make grounded aircraft more vulnerable to sudden attack," SAC adopted a rearward basing concept to maximize its warning time (Hopkins and Goldberg, 1986, p. 105). SAC's B-52s, therefore, were moved from northern bases, such as Dow in Maine, Glasgow in Montana, Larson in Washington, and Lincoln in Nebraska to southern bases, including Barksdale in Louisiana, Carswell in Texas, and March in Southern California (Marion, 2004, pp. 107–110). As a more affordable alternative to wholesale dispersal, in 1969 SAC began deploying its alert force to satellite bases—CONUS bases that were not owned by SAC—to complicate the Soviet's targeting and reduce the time required to launch the bombers (Headquarters Strategic Air Command, Office of the Historian, 1970, p. 19).

Concerns about vulnerability are not limited to the Cold War. Today, for example, China's increased ability to hold forward-based USAF forces at risk has led DoD to seek a more "geographically distributed" posture in the Asia-Pacific region (Gates, 2010). For the Air Force, this means seeking additional access for rotational forces in Australia and Southeast Asia (Defense Writers Group, 2013).

Preferred Defense Strategy of Administration in Office

Although U.S. officials often have the same goal, namely protecting U.S. security and interests, they may hold different views on how the military can best achieve these ends. Consequently, an additional strategic factor that influences basing decisions is the administration's preferred defense strategy, or "the link between military and political ends, the scheme for how to make one produce the other" (Betts, 2000, p. 5). Because a defense strategy may prioritize certain threats, military capabilities, or specific regions, the particular strategy adopted by an administration affects the shape of USAF posture.

During the Cold War, for instance, all U.S. presidents agreed that the Soviet Union posed a threat to the United States and therefore must be contained. Yet the Truman, Eisenhower, and Kennedy administrations differed in how they used U.S. forces to deter Soviet aggression. After facing Soviet-backed communist aggression on the Korean Peninsula, the Truman administration adopted an expansive strategy (articulated in NSC-68) of countering communist aggression across the globe. This included rolling back communist expansion in Korea while deterring the outbreak of a general war in Europe. The adoption of NSC-68 led to a dramatic expansion of USAF's network of bases at home and abroad. Because of the magnitude of the threat that the United States faced, NSC-68 called for comprehensive rearmament. In other words, NSC-68 did not prioritize one mission or subset of missions over the others. For the Air Force, this meant that while the strategic nuclear mission remained the foundation of the United States' deterrent, nuclear weapons alone were not believed to be capable of winning a protracted war against the Soviet Union (Millett and Maslowski, 1994, p. 516). Instead, the Truman administration believed that the United States also needed formidable conventional forces, including large numbers of tactical aircraft (Lemmer, 1974, p. 141; Millett and Maslowski, 1994, pp. 517–518).

President Dwight D. Eisenhower questioned the assumption that underlined the Truman administration's vast rearmament program, namely that resources would expand to meet security requirements. Consequently, the Eisenhower administration's New Look strategy stressed that containment needed to be affordable, because a strong free market economy was the foundation of the United States' military strength (Huntington, 1961, pp. 64–84; Gaddis, 2005, pp. 130–132). Eisenhower feared that unrestrained defense spending would undermine the U.S. economy and maintained that the United States needed "security with solvency" (Millett and Maslowski, 1994, p. 534). The New Look strategy, therefore, emphasized the deterrent value of nuclear weapons and continental defenses, while deemphasizing conventional forces (Huntington, 1961, p. 78). According to this logic, deterrence rested primarily on the credible threat to massively retaliate to any aggression with nuclear weapons coupled with strengthened air defenses. Because of this focus on nuclear retaliation, SAC and to a lesser extent Air Defense Command received the preponderance of the USAF's budget and base structure. To reduce military expenditures while expanding nuclear and air defense forces, the Eisenhower administration cut the budget of other USAF commands, also leading to reductions in their base structure (Haulman, 2004, p. 75).

This changed when John F. Kennedy entered the White House in 1961. Kennedy and his Secretary of Defense, Robert McNamara, believed that there were many circumstances where it was not appropriate to employ nuclear forces and, therefore, that the Eisenhower administration's threat of massive retaliation lacked credibility (Kaplan, Landa, and Drea, 2006, pp. 293–294). Believing that a strong deterrent rested on the threat of a calibrated and proportional response, the Kennedy administration shifted to a defense strategy of flexible response (Millett and Maslowski, 1994, p. 553; Gaddis, 2005, pp. 214–215). Flexible response's emphasis on conventional forces was reflected in how the budget was allocated among the USAF commands. The Kennedy administration ended SAC's ascendency and insisted that the command rely more heavily on missiles instead of bombers. As a result, SAC's vast network of CONUS bases declined by nearly 40 percent, while mobility air forces, which were needed to deploy ground forces, and tactical air forces, which supported ground units, gained new bases.[7] For instance between 1961 and 1972 Tactical Air Command gained nine additional CONUS bases (Marion, 2004, p. 122). The Military Air Transport Service and its successor Military Airlift Command obtained only two additional bases between 1961 and 1974, but the Air Force's airlift capacity significantly expanded due to the development of larger, more capable intercontinental airlifters (Marion, 2004, p. 103; Millett and Maslowski, 1994, p. 560).

Long-Term Strategic Influences

Maintain a Flexible Posture

Because the international environment can rapidly and unexpectedly change, a strategic factor that influences USAF basing decisions is a desire to preserve flexibility. In other words, the Air Force wants to maintain a base structure that can adapt to unforeseen demands. Flexibility is important because making changes to the Air Force's posture at home and abroad is a time-consuming and expensive process (Calder, 2007, p. 35; Schlossberg, 2012, pp. 3–5). At home, the Air Force must engage in lengthy and comprehensive environmental studies and secure resources from Congress for military construction. The overseas process is even more complicated because the United States must also negotiate with a host nation and navigate its domestic political processes. Flexibility may involve a number of factors, including an expandable base structure, protecting difficult to reconstitute assets, geographic balance, and prioritizing bases that provide reach into multiple regions.

To be capable of accommodating new demands, a base structure may need some excess capacity in the form of standby or inactive bases. Alternatively, flexibility may be embedded in a base structure if the current bases can be enlarged to expand existing activities or add additional missions. As the Air Force reduced its CONUS and OCONUS base structure in the 1990s, it sought to maintain a posture that enabled to it adjust to changing circumstances (DoD, 1978, p. 13; Sahaida, 2004, p. 157). During the 1995 round of base closures, 12 of the tactical aircraft bases that DoD retained had the potential to accommodate 378 additional aircraft at relatively little cost (DoD, 1999, p. 33). Consequently, the Air Force concluded that its CONUS base structure "retain[ed] the flexibility to absorb overseas force structure, provide surge capability, and accommodate changes in the strategic threat" (Department of the Air Force, 1995, p. 14).

Similarly, the Bush administration's 2004 Global Defense Posture Review (GDPR) sought to enable the United States to deal with an extremely fluid and uncertain environment characterized by asymmetric threats. Because the Bush administration believed that the "principal

---

[7] Between 1961 and 1969, the number of major SAC bases in the United States fell from 46 to 28 (Marion, 2004, p. 107).

characteristic of security environment" was "uncertainty," the 2004 GDPR proposed a number of changes to improve the agility of U.S. armed forces (DoD, 2004, pp. 9–15; JCS, 2004, p. 7). In particular, the GDPR sought to eliminate static formations in Europe and Asia by moving away from large main operating bases (MOBs) in favor of access to cold and warm facilities that could be scaled up or scaled down as needed (DoD, 2004, p. 10; Henry, 2006, p. 38). A nimble posture consisting primarily of expandable facilities would allow U.S. forces "to reach any potential crisis spot quickly" (Henry, 2006, p. 39).

At other times, the desire to preserve geographic balance has shaped USAF basing decisions. Concentrating bases limits the reach of USAF aircraft. Maintaining a network of geographically dispersed bases that offer adequate coverage of multiple regions guarantees that the Air Force can project power globally. AMC, for example, seeks to maintain an en route network across the world. Thus, even though it identified Southwest Asia, Southeast Asia, Korea, Africa, and the Black Sea as its focus, AMC seeks to improve its en route facilities in Latin America and to maintain its European locations so that it has "global coverage" (AMC, 2010, pp. 30–31). At home, the Air Force's mobility forces have retained terminals on each coast to facilitate deployments across the globe.

An additional way of creating an agile force posture is obtaining bases that provide reach into multiple regions. For instance, during and immediately after World War II, the Joint Chiefs of Staff developed basing plans for the postwar world. The Joint Chiefs had not identified any potential adversary, yet they sought to create a network of overseas bases along the perimeters of the European and Asian continents that would allow the United States "to rapidly deploy forces in any desired direction" (JCS, 1945). This posture of perimeter defense-in-depth emphasized offshore locations that could strike many directions and basing rights instead of large fixed garrisons.

Protect Unique Assets

As the United States' network of bases at home and abroad has contracted, DoD in general, and the Air Force in particular, have sought to protect particular unique assets that might be needed in the future and are difficult to recreate. While many physical structures (such as runways and buildings) can be easily rebuilt, other military possessions may be impossible—or at least very difficult and expensive—to reconstitute. The latter category includes things that require certain topographical or geophysical characteristics that are relatively rare in today's world, including "large, contiguous, unencumbered areas with specific characteristics needed to fulfill a military requirement" (DoD, 1999, p. 4). Examples of difficult-to-reconstitute assets are deep-water ports, large maneuver areas, airspace, and air training ranges. Nellis Air Force Base, for example, was excused from the 1995 BRAC because it offers an "irreplaceable" and "extensive" range complex (Department of the Air Force, 1995, pp. 23–24).

Among these difficult-to-reconstitute assets, the Air Force has prioritized protecting the CONUS airspace that it controls. When Reese Air Force Base was selected for closure in 1995, the Air Force transferred the control of its airspace to nearby bases that remained active (DoD, 1999, p. 14). The map on the left of Figure D.1 illustrates the special use airspace in the United States as of 1999, while the map on the right depicts the military airspace that was relinquished through the 1995 round of BRAC (DoD, 1999, p. 16). The pictures clearly demonstrate that while closing surplus ground facilities, the Air Force and the Navy have preserved nearly all of the airspace under their jurisdiction.

Figure D.1
Special Use Airspace in 1999 (left) and Airspace Ceded as a Part of BRAC Through 1995 (right)

At other times, the Air Force exempts bases that are "unique," or "mission essential," from the BRAC process. Some space ground stations, for example, need to be at specific locations to provide continuous coverage of satellites, or to provide access to the proper space orbit (Sturm, 1969, p. 93). Other facilities may offer a unique and strategically valuable geography. Andersen Air Force Base on Guam, for instance, was excluded from BRAC because it offers "an irreplaceable resource for overseas contingencies" (Department of the Air Force, 1995, pp. 23–24). Other protected assets include live training ranges, hazardous material storage, and weapon and system testing locations (DoD, 1999, p. 13).

Support Planned Mission

One of the primary strategic factors that influences basing decisions is the ability of a location to support the planned mission. The suitability of an existing or potential base site depends on two separate factors: geophysical characteristics and existing capacity. There are certain geophysical requirements for all airfields, namely level, well-drained parcels of unobstructed land that provide adequate space for a runway with safe approach and departure routes (USAF, 1963; Brown, 1990, p. 2; Marion, 2004, p. 103). Early aircraft needed only a few hundred acres of clear land to serve as an airfield, but over time, as aircraft have grown larger and heavier, the space required for an airfield has significantly increased (Brown, 1990, p. 2).

Additional necessary characteristics depend on the specific weapon system stationed at the base. For example, it is essential that ballistic missile bases have a particular soil composition and enough room to disperse the launch sites within the base. For tactical fighter airbases, fighter pilots must be able to regularly train for combat. Consequently, one of the most important factors in choosing where to station a fighter wing is the base's proximity to training ranges. Because of the hazards associated with combat training, suitable locations need large tracts of undeveloped land that have year-round good weather and uncongested airspace. Locations with well-developed transportation networks, which often are in urban areas, are preferred for intertheater airlift bases to facilitate the transloading of cargo. For example, the Air Force has favored bases along the U.S. coasts to maximize the unrefueled range and cargo payload capacity of its intercontinental airlifters (Sahaida, 2004, p. 173).

In addition to the geophysical characteristics, another factor that influences the ability of an existing base to support its assigned mission is its infrastructural capacity (Sahaida, 2004, pp. 157, 162). In particular, the Air Force weighs whether an existing base has sufficient infrastructure in good condition to support its mission. Old or deteriorating infrastructure demands frequent maintenance, major repairs, and perhaps even extensive upgrades, which increase the operating costs of a base. The Air Force can improve runways, ramps, hangars, and other supporting facilities, but these types of renovations are often very expensive. Therefore, when deciding which bases it should retain or use for new weapon systems, the Air Force considers whether there is adequate infrastructure at a site. For instance, as a part of the 1988 BRAC process, the Air Force recommended the closure of Norton Air Force Base over other airlift bases because of the poor quality of its warehouses and its lack of family housing and support facilities (Mayer, 1999). Also, when the Air Force seeks new bases, it has typically preferred locations that have significant infrastructural capacity. While under the command of General Curtis LeMay, SAC was notorious for using CONUS bases with the best infrastructure from other commands, while offloading its worst facilities (Haulman, 2004, p. 57).

Similarly, the Air Force has prioritized maintaining access to the most capable airbases overseas; however, at times this has been difficult and costly, as host nations can leverage this dependence during base rights negotiations. In Saudi Arabia, for example, the United States built Dhahran Airport in 1946 to facilitate the movement of forces between the European and Pacific theaters. As relations with the Soviet Union deteriorated, Dhahran became an important strategic asset because it was capable of supporting U.S. bomber operations. To retain basing rights at Dhahran, however, U.S. officials engaged in years of negotiations with the Saudis and secured only a short five-year lease in return for considerable financial and military aid (Gormly, 1994, pp. 203–204).

Extent of Strategy Influence on USAF Basing Decisions

This section focuses solely on strategic considerations and explores when and how strategy influences domestic basing decisions. It is difficult, or even impossible, to determine with any precision the degree to which strategic versus nonstrategic factors dictate basing outcomes—

especially across time. For much of the Air Force's history, it did not have a formalized basing process, but instead made decisions on an ad hoc basis. Consequently, many basing recommendations are not well documented.[8] Furthermore, because policymakers often claim that strategic calculations drove their decisions, it can be difficult to distinguish rhetoric from reality.[9]

Nevertheless, it is possible to conclude that strategy plays a role in the Air Force's CONUS and OCONUS basing decisions irrespective of the level of threat. Moreover, threat perceptions have influenced the *type* of strategic factors that have been most important. When the threat level is thought to be high, there is an urgent need to act, and *short-term* strategic factors tend to guide basing decisions. In contrast, when there is little to no threat, *long-term* strategic considerations play a less visible but important role in basing decisions. This distinction is not meant to be mutually exclusive. At all times, both long-term and short-term strategic and nonstrategic factors play a role in basing decisions. Nevertheless, an examination of the historical record reveals that at particular times certain strategic factors tend to dominate over others. Moreover, the role that strategy plays in CONUS and OCONUS basing decisions is remarkably similar. However, because threats can be geographically differentiated, since the end of the Cold War domestic basing decisions have occurred within the context of a relatively low threat environment, while at times overseas the level of threat has spiked.

The United States' experience during both world wars illustrates the importance of focusing on short-term concerns when vital U.S. interests are in peril. In April 1917, "practically no military aviation infrastructure existed in the country" (Warnock, 2004). As a result, the Army Air Service built nearly 105 airbases, but because the threat was contained to the European theater, there was no need to create air defenses. Instead, the airbases built during World War I were situated to facilitate the deployment of the American Expeditionary Forces to Europe (Warnock, 2004, p. 12). This base structure was determined to be inadequate as fears of an enemy attack on the United States increased in the late 1930s. The Army Air Corps Board even concluded that "in the past Air Corps stations have not been located solely in accordance with tactical and strategic requirements" (Futrell, 1951, p. 3). Yet this is an incorrect assessment. The airbases established during World War I were built primarily with immediate strategic requirements in mind, and these conditions differed significantly from the conditions during World War II. Beginning in 1939, the AAC constructed new airbases in the United States and Central and Southern America to remedy the deficiencies of its legacy posture and enable it to implement the President's strategy of hemispheric defensive (Pettyjohn, 2012, pp. 39–48; Warnock, 2004, pp. 22–23). This initiative included establishing airbases to defend critical parts of the nation that were currently vulnerable. Moreover, because of concerns about an attack on the East Coast, the AAC chose locations away from the coastlines to complicate enemy aircraft's

---

[8] Even in the 1990s, there have been questions about USAF BRAC decisions and the lack of documentation. See GAO, 1993, p. 5.

[9] Moreover, criteria can be applied inconsistently—either intentionally or unintentionally. See GAO, 1991, p. 3.

targeting problem. The Japanese attack on Pearl Harbor further amplified the AAC's concerns about airbase vulnerability, resulting in an order that new airbases be dispersed and built at least 350 miles inland (Futrell, 1951, p. 77).

At times, a focus on current exigencies has led to the creation of a posture that quickly became obsolete. During World War II, for example, the Army Air Force (AAF) relied on the capabilities of its current aircraft for all basing decisions, even though new aircraft were scheduled to be operational in the near future. This was a sensible but inefficient approach, as technological improvements often outstripped the pace of base construction. As a result, the AAF built many airfields that were not needed when planes with longer ranges entered into service. In the Pacific, the United States had planned to launch a bombing campaign against Japanese cities and therefore sought bases that would allow its medium bombers, such as the B-25, to strike targets in Japan. Yet by the time that the United States' island-hopping campaign had secured airbases within the B-25's range, longer-range B-24 and B-29 aircraft had entered the inventory and formed the core of air fleet used in the air campaign against Japan (Blaker, 1990, p. 27).

Similarly, as General Curtis LeMay noted, SAC built "bases in relation to specific emergency war plans and specific operations orders instead of developing them in accordance with intermediate or long range plans" (U.S. Air Force, Directorate of Information, History and Research Division, 1962, p. 4). Consequently, SAC's basing requirements dramatically changed as a result of the development of more capable bombers, the introduction of reliable aerial refueling, and the growing ability of the Soviet Union to strike SAC's overseas bases. For example, because of their limited (3,700 mile) ranges, SAC's B-29 bombers had to deploy overseas for strike missions, which were called Reflex operations (U.S. Air Force, Directorate of Information, History and Research Division, 1962, p. 1). To simplify their deployment, B-29s were based near atomic stockpiles in the U.S. southwest. Other SAC bases, such as MacDill AFB in Florida and Hunter AFB in Georgia, were used as transit stops by the bombers on their way to forward operating bases in Europe and North Africa (Haulman, 2004, p. 58). By 1948, however, SAC acquired its first intercontinental bomber, the B-36 Peacemaker (with a 10,000 mile unrefueled range), enabling it to launch attacks on targets in the Soviet Union from CONUS. While the B-36 could not reach the Soviet Union from many of SAC's current southwestern bases, it could from locations in the northeastern United States. As a result, SAC began to build Limestone (later Loring) airbase in northeastern Maine in 1947 (Mueller, 1989, p. 327).

In 1953, a RAND analysis led by Albert Wohlstetter prompted wholesale changes to SAC's global posture and its concept of operations. Prior to this analysis, SAC, which had always emphasized offensive operations, had never given serious thought to the vulnerability of its bomber force. Yet Wohlstetter's work revealed that the United States' nuclear deterrent rested on a "house of cards," as SAC's overseas bases were vulnerable to a Soviet first strike (Kaplan, 1983, p. 98). RAND concluded that 68 percent of SAC's foreign bases were within the range of

Soviet short-range jet bombers and that all of its OCONUS bases were within the unrefueled range of Soviet medium bombers (Wohlstetter et al., 1954, p. 237). Concerns about the survivability of SAC's overseas bases expanded when the Soviet Union successfully tested its first hydrogen bomb in August 1953.

Because of these revelations, the Air Force began to experiment, in an exercise called Full House, with its new B-47 bombers, which had a limited (2,000 nautical mile) range, but were capable of in-flight refueling, to see whether they could successfully launch a bombing strike on the Soviet Union from CONUS. Full House demonstrated that SAC could carry out strikes from CONUS and, therefore, that it needed overseas bases only for tankers and post-strike recovery (Strum, 1969, p. 23). Consequently, SAC ended its Reflex operations and began an extensive base construction program in the northeastern United States, Newfoundland, and Canada (U.S. Air Force, Directorate of Information, History and Research Division, 1962, p. 5; Mueller, 1989, pp. 467, 475–476). In the United States, SAC made improvements to an inactive base in Pearse, New Hampshire, constructed a new airbase in Plattsburgh, New York, and acquired Westover AFB in Massachusetts from the Military Air Transport Service (Mueller, 1989, pp. 467, 475–476). The Full House initiative also called for the development of 21 Royal Canadian Air Force (RCAF) bases in Canada.[10]

Yet SAC's posture was again made obsolete by technological advances, in particular the Soviet Union's development of ICBMs, which called into question the survivability of northern airbases. While improvements to Canadian airbases were dramatically scaled back due to budgetary concerns, it still took almost three years for the Canadian and U.S. governments to accept a modified plan to upgrade four airbases. By the time that the construction at the RCAF bases was completed in 1960, the Soviet Union had already successfully tested an ICBM, undermining the viability of the original Full House concept of operations (Sturm, 1969, p. 24). Moreover, in 1963 SAC was completing its transition from the B-47 to the B-52 bomber, which reduced its requirement for northern tanker bases and ultimately prompted the Air Force to withdraw its KC-97 tankers and close its last RCAF base in 1964 (Sturm, 1969 pp. 52–53). Because the long-range B-52 bomber (which had an unrefueled radius of 4,000 nautical miles that could be extended through air refueling) began to enter into service in 1955, SAC could reach most targets in the Soviet Union from anywhere in CONUS. Therefore, the B-52s were stationed at the former B-36 bases in the southwest to maximize their warning times, allowing Secretary McNamara to close nine former B-47 bases, which were primarily in the northeastern part of the United States (USAF, 1963, p. 4; Sturm, 1969, pp. 107, 110). This shift was also a product of the Kennedy administration's strategy of flexible response, which emphasized the importance of conventional forces in addition to nuclear forces for deterrence. While flexible response necessitated a costly buildup of conventional U.S. forces, the Kennedy and Johnson

---

[10] At that time, SAC only had one tanker base in Canada, at Goose Bay, which was a holdover from the Bases for Destroyers Agreement (Sturm, 1969, p. 23).

administrations simultaneously sought to control defense expenditures by decreasing SAC's reliance on bombers in favor of ICBMs, which were seen as more cost-efficient, dependable, and survivable (Strum, 1969, p. 113). As this discussion makes clear, the rapid pace of technological developments and changing strategies in the early years of the Cold War generated a shifting set of basing requirements.

More recently, the level of threat at home and abroad has diverged. The dissolution of the Soviet Union in 1991 dramatically reduced the probability of a superpower confrontation, and since that time existential threats to the U.S. homeland have disappeared. By contrast, threats have continued to emerge overseas, creating new basing requirements. For example, in 1990, the United States deployed hundreds of thousands of troops to the Persian Gulf to deter Iraq from attacking Saudi Arabia. U.S. forces then forcibly expelled Iraqi troops from Kuwait, because U.S. policymakers believed that the potential for Saddam Hussein to control a large portion of the world's oil supply posed a serious threat to U.S. interests (Brands, 2004, pp. 117–118). Even after defeating the Iraqi forces and driving them back into their country, U.S. officials believed that Saddam Hussein remained a serious concern and therefore sought to undermine his ability to threaten his neighbors by establishing an economic sanctions regime and two no-fly zones.[11] At the time, however, the United States had only limited peacetime access to a few airbases in Oman, which were too remote for enforcing a no-fly zone over Iraq. Consequently, the United States acquired the right to use closer airbases in Saudi Arabia, the United Arab Emirates, Kuwait, and Qatar (Cordesman, 1998).

While the 9/11 terrorist attack struck the United States at home, the existing USAF base structure remained sufficient to provide for the increasingly important mission of homeland defense. NORAD had previously focused on external approaches to the U.S. homeland, but after 9/11 its mandate expanded to the United States' interior airspace as well. Since 2001, as a part of Operation Noble Eagle, the Air Force has expanded the number of defense combat air patrols (CAPs) over key U.S. cities and increased the number of NORAD sites that are on strip alert. Yet the vast majority of these duties have been carried out by the Air National Guard, using its existing force structure and posture (Hebert, 2005).

---

[11] For more on this period, see Knights, 2005, Chapters 5–8.

Appendix E: Basing Decisionmaking Processes of the U.S. Army and U.S. Navy

The U.S. Army and U.S. Navy provide interesting basing decision alternatives. In examining the basing decisionmaking processes in these services, we reviewed service guidance documents, decision packages for specific basing decisions, and supporting documentation, such as environmental analyses.[1] We also examined relevant studies by other government agencies, such as the U.S. Government Accountability Office (GAO) and the Congressional Research Service, and interviewed selected officials involved with basing decisions. Our purpose for a high-level examination of the basing process in the Army and the Navy was to reveal major differences in the process and add context to the discussion around basing, rather than to conduct a detailed comparison of basing processes across the services.

Army Stationing Decisions

Authorities and Strategic Guidance

The Army term for basing or beddown decisions is *stationing*. Decision authority for stationing actions varies with the size of the unit affected. The Secretary of Defense approves stationing changes for Army divisions, the Secretary of the Army does so for brigades, and the Assistant Secretary of the Army has authority for BRAC-related discretionary actions.[2] Most other stationing decisions are delegated to the Deputy Chief of the Army Staff, G-3/5/7, including units at battalion level and below (AR 5-10, 2010, p.10).

This differs from USAF practice, in which squadron-level basing decisions (equivalent to battalions, in that both units are commanded by O-5s) are made by the SecAF. It also marks a difference with the Navy, where homeporting decisions for any ship (including destroyers, also commanded by O-5s) are made by the Secretary of the Navy.

Although this difference appears significant, it is somewhat less so in actual practice. Many stationing decisions made by the G-3/5/7 simply implement force posture moves that are at least implied, and sometimes explicitly directed, by strategic guidance. For example, a recent decision to move an aviation battalion to Joint Base Lewis-McChord was an implementing action for a larger initiative to establish a combat aviation brigade there. That initiative was approved by the Vice Chief of Staff of the Army in 2011 and was itself part of the larger Army Transformation

---

[1] The basing process for the U.S. Marine Corps is not explicitly reviewed in this appendix.

[2] BRAC discretionary actions are decisions regarding minor tenants at installations slated for closure under BRAC. BRAC decisions typically identify new stations for major tenants and leave the disposition of smaller tenants to the services' discretion.

plan approved by the Secretary of the Army. The Army Transformation plan is intended to migrate the Army from a division- to a brigade-centric force, providing more useful force packages that are easier to deploy, especially for small- and medium-sized contingencies. The plan includes a number of reorganizations, which necessitate rebasing many units. Thus, while the G-3/5/7 may appear to have some of the same authorities as the other services' secretaries, in many cases this authority is limited to approving the execution of moves already directed by senior leaders' guidance.[3]

Process and Supporting Analyses

Proposals for stationing actions are referred to as stationing packages. Stationing packages are submitted to the Army staff by commands with a direct administrative reporting chain to Army Headquarters.[4] This closely parallels the Air Force, whose MAJCOMs are responsible for submitting basing proposals and include many similar commands. Stationing packages have several components. The most important of these are the stationing summary, environmental documentation, facility requirements summary, and information for members of Congress (IMC).[5] Besides summarizing the contents of the entire package, the stationing summary includes an analysis of the military value of the proposed action.

Military Value Analyses

Stationing packages must include consideration of operational issues, but Army guidance does not prescribe specific considerations. Instead, it acknowledges that operational issues are likely to differ from action to action, and requires that stationing summaries describe the proposed action's impact on strategy, operations, and (at an unclassified level) on plans (AR 5-10, 2010, p. 12).

The Army also mandates that stationing actions consider alternatives to the proposed stationing. There is no requirement for a specific number of alternatives that must be considered, but at a minimum, the options must include a "no-action" alternative. If multiple alternatives are studied, the package must explain what screening and evaluation criteria were used and how the

---

[3] Other stationing decisions made by the G-3/5/7 may involve offices rather than tactical units. Two recent examples include the relocation of the Army Marketing Research Group from Fort Knox, Kentucky, to the Washington, D.C., area, and the Army Intermodal Distribution Platform Management Office from Tobyhanna Army Depot, Pennsylvania, to Scott Air Force Base, Illinois.

[4] These commands fall into three categories: Army commands, such as Army Materiel Command; Army service component commands, such as U.S. Army Europe; and direct reporting units, such as the Army Corps of Engineers and Army Medical Command.

[5] Other elements of the stationing package include the manpower migration diagram (a document showing changes in the size or chain of command of units affected by the stationing action), civilian employee impacts (a document detailing the cost or savings of any changes in civilian employees caused by the action), public notification documents, and community impact analyses (where applicable).

criteria support the proposed action. Stationing packages must also consider the ability to support the "training density" at the gaining installation (AR 5-10, 2010, p. 13).

Just as strategic guidance dictates certain stationing actions, it may also underpin the military value of such an action. This has been the case in several recent stationing actions directed by the Army Transformation plan. This plan, as noted above, is intended to result in a more effective Army. It is therefore typically cited as providing the military value of moves that it directs. This was the case in the relocation of aviation units to Joint Base Lewis-McChord, mentioned above. This move was an element of the Army Transformation plan's goal of establishing a combat aviation brigade at the base. The stationing package's military value discussion referred to this element of the plan (HQDA, 2012).

Facility and Capacity Analyses

Army stationing planners obtain facilities data from three sources: centralized information systems and databases, requests to individual facilities, and site visits. The primary information system for facilities planning is the Real Property Planning and Analysis System. This system provides analyses of facility assets against allowances, evaluates the impact of candidate stationing actions, and validates construction programs (AR 5-10, 2010, pp. 11–12).[6] There are also at least seven Army databases with additional information on facility capacities, capabilities, and requirements. These databases provide information on the following (from AR 5-10, 2010, pp. 11–12):

1. current and future-year authorized civilian and military populations, by installation (Army Stationing and Installation Plan [ASIP])
2. real property facility data (Integrated Facilities System)
3. allowances, personnel, and equipment lists for tenant units, by installation (Facility Planning System)
4. facility conditions (installation status reports)
5. directed or major weapon-specific facility considerations (such as logistics, maintenance, training, special physical security, or safety needs) (Support Facility Annex)
6. training facilities data (active and inactive range inventories)
7. environmental cleanup requirements, by installation (Army Environmental Database).

A second step in obtaining facility information is through data calls to installations. These calls are routed through Army Installations Management Command, which vets the requests, forwards them to installations for response, and vets the responses before returning them to the initiating Army command/Army service component command/direct reporting unit. Army Installations Management Command also vets draft final stationing packages before they are formally submitted to Headquarters Department of the Army.

---

[6] GAO has raised some concerns about the Army's management information systems and databases. In 2010, GAO found cases in which some of these systems held inaccurate information or returned unusual results. GAO cited a case in which one system reported a requirement for 74 baseball fields at Fort Bragg (GAO, 2010b).

Site surveys are a third means of obtaining facility data. Army guidance anticipates that site surveys will be conducted after the first two steps have been taken. The data from the prior steps may limit the number of candidate installations, thus reducing the number of site surveys required (AR 5-10, 2010, p.7).

Environmental Analyses

Army guidance specifically requires that stationing actions comply with NEPA. As noted earlier, the Army does not require that any specific minimum number of alternatives be studied, other than a "no-action" alternative (AR 5-10, 2010, pp. 12–13). Army guidance also does not discuss whether any alternatives can be eliminated based on non-environmental factors (such as military value and facilities requirements) before determining which alternatives will be considered in the environmental analysis. In recent years, the Army has completed broad environmental impact statements for multiple potential stationing actions before deciding whether to implement them.

If not included in NEPA documentation, Army stationing packages must include a separate community impact analysis. Community impact analyses contain estimates of the stationing action's impact on the population and economy of the surrounding communities, including businesses, schools, housing, and public services. A model developed by the Army Construction Engineering Research Lab, the Economic Impact Forecast System, is available to support these analyses (AR 5-10, 2010, p. 13).

Congressional Notification and Timelines

Information for members of Congress (IMC) is another component of stationing packages. IMC is typically a single-page summary explaining the "who, what, where, when, and why" of the stationing action. The Army's Chief of Legislative Liaison determines which members of Congress will receive this information and obtains Secretary of the Army approval regarding what IMC will be shared (AR 5-10, 2010, p. 13).

The Army informs Congress that a stationing study has begun only in exceptional, "highly politically sensitive" cases. Such cases require approval from senior Army leadership and are usually decided in initial briefings to such leaders.[7] This appears to contrast with the Air Force approach, where the service routinely shares information with Congress near the beginning of the analysis, during the decision process, and after a decision is reached.

Army guidance directs that stationing actions requiring military construction at the gaining installation be submitted to the Army staff at least five years before the proposed effective date. This allows time for approvals and coordination, with a further goal of having all designs complete and contracts awarded three years before the effective date. For actions not requiring

---

[7] The Office of the Secretary of Defense must be notified before Congress in cases where a stationing action will close or substantially reduce the population of an installation, result in the release of 50 or more civilian personnel from government employment, or substantially reduce contract operations or employment involving 100 or more people (see Department of Defense Directive 5410.10, 1960).

military construction, the Army requires that stationing packages be submitted at least 18 months before the proposed effective date (AR 5-10, 2010, pp. 7–8).

In practice, some recent stationing actions have not met these goals. In a 2010 report, GAO found that some actions requiring military construction were submitted for approval less than one year before the effective date. Several of these moves went forward, with units occupying trailers and other temporary facilities at their new stations. In some cases, even temporary facilities were unavailable, and newly arrived units were forced to occupy spaces vacated by deployed personnel. In some situations, this required that the new tenants move from building to building as deployed soldiers returned to home station (GAO, 2010b).

Navy Homeporting Decisions

Authorities and Strategic Guidance

The U.S. Navy term for basing decisions is *homeporting*. The Secretary of the Navy is authorized to make decisions regarding homeports for new ships, as well as any temporary homeport change exceeding nine months' duration. This authority is delegated for temporary homeport changes lasting less than nine months. Temporary homeport changes typically occur for maintenance.

Strategic Guidance

The Navy uses a variety of strategic guidance documents to allocate ships to the Atlantic and Pacific Fleets. This allocation, which the Navy terms its *strategic laydown*, uses input from the most recent Quadrennial Defense Review, Navy strategy and vision documents, and other sources. The resulting Strategic Laydown Plan is updated annually (OPNAVINST 5400.44a, 2011b, p. 3–2; GAO, 2010a, pp. 30–31). Annual updates may modify the allocation of ships between the fleets based on changes in force structure, the international security environment, or shipbuilding or aviation procurement programs (OPNAVINST 5400.44a, 2011b, p. 3–2).

After the determining the fleet in which a ship or ships will be homeported, the Navy's decision process focuses on *strategic dispersal*. This term refers to the distribution of ships within a fleet by homeport (GAO, 2010a, p. 31). The Navy's type commands for air, surface, and submarine forces use the strategic laydown plans to narrow homeport options for new ships. They then consider the new ships' manning, training, maintenance, and other requirements. If environmental documentation is not required, the type commands recommend specific ports for new assets. If environmental documentation is required, the NEPA process begins and homeport decisions are deferred until it is complete. The Navy's approach to NEPA documentation is discussed in more detail below (OPNAVINST 5400.44a, 2011b, p. 3-3).

Strategic dispersal has been an issue of historical importance to the Navy. As it grew toward 600 ships in the 1980s, the Navy pursued a Strategic Homeporting initiative that sought to increase the number of ports capable of hosting battle groups and smaller formations. The Navy advocated Strategic Homeporting based on the vulnerability of a concentrated fleet to small numbers of Soviet strikes (O'Rourke, 1987).[8] Dispersal remains a significant aspect of military value for the Navy today. This is seen in part in the Mayport decision discussed below.

Process and Supporting Analyses

Military Value Analyses

The Navy does not dictate specific measures of military value for homeporting requests to consider. Guidance is limited to directing that the command requesting the homeporting action state the reason for the request (OPNAVINST 5400.44a, 2011b, pp. 1–17). In a recent decision to move an aircraft carrier from Hampton Roads, Virginia, to Mayport, Florida, the Navy cited three main operational benefits (U.S. Department of Defense and Department of the Navy, 2009). The first was reduced steaming time to the open ocean to approximately one hour in Mayport as compared with 3.5 hours in Hampton Roads (GAO, 2010a, pp. 30–31). The second operational benefit cited in the Mayport decision was lessened risk from natural and manmade disasters. The Navy noted a lower risk of terror attack at Mayport compared with Hampton Roads. It also cited the possibility of losing access to the open ocean at Hampton Roads due to terror attacks or natural disasters affecting choke points between the carrier piers and the ocean. The third operational benefit relates to dispersal. The Navy cited the risk of having all of the Atlantic Fleet's nuclear carrier-trained crews and support infrastructure concentrated at Hampton Roads. Moving a carrier, its crew, and support infrastructure to Mayport reduced this risk (GAO, 2010a, pp. 30–31).

The Navy does not require that homeporting actions consider a specific number of alternatives. In deciding whether to homeport an aircraft carrier at Mayport, the Navy considered 13 alternatives, and all but one (a "no-action" alternative) involved homeporting different mixes of ships at Mayport. No other ports were considered. The Navy reportedly "worked on the assumption that it would not establish a new carrier homeport but upgrade an existing" one to support a nuclear-powered ship (GAO, 2010a, p. 31).

Facilities Issues

Navy homeporting requests include two types of information with respect to facilities: buildings and land and personnel impacts. Data concerning buildings and land include the square footage occupied by the affected command(s), the value of the buildings and any stationary equipment within them, and the cost of maintaining the buildings and equipment over a three-year period. If land and/or buildings must be acquired at the gaining installation, homeport requests must include acquisition cost and further detail whether the buildings are to be permanent, semipermanent, or temporary (OPNAVINST 5400.44a, 2011b).

---

[8] See also O'Rourke, 1990.

Environmental Impact

The Navy explicitly requires that homeporting actions comply with NEPA. As noted earlier, the Navy does not require that any minimum number of alternatives be considered beyond a no-action option. The Navy does, however, highlight the importance of explaining what alternatives were considered in terms of "what requirements are essential to achieving the proposed action's purpose" and what alternatives would achieve it. This is especially important in the case of Environmental Impact Statements, which should include a "description of proposed action and alternatives (DOPAA) (OPNAVINST 5090.1C, 2011a, p. 5–40). Where screening criteria such as operational or location needs are used, they must be identified and explained (OPNAVINST, 2011a, pp. 5-52–5-53).

Congressional Notification and Timelines

Navy guidance requires that homeporting requests identify members of Congress with districts affected by the proposed action. There is no requirement to include a draft summary of information to be provided to Congress. Navy guidance acknowledges the possibility that a homeporting action will have been discussed with interested members of Congress before formal submission to the Office of the Chief of Naval Operations. In such cases, the homeporting request must indicate which members of Congress were advised of the potential action, and the forum or venue in which the exchange took place.

Summary

Strategic guidance strongly affects basing decisions for the Army and Navy. In the Army, such guidance may explicitly define stationing actions. The Deputy Chief of Staff G-3/5/7, although nominally empowered to base battalion and smaller-sized units, is often implementing strategic guidance when doing so. In the Navy, the division of ships between the Atlantic and Pacific Fleets is determined by strategic guidance, and the Navy also perceives a strategic need to disperse the fleet to the extent possible. This aspect of strategic guidance may also influence many homeporting decisions.

Strategic guidance may also describe the military value of many basing decisions in both services. The Army's Transformation Plan is a strategic initiative designed to give the entire service more military value. It directs (either implicitly or explicitly) a number of stationing actions in pursuit of this goal. The Navy's Strategic Homeporting initiative of the late Cold War era is another historical example of a strategy that defined the military value of a number of homeporting decisions.

Neither the Army nor the Navy requires that a specific number of basing alternatives be considered in environmental analyses, beyond a "no-action" alternative. Navy guidance suggests the possibility of ruling out some options based on non-environmental factors, before deciding which alternatives will be considered in an environmental impact statement.

Finally, both the Army and the Navy have a default position of sharing basing decisions with Congress after they are made. Each service recognizes that there may be exceptions to this rule, but neither plans for the series of engagements with Congress that the Air Force does in making its basing decisions.

Table E.1 summarizes key aspects of the USAF basing decisionmaking process compared with those of the Army and the Navy.

Table E.1
Comparison of Service Basing Decisionmaking Processes

Key Aspect	Air Force	Army	Navy
Decision Authority			
O-6-level commands	SecAF	Secretary of the Army	Secretary of the Navy
O-5-level commands	SecAF	Deputy Chief of Staff G-3/5/7a	Secretary of the Navy
Supporting analyses			
Military value	Considered by BRRP and USAF leadership	May be inherent in strategic guidance	May be inherent in strategic guidance
Environmental issues	Included as criteria in strategic basing process	May encompass multiple potential moves, where envisioned by strategic guidance	Alternatives may be eliminated based on requirements to achieve proposed action's purpose.
Congressional notification	Multiple points	Post-decision[b]	Post-decision[c]
Timeline for proposal submission (months before execution)	Varies, proponent MAJCOM is responsible	60 (if military construction) 18 (no military construction)	14

[a] Often implementing stationing actions decided or implied by higher-level strategic guidance.
[b] Exceptions in highly sensitive cases, with SA approval.
[c] Navy guidance acknowledges possibility that some information will have been shared with Congress before the final decision.

References

AFI—*See* Air Force Instruction.

Air Force Instruction 10-503, *Strategic Basing*, Washington, D.C.: Department of the Air Force, September 27, 2010. As of November 11, 2015:
http://www.acq.osd.mil/dpap/ccap/cc/jcchb/Files/FormsPubsRegs/Pubs/afi10-503.pdf

Air Mobility Command, *Air Mobility Command Global En Route Strategy White Paper*, version 7.2.1, July 14, 2010.

Alaska Delegation, letter to Secretary Donley and General Schwartz, March 7, 2012. As of November 11, 2015:
http://www.murkowski.senate.gov/public/?a=Files.Serve&File_id=a1bdf937-34dd-4b9a-b6a8-181c7037bd9a

AMC—*See* Air Mobility Command.

AR—*See* Army Regulation.

Army Regulation 5-10, *Stationing*, Washington, D.C.: Headquarters Department of the Army, August 20, 2010.

Bender, Bryan, "As Jets Seem Bound for Vt., Questions of Political Influence Arise," *The Boston Globe,* April 14, 2013. As of November 11, 2015:
http://www.bostonglobe.com/news/nation/2013/04/13/selection-vermont-guard-base-for-jets-was-based-flawed-data-raising-questions-political-influence/pmhiPtI1BPWxwC3yK1adAL/story.html

Benson, Lawrence R., *USAF Aircraft Basing in Europe, North Africa, and the Middle East, 1945–1980*, Ramstein Air Base, Germany: Office of History, Headquarters United States Air Forces in Europe, 1982.

Betts, Richard K., "Is Strategy an Illusion?" *International Security*, Vol. 25, No. 2, 2000.

Blaker, James R., *The United States Overseas Basing: An Anatomy of the Dilemma*, New York: Praeger, 1990.

Brands, H. W., "George Bush and the Gulf War of 1991," *Presidential Studies Quarterly*, Vol. 34, No. 1, March 2004.

Brown, Jerold E., *Where Eagles Land: Planning and Development of U.S. Army Airfields, 1910–1941*, New York: Praeger, 1990.

Calder, Kent E., *Embattled Garrisons: Competitive Base Politics and American Globalism*, Princeton, N.J.: Princeton University Press, 2007.

Chief of Naval Operations Instruction 5090.1C, *Environmental Readiness Program Manual*, Washington, D.C.: Office of the Chief of Naval Operations, July 18, 2011a.

Chief of Naval Operations Instruction 5400.44a, Washington, D.C.: Office of the Chief of Naval Operations, *Navy Organization Change Manual*, October 13, 2011b.

Clinton, Hillary, "America's Pacific Century," *Foreign Policy*, November 2011.

Conger, John, "Is Base Realignment and Closure (BRAC) Appropriate at this Time?" testimony before the Subcommittee on Readiness of the Committee on Armed Services, House of Representatives, H.A.S.C. No. 113–18, Washington, D.C., March 14, 2013. As of November 11, 2015:
http://www.gpo.gov/fdsys/pkg/CHRG-113hhrg80188/html/CHRG-113hhrg80188.htm

Conn, Stetson, Rose C. Engelman, and Byron Fairchild, *Guarding the United States and Its Outposts*, Washington, D.C.: U.S. Army Center of Military History, 1989.

Conn, Stetson, and Byron Fairchild, *The Framework of Hemisphere Defense*, Washington, D.C.: U.S. Army Center of Military History, 1989.

Cordesman, Anthony H., *CENTCOM and Its Area of Operations: Cooperation, Burden Sharing, Arms Sales, and Analysis by Country and Subregion*, Washington, D.C.: Center for Strategic and International Studies, April 1998.

Defense Writers Group, interview of General Herbert J. "Hawk" Carlisle, commander of Pacific Air Forces to the Defense Writers Group, transcript, July 29, 2013. As of November 11, 2015:
http://www.airforcemag.com/DWG/Documents/2013/July%202013/072913Carlisle.pdf

Department of Defense Directive 5410.10, *Coordination and Clearance of Announcements of Personnel Reductions, Closures of Installations, and Reductions of Contract Operations Within the United States*, February 2, 1960 (incorporating Change 1, July 22, 1968).

"Department of Defense Draft Selection Criteria for Closing and Realigning Military Installations Inside the United States," *Federal Register*, Vol. 68, No. 246, December 23, 2003, pp. 74221–74222.

Department of the Air Force, *Department of the Air Force Analyses and Recommendations*, 1960 Base Closure and Realignment Report to the Commission, Vol. V, February 1995.

———, *Analysis and Recommendations BRAC 2005*, Department of Defense Report to the Defense Base Closure and Realignment Commission, Volume V, Part 1, May 2005.

DoD—*See* U.S. Department of Defense.

DWG—*See* Defense Writers Group.

Elith, Jane, John R. Leathwick, and Trevor Hastie. "A Working Guide to Boosted Regression Trees," *Journal of Animal Ecology*, Vol, 77, No. 4, 2008, pp. 802–813.

Flinn, Michael H., "Air Force BRAC Recommendations for Consolidating C-130s: A BRAC Commission Perspective," *Air & Space Power Journal*, September 2006, pp. 1–11.

Friedman, J.H. "Greedy Function Approximation: A Gradient Boosting Machine," *Annals of Statistics*, Vol. 29, No. 5, 2001, pp. 1189–1232.

Futrell, Robert Frank, *Development of AAF Base Facilities in the United States, 1939–1945*, USAF Historical Study No. 69, Montgomery, Ala.: Air Force Historical Research Agency, Maxwell Air Force Base, 1951.

GAO—*See* U.S. Government Accountability Office.

Gaddis, John Lewis, *Strategies of Containment: A Critical Appraisal of American National Security Policy During the Cold War*, New York: Oxford University Press, 2005.

Gates, Robert M., *Remarks as Delivered by Secretary of Defense Robert M. Gates*, International Institute for Strategic Studies (Shangri-La-Asia Security), Shagri-La Hotel, Singapore, Saturday, June 5, 2010.

Goren, Lilly J., *The Politics of Military Base Closings: Not in My District*, New York: Peter Lang, 2003.

Gormly, James L., "Keeping the Door Open in Saudi Arabia: The United States and Dhahran Airfield, 1945–46," *Diplomatic History*, Vol. 4, No. 2, 1994, pp. 203–204.

Hammack, Katherine, "Is Base Realignment and Closure (BRAC) Appropriate at this Time?" testimony before the Subcommittee on Readiness of the Committee on Armed Services, House of Representatives, H.A.S.C. No. 113-18, Washington, D.C., March 14, 2013. As of November 11, 2015:
http://www.gpo.gov/fdsys/pkg/CHRG-113hhrg80188/html/CHRG-113hhrg80188.htm

Haulman, Daniel L., "Air Force Bases, 1947–1960," in Frederick J. Shaw, ed., *Locating Air Force Base Sites: History's Legacy*, Washington, D.C.: Air Force History and Museums Program, United States Air Force, 2004.

Headquarters, Department of the Army, *Stationing Package for FY 12–13 Relocation of 1st Bn, 229th Air Cavalry Squadron and D Co., 123rd Aviation Regiment to Joint Base Lewis-McChord*, Washington, D.C., May 20, 2012.

Headquarters Strategic Air Command, Office of the Historian, *History of Strategic Air Command FY1969*, Historical Study No 116, Vol. I Narrative, March 1970.

Hebert, Adam J., "Noble Eagle Without End," *Air Force Magazine*, February 2005.

Heinrich, Bernd, Marcus Kaiser, and Mathias Klier, "How to Measure Data Quality? A Metric Based Approach," in S. Rivard and J. Webster, eds., *Proceedings of the 28th International Conference on Information Systems (ICIS)*, Montreal (Canada): Association for Information Systems, December 2007. As of November 11, 2015: http://www.wi-if.de/paperliste/paper/wi-205.pdf

Henry, Ryan, "Transforming the U.S. Global Defense Posture," in Lord Carnes, ed., *Reposturing the Force: U.S. Overseas Presence in the Twenty-First Century*, Newport, R.I.: Naval War College Press, 2006.

Hopkins, J. C., and Sheldon A. Goldberg, *The Development of Strategic Air Command 1946–1986*, Offutt Air Force Base, Nebraska: Office of the Historian, Headquarters Strategic Air Command, 1986.

HQDA—*See* Headquarters Department of the Army.

JCS—*See* Joint Chiefs of Staff.

Joint Chiefs of Staff, *Enclosure C: Overall Examination of United States Requirements for Military Bases and Base Rights*, JCS 570/40, RG 218, Combined Chiefs of Staff Series 360 (19-9-42), November 7, 1945.

——, *The National Military Strategy of the United States of America: A Strategy for Today; a Vision for Tomorrow*, Washington, D.C.: Office of the Chairman of the Joint Chiefs of Staff, 2004.

Kaplan, Fred, *The Wizards of Armageddon*, Stanford, Calif.: Stanford University Press, 1983.

Kaplan, Lawrence S., Ronald D. Landa, and Edward J. Drea, *History of the Office of the Secretary of Defense, Volume V: The McNamara Ascendancy, 1961–1965*, Washington, D.C.: Historical Office, Office of the Secretary of Defense, 2006.

Keller, Kirsten, Sean Robson, Kevin O'Neill, Paul D. Emslie, Lane F. Burgette, Lisa M. Harrington, Dennis Curran, *Promoting Airmen with the Potential to Lead: A Study of the Air Force Master Sergeant Promotion System*, Santa Monica, Calif.: RAND Corporation, RR-581-AF, 2014. As of November 11, 2015: http://www.rand.org/pubs/research_reports/RR581.html

Kennedy, Paul, *Grand Strategies in War and Peace*, New Haven, Conn.: Yale University Press, 1992.

Knights, Michael, *Cradle of Conflict: Iraq and the Birth of Modern U.S. Military Power*, Annapolis, Md.: Naval Institute Press, 2005.

Kreis, John F., *Air Warfare and Air Base Air Defense*, Washington, D.C.: Office of Air Force History, United States Air Force, 1988.

Leighton, Richard M., *History of the Office of the Secretary of Defense Vol. III, Strategy, Money, and the New Look, 1953–1956*, Washington, D.C.: Historical Office of the Secretary of Defense, 2001.

Lemmer, George F., "Bases," in Alfred Goldberg, ed., *A History of the United States Air Force*, New York: Arno Press, 1974.

Manyin, Mark E., Stephen Daggett, Ben Dolven, Susan V. Lawrence, Michael F. Martin, Ronald O'Rourke, and Bruce Vaughn, *Pivot to the Pacific? The Obama Administration's "Rebalancing" Toward Asia*, Washington, D.C.: Congressional Research Service, March 28, 2012.

Marion, Forrest L., "Retrenchment, Consolidation, and Stabilization, 1961–1987," in Frederick J. Shaw, ed., *Locating Air Force Base Sites: History's Legacy*, Washington, D.C.: Air Force History and Museums Program, United States Air Force, 2004.

Mayer, Kenneth R., "The Limits of Delegation: The Rise and Fall of BRAC," *Regulation,* Vol. 22, No. 3, 1999, p. 33.

Millett, Allan R., and Peter Maslowski, *For the Common Defense: A Military History of the United States of America*, New York: The Free Press, 1994.

Mueller, Robert, *Air Force Bases Volume I: Active Air Force Bases Within the United States of America on 17 September 1982*, Washington, D.C.: Office of Air Force History, 1989.

"Murkowski Gets Transparency Pledge from Next USAF Chief of Staff," *Alaska Monthly*, July 2012. As of December 15, 2015:
http://www.akbizmag.com/Alaska-Business-Monthly/July-2012/Murkowski-Gets-Transparency-Pledge-from-Next-USAF-Chief-of-Staff/

NSC-68—*See* National Security Council.

National Security Council, "NSC 68: United States Objectives and Programs for National Security (April 14, 1950)," *Naval War College Review*, Vol. 28, May–June 1975, pp. 51–108. As of November 11, 2015:
https://www.mtholyoke.edu/acad/intrel/nsc-68/nsc68-1.htm

O'Hanlon, Michael and Bruce Riedel, "Why the United States Should Open More Bases in the Middle East," *Foreign Affairs*, July 2, 2013.

O'Rourke, Ronald, *The Navy's Strategic Homeporting Program: Issues for Congress*, Washington, D.C.: Congressional Research Service, June 11, 1987.

——, *Strategic Homeporting Reconsidered*, Washington, D.C.: Congressional Research Service, December 20, 1990.

OPNAVINST—*See* Chief of Naval Operations Instruction.

Pettyjohn, Stacie L., *U.S. Global Defense Posture, 1783–2011*, Santa Monica, Calif.: RAND Corporation, MG-1244-AF, 2012. As of November 11, 2015: http://www.rand.org/pubs/monographs/MG1244.html

Pettyjohn, Stacie L., and Evan Montgomery, "By Land and by Sea," *Foreign Affairs*, July 19, 2013.

Pohlmeier, Mark, "Air Force Strategic Basing Division," AF/A8PB, briefing, November 2012.

Public Law 91-190, National Environmental Policy Act, January 1, 1970.

Public Law 100-526, Defense Authorization Amendments and Base Closure and Realignment Act, October 24, 1988.

Public Law 101-510, Defense Base Realignment and Closure Act of 1990, November 5, 1990.

Reynolds, David, *From Munich to Pearl Harbor: Roosevelt's America and the Origins of the Second World War*, Chicago, Ill.: Ivan R. Dee, 2001.

Ridgeway, G., "Generalized Boosted Models: A Guide to the GBM Package," September 21, 2009. As of November 9, 2014:
https://r-forge.r-project.org/scm/viewvc.php/*checkout*/pkg/inst/doc/gbm.pdf?revision=18&root=gbm&path rev=22

Ross, Robert S., "The Problem with the Pivot: Obama's New Asia Policy Is Unnecessary and Counterproductive," *Foreign Affairs*, November/December 2012.

Saab, Bilal Y., "Asia Pivot, Step One: Ease Gulf Worries," *National Interest*, June 20, 2013.

Sahaida, Jeffrey P., "Reorganization After the Cold War, 1988–2003," in Frederick J. Shaw, ed., *Locating Air Force Base Sites: History's Legacy*, Washington, D.C.: Air Force History and Museums Program, United States Air Force, 2004.

Schlossberg, George, "How Congress Cleared the Bases: A Legislative History of BRAC," *Journal of Defense Communities,* Vol. 1, 2012, pp. 1–12.

Shaw, Frederick J., ed., *Locating Air Force Base Sites: History's Legacy*, Washington, D.C.: Air Force History and Museums Program, United States Air Force, 2004.

Sorenson, David S., *Shutting Down the Cold War: The Politics of Military Base Closure*, New York: Palgrave MacMillian, 1998.

——, *Military Base Closure: A Reference Handbook*, Westport, Conn.: Praeger Security International, 2007.

Sturm, Thomas, *USAF Overseas Forces and Bases: 1947–1967*, Washington, D.C.: Office of Air Force History, 1969.

Sutter, Robert G., Michael E. Brown, and Timothy J. A. Adamson, with Mike M. Mochizuki, and Deepa Ollapally, *Balancing Acts: The U.S. Rebalance and Asia-Pacific Stability*, Washington, D.C.: Elliott School of International Affairs, George Washington University, August 2013. As of November 11, 2015:
http://www2.gwu.edu/~sigur/assets/docs/BalancingActs_Compiled1.pdf

USAF—*See* U.S. Air Force.

USAFE—*See* U.S. Air Forces in Europe.

U.S. Air Force, Directorate of Information, History and Research Division, *Overseas Bases: A Military and Political Evaluation*, April 2, 1962.

———, *Proposal to Relocate the 18th Aggressor Squadron (18 AGRS) from Eielson Air Force Base (EAFB), Alaska to Joint Base Elmendorf-Richardson (JBER), Alaska and to Right-Size the Remaining Wing/Overhead/Base Operating Support at EAFB, Alaska*, Draft Environmental Impact Statement, Hickam Air Force Base, HI: Air Force Civil Engineer Center Pacific Air Forces, May 2013a.

———, "Air Force Announces Aggressors Will Remain at Eielson," News post, Eielson Air Force Base Web site, October 2, 2013b. As of January 7, 2013:
http://www.eielson.af.mil/news/story.asp?id=123365696

U.S. Air Force, *Ideal Base Study*, 1963.

U.S. Air Forces in Europe, *Short History and Chronology of the USAF in the United Kingdom*, Historical Division Office of Information Third Air Force, United States Air Forces in Europe, May 1967.

U.S. Department of Defense, *Department of Defense Base Realignment Policy*, Hearing Before the Subcommittee on Military Construction and Stockpiles of the Committee on Armed Services, United States Senate, Ninety-Fifth Congress, Second Session, Washington, DC: U.S. Government Printing Office, August 4 1978.

———, *Report on the Effect of Base Closures on Future Mobilization Options*, Washington, D.C.: Office of the Deputy Under Secretary of Defense (Installations), November 10, 1999.

———, *Strengthening U.S. Global Defense Posture Report to Congress*, Washington, D.C., September 2004.

———, *National Defense Strategy*, Washington, D.C., June 2008.

———, *Sustaining U.S. Global Leadership: Priorities for 21st Century Defense*, Washington, D.C., January 2012.

U.S. Department of Defense and Department of the Navy, *Record of Decision for Homeporting of Additional Surface Ships and Naval Station Mayport, Florida*, Washington, D.C., January 14, 2009.

U.S. Government Accountability Office, *Military Bases: Observations on Analyses Supporting Proposed Closures and Realignments*, GAO/NSIAD-91-224, Washington, D.C., May 1991.

———, *Military Bases: Analysis of DoD's Recommendations and Selection Process for Closures and Realignments*, GAO/ NSIAD-93-173, Washington, D.C., April 1993.

———, *Military Base Closures: Observations on Prior and Current BRAC Rounds*, Statement Before the Defense Base Closure and Realignment Commission, GAO-05-614, Washington, D.C., May 3, 2005.

———, *Military Base Realignments and Closures: Observations Related to the 2005 Round*, Letter to Congress, GAO-07-1203R, Washington, D.C., September 6, 2007.

———, *Military Base Realignments and Closures: Transportation Impact of Personnel Increases Will be Significant, but Long-Term Costs Are Uncertain and Direct Federal Support Is Limited*, Report to Congressional Committees, GAO-09-750, Washington, D.C., September 9, 2009a.

———, *Military Base Realignments and Closures: Estimated Costs Have Increased While Savings Estimates Have Decreased Since Fiscal Year 2009*, Letter to Chairmen and Ranking Members of Congressional Committees, GAO-10-98R, Washington, D.C., November 13, 2009b.

———, *Defense Infrastructure: Opportunities Exist to Improve the Navy's Basing Decision Process and DOD Oversight*, GAO-10-482, Washington, D.C., May 2010a.

———, *Defense Infrastructure: Army Needs to Improve its Facility Planning Systems to Better Support Installations Experiencing Significant Growth*, GAO-10-602, Washington, D.C., June 24, 2010b.

———, *Military Base Realignments and Closures: DOD Is Taking Steps to Mitigate Challenges but Is Not Fully Reporting Some Additional Costs*, Letter to Congressional Committees, GAO-10-725R, Washington, D.C., July 21, 2010c.

———, *Military Base Realignments and Closures: Key Factors Contributing to BRAC 2005 Results*, Testimony Before the Subcommittee on Readiness, Committee on Armed Services, House of Representatives, GAO-12-513T, Washington, D.C., March 8, 2012.

———, *Military Bases: Opportunities Exist to Improve Future Base Realignment and Closure Rounds*, Report to Congressional Committees, GAO-13-149, Washington, D.C., March 7, 2013a.

Warnock, A. Timothy, "Locating Army Air Installations, 1907–1947," in Frederick J. Shaw, ed., *Locating Air Force Base Sites: History's Legacy*, Washington, D.C.: Air Force History and Museums Program, United States Air Force, 2004.

Weathers, Bynum E., Jr., *Acquisition of Air Bases in Latin America, June 1939–June 1943*, Maxwell Air Force Base, Ala.: U.S. Air Force Historical Division, Research Studies Institute, 1943.

The White House, *National Security Strategy*, Washington, D.C., May 2010.

Wohlstetter, A. J., Fred S. Hoffman, R. J. Lutz, and Henry S. Rowen, *Selection and Use of Strategic Air Bases*, R-266, Santa Monica, Calif.: RAND Corporation, 1954. As of November 11, 2015:
http://www.rand.org/pubs/reports/R0266.html

World Bank East Asia and Pacific Regional Report

Toward Gender Equality in East Asia and the Pacific

A Companion to the
World Development Report

THE WORLD BANK
Washington, D.C.

Contents

Boxes

Tables

Foreword

Over the past few decades, the East Asia and Pacific region has been the most economically dynamic region in the world. In most countries in the region, incomes have grown dramatically, and with that growth, absolute poverty has declined rapidly. Most of the region's economies have also shifted away from agriculture and toward manufacturing and services. Rapid growth, structural transformation, and poverty reduction have been accompanied by progress toward gender equality in several key areas. Economic development has led to the closing of gender gaps in school enrollments and a decline in maternal mortality rates: girls in the region as a whole now enroll in secondary schools at a higher rate than boys, and maternal mortality has fallen by half over the past 20 years. Access to economic opportunities has also increased, particularly among younger, better educated women. In many ways, women in East Asian and Pacific countries are better positioned than ever before to participate in, contribute to, and benefit from development.

Yet, the experience of the region illustrates also how growth and economic development are not enough to attain gender equality in all its dimensions. Women still have less access than men to a range of productive assets and services, including land, financial capital, agricultural extension services, and new information technologies. Substantial employment segregation by gender remains. Women are less likely than men to work in formal sector jobs and more likely to work in poorly remunerated occupations and enterprises. Despite the closing of education gaps, women still earn less than men for similar work all across the region. Women in East Asian and Pacific countries still have a weaker voice and less influence than men, whether within the household, in the private sector, in civil society, or in politics. And women across the region remain vulnerable to gender-based violence, often at the hand of an intimate partner.

The main message of this book, *Toward Gender Equality in East Asia and the Pacific*—a regional companion to the *World Development Report 2012: Gender Equality and Development*—is that policy makers in the region need to understand why progress in closing gender gaps has been mixed and to implement corrective policies where gaps remain persistent. The reason is that gender equality is both an important development objective in its own right as well as good development policy. A growing body of evidence shows that promoting gender equality in access to productive resources, economic opportunity, and voice can contribute to higher economic productivity, improve the economic prospects and wellbeing of the next generation, and lead to more effective development policy making. Yet, gender equality in many areas does not happen automatically. Thus, gender-aware public policy is required if countries are to achieve both gender equality and more rapid development.

As a regional report, *Toward Gender Equality in East Asia and the Pacific* focuses on issues that are particularly pertinent to the region. Among other things, the report examines the gender dimensions of several emerging trends in the region—increased global economic integration, the rising use of information and communication technologies, migration, urbanization, and rapid population aging—all of which are generating new opportunities, but also new risks, for promoting gender equality. The report also contributes to the development of new data and evidence on gender and development, significantly strengthening the ability of countries to formulate evidence-based policy in this area.

Drawing on this evidence base, the report identifies four priority areas for

public action in the countries of East Asia and the Pacific. First, promoting gender equality in human development remains important where gender gaps in education and health outcomes remain large. Second, policies to close gender gaps in economic opportunity have a critical role. Such measures are often warranted on both equity and efficiency grounds. Third, initiatives to strengthen women's voice and influence—and to protect them from violence—are also called for across the region. Strengthening women's agency will enhance the quality of development decision making and, thus, development broadly. And, finally, public policy can foster the opportunities and manage the risks associated with emerging trends in the region; taking a gender-aware approach to policy making in this area will lead to better gender—and development—outcomes.

This report shows that in East Asia and the Pacific, as in other parts of the world, gender equality is both the right development objective as well as good development policy.

Pamela Cox
Vice President
East Asia and Pacific Region
The World Bank

Acknowledgments

This study has been prepared by a multisectoral, multidisciplinary team led by Andrew Mason under the guidance of Bert Hofman and Sudhir Shetty. The report was written by a core team comprising Reena Badiani, Trang Van Nguyen, Katherine Patrick, Ximena Del Carpio, and Andrew Mason, with significant contributions from Jennifer Golan. Patricia Fernandes, Anne Kuriakose, Rea Chiongson, and Daniel Mont also provided substantive inputs. New data analysis for the report was carried out by Juan Feng, Reno Dewina, and Flora Nankhuni, using the East Asia and Pacific Region's Poverty Monitoring Database. Background papers for the report were written by Sarah Iqbal, Nayda Almodovar Reteguis, Yasmin Klaudia bin Humam, Josefina Posadas, Reena Badiani, John Rang, Benedikte Bjerge, Chris Sakellariou, and Dongxiao Liu. Qualitative studies on women's economic decision making in Fiji, Indonesia, Papua New Guinea, and Vietnam were coordinated by Carolyn Turk and Patti L. Petesch as part of a 22-country research effort carried out under the auspices of the *World Development Report 2012* on gender equality and development. Production support was provided by Lynn Yeargin, Cathryn Summers, and Mildred Gonsalvez.

The team is grateful for the ongoing support and guidance provided by members of the East Asia and Pacific Region's management, including James Adams, Pamela Cox, Vikram Nehru, John Roome, Emmanuel Jimenez, Linda Van Gelder, Magda Lovei, Xiaoqing Yu, Annette Dixon, Victoria Kwakwa, and Coralie Gevers. The team also appreciates the support of the members of the Region's Gender Practice Group, past and present, including Nina Bhatt, Helene Carlsson Rex, Markus Kostner, Eduardo Velez, and Lester Dally. Furthermore, the team benefitted from the ongoing interaction, coordination, and support of the *World Development Report 2012* team, namely Ana Revenga, Sudhir Shetty, Ana Maria Munoz, Carolina Sanchez-Paramo, Luis Benveniste, Markus Goldstein, Jishnu Das, and Aline Coudouel.

During the report's preparation, the team received helpful comments from four peer reviewers: Gillian Brown, Jeni Klugman, Pierella Paci, and Martin Rama. Several country teams, represented by their Country Gender Coordinators (including Laura Bailey, Edith Bowles, Yulia Immajati, Vanna Nil, Solvita Klapare, Erdene Ochir Badarch, and Pamornrat Tansanguanwong), also provided valuable feedback at various stages.

In addition, the team received constructive comments and inputs from Keith Bell, Shubham Chaudhury, Qimiao Fan, Mary Hallward Dreimeier, Mathew Verghis, Robert Jauncey, Gladys Lopez-Acevedo, David Newhouse, Carmen Niethammer, Bob Rijkers, Sevi Simavi, Monica das Gupta, Isabella Micali Drossos, Thuy Thi Thu Nguyen, Tehmina Khan, and Carlos Sobrado. Mark Ingebretsen, Patricia Katayama, and Andrés Meneses, from the Office of the Publisher, provided excellent support in the design and publication of the report. Mohamad Al-Arief and Carl Hanlon provided invaluable guidance on a dissemination and communication strategy.

The team benefited from early consultations with policy makers, civil society representatives, academics, and development partners in Indonesia, Mongolia, Thailand, and Vietnam. It also gained from feedback on preliminary findings and messages from the participants of the International Association for Feminist Economics, held in June 2011 in Hangzhou, China; from country team colleagues in Cambodia, the Lao People's Democratic Republic, and Thailand who participated in a video seminar in September 2011; from participants in the Asia-Pacific Economic Cooperation (APEC) Women and the Economy Summit held in San Francisco in September 2011; and from colleagues, counterparts, and other stakeholders who participated in events in Australia and New Zealand in November 2011; Japan, Indonesia, and Vietnam in December 2011; and in Papua New Guinea, the Solomon Islands, and Fiji in March 2012.

Generous financial support for the preparation and dissemination of this report was provided by AusAID.

Abbreviations

ADAPT	An Giang Dong Thap Alliance for the Prevention against Trafficking
ADB	Asian Development Bank
AED	Academy for Education Development
AGI	Adolescent Girls Initiative
ALMPs	active labor market policies
ASEAN	Association of South East Asian Nationals
AusAID	Australian Agency for International Development
BFC	Better Factories Cambodia
BMI	body mass index
BPS	Badan Pusat Statistik
BREAD	Bureau for Research and Economic Analysis of Development
CAPWIP	Center for Asia-Pacific Women in Politics
CAREM-Asia	Coordination of Action Research on AIDS, Mobility-Asia
CCT	conditional cash transfers
CDD	community-driven development
CEDAW	Convention on the Elimination of All Forms of Discrimination against Women
CGA	Country Gender Assessment
CIDA	Canadian International Development Agency
CWCC	Cambodian Women's Crisis Centre
CWDI	Corporate Women Directors International
DFID	Department for International Development (United Kingdom)
DHS	Demographic and Health Surveys
EAP	East Asia and Pacific
EC	European Commission
ECA	Europe and Central Asia
ECPAT	End Child Prostitution in Asian Tourism
EFA	education for all
EPZ	Export Processing Zone
FAO	Food and Agriculture Organization of the United Nations
FCND	Food Consumption and Nutrition Division

FODE	Flexible Open and Distance Education
GBV	gender-based violence
GDC	Gender and Development for Cambodia
GDP	gross domestic product
GEM	gender equity model
GNP	Grand National Party
G-PSF	Government-Private Sector Forum
GRID	Global Resource Information Data Base
HEF	Health Equity Fund
HNP	Health Nutrition and Population
HSI	Hang Seng Index
ICT	information and communication technology
ICRW	International Center for Research on Women
IDEA	International Development Evaluation Association
IFC	International Finance Corporation
ILO	International Labour Office
INSTRAW	United Nations International Research and Training Institute for the Advancement of Women
IOM	International Organization for Migration
ISCO	International Standard Classification of Occupations
ISD	Indices of Social Development
KDP	Kecamantan Development Program
KILM	Key Indicators of the Labour Market
LAC	Latin America and the Caribbean
Lao PDR	Lao People's Democratic Republic
MDGs	Millennium Development Goals
MENA	Middle East and North Africa
MIC	middle income countries
MMR	maternal mortality rate
MPDF	Mekong Project Development Facility
MSMEs	micro, small, and medium enterprises
NCRFW	National Commission on the Role of Filipino Women
NGOs	nongovernmental organizations
NIPH	National Institute of Public Health
NIS	National Institute of Statistics
NSD	National Statistics Directorate
NSO	National Statistics Office
OECD	Organisation for Economic Co-operation and Development
ORC	Opinion Research Corporation
PEKKA	Indonesia Women Headed Household Program
PISA	Programme for International Student Assessment
PPP	purchasing power parity
SAR	special administrative region
SciCon	Science Connections
SEZ	Special Economic Zone
SHG	self-help group
SIGI	Social Institutional and Gender Index
SMEs	small and medium enterprises
SPC	Secretariat of the Pacific Community
SRB	sex ratios at birth
STI	sexually-transmitted infections
SUSI	Survey of Cottage and Small-Scale Firms (*Survei Usaha Terintegrasi*)

ThaiHealth	Thai Health Promotion Foundation
TIMSS	Trends in International Mathematics and Science Study
UN	United Nations
UNDP	United Nation Development Programme
UNESCO	United Nations Educational, Scientific and Cultural Organization
UNFPA	United Nations Fund for Population Activities
UNICEF	United Nations Children's Fund
UNIFEM	United Nations Development Fund for Women
VCCI	Vietnam Chamber of Commerce and Industry
WBL	Women, Business, and the Law
WDR	*World Development Report*
WHO	World Health Organization

ISO 3166 country name abbreviations

AUS	Australia
CHN	China
FJI	Fiji
FSM	Federated States of Micronesia
HKG	Hong Kong SAR, China
IDN	Indonesia
JPN	Japan
KHM	Cambodia
KIR	Kiribati
KOR	Republic of Korea
LAO	Lao People's Democratic Republic
MHL	Marshall Islands
MMR	Myanmar
MNG	Mongolia
MYS	Malaysia
NZL	New Zealand
PHL	Philippines
PLW	Palau
PNG	Papua New Guinea
PRK	Democratic Republic of Korea
PYF	French Polynesia
SGP	Singapore
SLB	Solomon Islands
THA	Thailand
TMP	Timor-Leste
TON	Tonga
TUV	Tuvalu
TWN	Taiwan, China
VNM	Vietnam
VUT	Vanuatu
WSM	Samoa

Overview

In recent decades, women across the globe have made positive strides toward gender equality. Literacy rates for young women and girls are higher than ever before, while gender gaps in primary education have closed in almost all countries. In the last three decades, over half a billion women have joined the world's labor force (World Bank 2011c). Progress toward gender equality in East Asia and the Pacific has been similarly noteworthy. Most countries in the region have either reached or surpassed gender parity in education enrollments. Health outcomes for both women and men have improved significantly. Female labor force participation rates in the region are relatively high. Yet, despite considerable progress in this economically dynamic region, gender disparities persist in a number of important areas—particularly in access to economic opportunity and in voice and influence in society. For policy makers in East Asian and Pacific countries, closing these gender gaps represents an important challenge to achieving more inclusive and effective development.

The East Asia and Pacific Region's significant economic growth, structural transformation, and poverty reduction in the last few decades *have been associated with reduced gender inequalities in several dimensions.* The region grew at 7 percent on average between 2000 and 2008 (figure O.1), the structure of the region's economies has shifted away from agriculture toward manufacturing and services, and extreme poverty has fallen dramatically. Indeed, the share of the region's population living on less than US$1.25 a day has declined by more than 50 percent since 1990—from among the highest rates of poverty in the world to among the lowest (figure O.2). Growth, structural transformation, and poverty reduction have been accompanied by considerable progress toward gender equality in several key areas, particularly education and health. Many countries in the region have experienced closing gender gaps in school enrollments and declining maternal and child mortality rates.

But growth and development have not been enough to attain gender equality in all its dimensions. Women still have less access than men to a range of productive assets and services, including land, financial capital, agricultural extension services, and new information technologies. Substantial employment segregation by gender remains. Women are less likely than men to work in

1

FIGURE O.1   The East Asia and Pacific region has experienced rapid economic growth

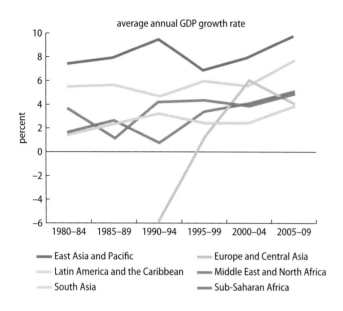

Source: World Development Indicators (WDI) database.

FIGURE O.2   Poverty reduction in the East Asia and Pacific region has been impressive

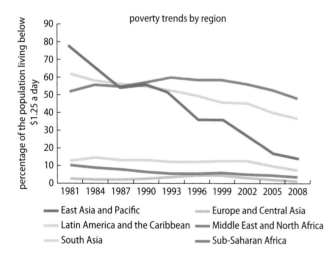

Source: PovcalNet.

formal sector jobs and more likely to work in poorly remunerated occupations and enterprises. And despite closing of education gaps, women continue to be paid less than men for similar work. Women in East Asian and

Pacific countries still have a weaker voice and less influence than men, whether in household decision making, in the private sector, in civil society, or in politics. Moreover, women across the region remain vulnerable to gender-based violence, often at the hand of an intimate partner.

This report clarifies empirically the relationship between gender and development and outlines an agenda for public action to promote gender equality in East Asian and Pacific countries. The report was written as a companion to the *World Development Report 2012: Gender Equality and Development* (World Bank 2011c) and is intended as a tool to help policy makers in the region promote both gender equality and more effective development. Following the *World Development Report 2012*, this report focuses on gender outcomes in three domains: (a) endowments—human and productive capital; (b) economic opportunity—participation and returns in the economy; and (c) agency—women's voice and influence in all facets of society.

The report makes several distinct contributions to policy makers' understanding of gender, development, and public policy in East Asian and Pacific countries.

- First, the analysis focuses on issues that are particularly relevant to the region. Compared with other developing regions, for example, female access to basic education is no longer a first-order concern in most East Asian and Pacific countries. Gender stereotyping and gender "streaming" in education still represent critical challenges, however, and thus receive particular emphasis in the report.
- Second, the report examines the gender dimensions of several emerging trends that are important to the region's development: increased global economic integration, rising use of information and communication technologies (ICTs), migration, urbanization, and rapid population aging. These trends have gender dimensions that

are not commonly accounted for by policy makers but that will generate a distinct set of challenges for promoting gender equality going forward.

- Third, the East Asia and Pacific region is vast and diverse, with important differences in economic and social characteristics that affect progress toward gender equality. The report accounts for intraregional diversity in a way that is not possible in a global report. Particular emphasis is placed, where possible, on the challenges faced by countries in the Pacific as distinct from those in East Asia.
- Finally, the report has undertaken extensive empirical analysis of gender equality using a newly created database of household surveys for the region. In doing so, the report has contributed significantly to the development of indicators and evidence on gender, development, and public policy that were not available previously.

Why does gender equality matter for development?

Gender equality matters intrinsically. Nobel prize–winning economist Amartya Sen transformed the discourse on development when he argued that development not only is about raising people's incomes or reducing poverty but rather involves a process of expanding freedoms equally for *all* people (Sen 1999).[1] Viewed from this perspective, gender equality is intrinsically valued. The near-universal ratification and adoption of the Convention on the Elimination of All Forms of Discrimination against Women (CEDAW)[2]—and the subsequent commitment of the international community to Millennium Development Goals 3 and 5—underscores a near-global consensus that gender equality and women's empowerment are development objectives in their own right.

Gender equality also matters for development. A growing body of empirical literature from around the world demonstrates that promoting gender equality is also good development policy, or as stated in the *World*

Development Report 2012 (World Bank 2011c, 3), "Gender equality is smart economics." Indeed, the literature shows that greater gender equality in endowments, access to economic opportunities, and agency can (a) contribute to higher productivity, income growth, and poverty reduction; (b) improve the opportunities and outcomes of the next generation; and (c) enhance development decision making. This section explores the evidence on these three pathways, in turn.

Gender equality can contribute to higher productivity and income growth

For households and economies to function at their full potential, resources, skills, and talent should be put to their most productive use. If societies allocate resources on the basis of one's gender, as opposed to one's skills and abilities, this allocation comes at a cost. Indeed, the economic costs of gender inequalities—whether caused by the persistence of traditional norms or by overt discrimination—can be considerable. A recent study commissioned for the *World Development Report 2012* found that in the East Asia and Pacific region, output per worker could be 7 to 18 percent higher across a range of countries if female entrepreneurs and workers were to work in the same sectors, types of jobs, and activities as men and to have the same access to productive resources (Cuberes and Teignier-Baqué 2011).

Evidence suggests that misallocation of female skills and talent commonly begins before women enter the labor force, when families and societies underinvest in girls' schooling. A number of cross-country studies have found a robust inverse relationship between the size of the gender gap in education and gross domestic product (GDP) growth, controlling for average education levels and other factors associated with economic growth (see, for example, Klasen 2002; Knowles, Lorgelly, and Owen 2002).[3] Moreover, to the extent that young women (or men) choose fields of study on the basis of their gender rather than their abilities, this too will exact costs not only on individuals' employment and earnings, but also on a country's economic productivity more broadly.

Gender inequalities in access to productive assets also have costs in terms of productivity and income. Microeconomic studies from a number of countries across developing regions show that female farmers and entrepreneurs are inherently no less productive than male farmers and entrepreneurs; rather, they tend to have less access to productive inputs.[4] A recent study by the Food and Agriculture Organization of the United Nations estimates that equalizing access to productive resources between female and male farmers could increase agricultural output in developing countries by 2.5 to 4.0 percent (FAO/Sida Partnership Cooperation 2010).

A number of studies show that gender-based violence also imposes significant costs on the economies of developing countries through lower worker productivity and incomes, lower human capital investments, and weaker accumulation of social capital (Morrison, Ellsberg, and Bott 2007). In addition to indirect costs, gender-based violence has large direct economic costs on society. A study in the United States found that the direct health care costs of intimate partner violence against adult women were more than $4 billion USD in 1995 (USCDC 2003). Reducing gender-based violence would thus have significant positive effects on the region's economies by reducing health care costs and increasing investments in women's human capital, female worker productivity, and women's accumulation of social capital.

As the global economy becomes more integrated, the productivity effects associated with greater gender equality are likely to be increasingly important to East Asian and Pacific countries. Recent studies on the relationship between gender and trade suggest that gender inequalities have become financially detrimental for countries in a world of open trade (Do, Levchenko, and Raddatz 2011). To participate effectively in

an increasingly competitive world, countries will need to harness their resources efficiently by improving opportunities for all and allocating labor on the basis of skill instead of gender. Gender inequality, whether in endowments, economic opportunities, or agency, reduces a country's ability to compete in this increasingly globalized economic environment (World Bank 2011c).

Promoting gender equality is also an investment in the next generation

A large body of cross-country and country-specific literature shows that healthier, better educated mothers have healthier, better educated children, which can be expected to positively affect children's future productivity and economic prospects. The effects begin even before childbirth. In Timor-Leste, highly educated mothers and those in the wealthiest households are more likely to have their babies delivered by skilled attendants than less educated mothers and those from poorer households (NSD, Ministry of Finance, and ICF Macro 2010). Similarly, Demographic and Health Survey data show that Cambodian women with little education are relatively less likely to receive prenatal care and assistance from trained health personnel during birth deliveries than women with more education (Johnson, Sao, and Hor 2000). A mother's health and nutrition status is also found to strongly affect children's physical health as well as cognitive and noncognitive abilities, which can have long-lasting developmental and societal consequences (Naudeau et al. 2011).

Higher labor force participation as well as income earned and assets held by women have also been shown to have positive effects on the next generation. In Indonesia, for example, women with a higher share of household assets before marriage tend to use more prenatal care and are more likely to have their births attended by skilled health care providers (Beegle, Frankenberg, and Thomas 2001). Similarly, in China, increasing adult female income by 10 percent of the average household income raised the fraction of surviving girls by 1 percentage point and increased years of schooling for both boys and girls. In contrast, a similar increase in male income reduced survival rates and educational attainment for girls with no impact on boys (Qian 2008). Studies from across developing and developed regions (for example, from places as diverse as Brazil, Ghana, South Africa, and the United States) show that income in the hands of women positively affects their female children's health (Duflo 2003; Thomas 1995); commonly, the marginal effects of income and assets in the hands of mothers are larger than effects of similar income and assets in the hands of fathers.

Reductions in gender-based violence through greater female agency can also have important intergenerational benefits. Several studies show that experiencing domestic violence between parents as a child contributes to a higher risk of both women experiencing domestic violence as adults and of men perpetrating violence against their spouses (Fehringer and Hindin 2009). In Timor-Leste, 56.4 percent of women who were victims of spousal violence had a father who beat their mother (NSD, Ministry of Finance, and ICF Macro 2010). In Cambodia, women who reported that their mothers experienced domestic violence were more likely to experience physical and psychological domestic violence as well (NIPH, NIS, and ORC Macro 2006). Efforts that increase women's safety and security and that reduce domestic violence can thus lead to lower intergenerational transmission of violence within families.

Strengthening women's voice can enhance the quality of development decision making

Several studies show that women and men have different policy preferences (Edlund and Pande 2001; Lott and Kenny 1999). Despite perceptions in some East Asian and Pacific countries that women do not make

as good leaders as men, studies suggest that capturing these gender-based differences in perspective can lead to not only more representative but also better decision making. Evidence from South Asia suggests that development policy making can benefit from greater gender equality in voice. As an example, a study of women elected to local government in India found that female leadership positively affected the provision of public goods at the local level in ways that better reflected both women's and men's preferences (Chattopadhyay and Duflo 2004). Similarly, studies from rural India and Nepal found that when women who were previously excluded from decisions about local natural resource management gained greater voice and influence, local conservation outcomes improved significantly (Agarwal 2010a, 2010b).

Women's collective agency can also be transformative, both for individuals and for society as a whole. For example, for a group of ethnic minority women in rural China, information sharing among them has helped empower them and raise their social standing in the Han-majority communities into which they married (Judd 2010). In a more formal setting, over the last 15 years, migrant domestic workers in Hong Kong SAR, China, have been engaged in civic action focused on local migrant workers' rights as well as international human rights (Constable 2009). These efforts have contributed to the enactment of laws that now provide migrant domestic workers in Hong Kong SAR, China, with some of the most comprehensive legal protections in the world.

Recent progress, pending challenges

Over the last few decades, most East Asian and Pacific countries made considerable progress toward gender equality in several dimensions. In other dimensions, gender disparities have been more persistent. This section reviews recent progress and pending challenges in the region, noting where economic growth and development have contributed to advances and where they have been insufficient.

Growth and development have been accompanied by reduced gender inequalities in several dimensions

Many gender gaps in education have closed. Over the last few decades, boys' and girls' schooling outcomes have converged at levels that are high by international standards. East Asia and the Pacific has performed better than other developing regions, in terms of both increasing female and male educational enrollments and raising the female-to-male enrollment ratio. In 2010, the region had the highest primary school ratio of female-to-male enrollments among all developing regions; at the secondary level, only Latin America and the Caribbean had a higher female-to-male enrollment ratio (figure O.3).

Key health outcomes have improved. During the past half century, the region has experienced significant advances in several health indicators. Fertility rates have

sharply declined, and under-five mortality rates have halved since 1990 for both boys and girls. Noteworthy gains have been made in birth attendance by health professionals. In addition, the East Asia and Pacific region has seen substantial declines in the maternal mortality rate, from approximately 200 deaths per 100,000 live births in 1990 to 100 in 2008 (figure O.4).

Gender gaps in labor force participation have narrowed. Female labor force participation in East Asian and Pacific countries is high by international standards (figure O.5), and among younger cohorts, female labor force participation has tended to rise over time. Moreover, as countries grow and develop, women are increasingly moving into jobs in the nonagricultural sector and are migrating to urban areas in search of better employment opportunities. Trends and patterns of labor force participation look similar to those observed in the United States and other countries of the Organisation for Economic Co-operation and Development (OECD) during their economic transformations.

In many ways, women in East Asia and the Pacific are better positioned today than ever before to participate in and contribute to their countries' development.

FIGURE O.3   Girls' secondary school enrollments have converged to those of boys

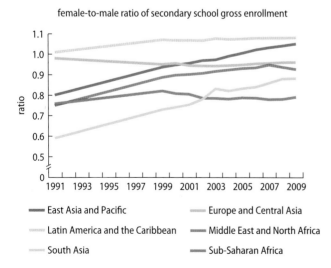

female-to-male ratio of secondary school gross enrollment

- East Asia and Pacific
- Latin America and the Caribbean
- South Asia
- Europe and Central Asia
- Middle East and North Africa
- Sub-Saharan Africa

Source: WDI database, 2011 data.

Despite progress, important challenges to promoting gender equality remain

Progress has been uneven across the region. Substantial variation remains across countries, both in overall enrollment rates and in female-to-male enrollment ratios. Countries such as Cambodia, the Lao People's Democratic Republic (Lao PDR), and Papua New Guinea still have relatively low enrollment levels and low female-to-male enrollment ratios, particularly at the secondary school level. Furthermore, although countries have experienced convergence in enrollment among the young, substantial gaps still remain in the educational endowments of adult populations.

Maternal mortality remains high in lower-income countries and in several Pacific countries (figure O.4). In Lao PDR, for example, maternal mortality rates were still more than 500 deaths per 100,000 births in 2008, among the highest rates in the world. Indonesia's maternal mortality rate remains high compared to other countries in the region at similar levels of development.

Substantial differences in labor force participation occur across countries in the region, even among countries with similar income levels. Relative to their income levels, countries such as China and Vietnam have substantially higher rates of female labor force participation than the world average, whereas participation is near the world average in countries such as Indonesia and the Philippines, and below average in countries such as Fiji and Malaysia (figure O.5).

Within countries, interactions between gender and other socioeconomic characteristics can often exacerbate disparities. Economically disadvantaged and minority populations often experience lower educational enrollments, for example. In Vietnam, school participation among 15- to 17-year-olds is substantially higher among the Kinh and Hoa (Chinese) majorities than among many of the 52 ethnic minority populations. Among the more economically disadvantaged and less well integrated Hmong, Dao, and Khmer minorities, far fewer girls attend school than boys (Baulch et al. 2002).[5] Geographic distance, or remoteness, can also serve to compound gender disadvantage. Women in remote rural areas commonly have limited access to health care, significantly raising the risks associated with pregnancy and childbirth. While Vietnam has experienced noteworthy declines in maternal mortality, on average, over the last decade, progress has been much slower in remote and ethnic minority regions (World Bank 2011b).

Some gender disparities fail to close— or close very slowly—with development

More than a million girls and women per year are "missing" in East Asia. Among

FIGURE O.4   Maternal mortality rates have declined in most countries in the region

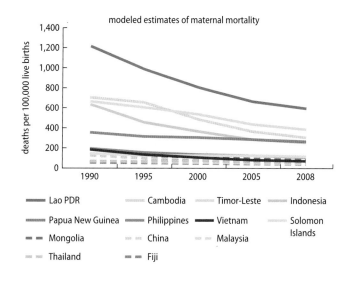

Source: WDI database, Gender Statistics, 2010 data.

FIGURE O.5   Female labor force participation is high by global standards but also varies substantially across the region

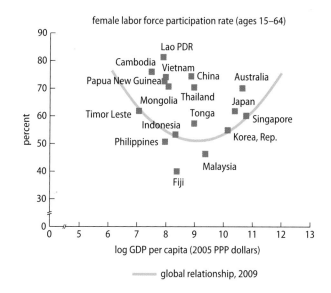

Sources: World Bank staff estimates using Key Indicators of Labour Market (KILM) labor force data (International Labour Organization) and purchasing power parity adjusted GDP per capita (in logs and at 2005 prices) from the Penn World Tables.
Note: GDP = gross domestic product, PPP = purchasing power parity. The data shown for each country are from 2009 and the estimated U-shaped relationship uses data from across the world.

the most concerning issues is that despite growth and development, the problem of missing girls remains significant. The term "missing women" was first coined

FIGURE O.6 Across the region, female-headed households own less land than male-headed households

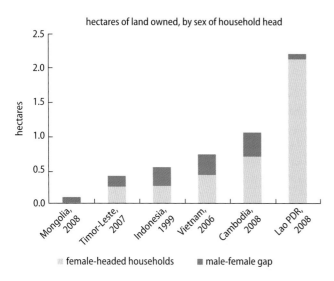

hectares of land owned, by sex of household head

female-headed households ■ male-female gap

Source: World Bank staff estimates using household income and expenditure surveys.

FIGURE O.7 Women in East Asia and the Pacific still earn less than men

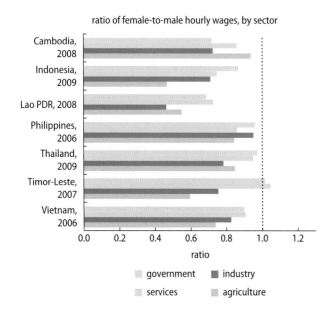

ratio of female-to-male hourly wages, by sector

government ■ industry
services agriculture

Source: World Bank staff estimates using household income and expenditure surveys.

by Amartya Sen (1999) to refer to the phenomenon that many low-income countries have far fewer women than men relative

to what is observed in developed countries. Sen argued that this imbalance in sex ratios reflected severe forms of gender bias in affected societies. Biological differences between males and females imply that approximately 105 boys are born for every 100 girls. Nonetheless, China, Vietnam, and until recently, the Republic of Korea have experienced substantial deviations from the biological norm, and the trend over time, particularly in China, has been alarming. In China, the number of girls who are missing per year at birth increased from 890,000 in 1990 to 1,092,000 in 2008. Missing girls as a fraction of the total number of female births increased from 8.6 percent in 1990 to 13.3 percent in 2008 (World Bank 2011c).

Gender disparities still exist in access to and control of productive resources. Gender disparities in access to and control of land and farm inputs are pervasive in the region despite growth and development. Women remain less likely to own land than men. And when women, or specifically, female-headed households, do own land, they typically have smaller holdings (figure O.6).[6] Female-headed households also tend to have poorer access to other productive inputs and support services, including livestock holdings and access to agricultural extension services.

Despite high labor force participation, important gender inequalities in economic opportunity remain. Women still earn less than men in nearly all sectors in all countries in the region (figure O.7). Gender wage gaps increase with age, reflecting in part lower levels of experience among women caused by workforce interruptions and reduced working hours during childbearing years, as well as gender disparities in education among older cohorts. Gender wage gaps in the region are also strongly influenced by occupational and sectoral segregation, mirroring patterns seen in the United States and other

OECD countries. Together, differences in education endowments, experience, and industrial and occupational segregation explain up to 30 percent of observed gender wage disparities in East Asian countries (Sakellariou 2011).

Gender wage gaps in the region are often greatest among men and women with relatively low education and skill levels. Several studies from East Asian and Pacific countries point toward "sticky floors," that is, wider wage gaps at the bottom than at the top of the earnings distribution (figure O.8). The finding of sticky floors contrasts with studies from OECD countries, which more commonly find "glass ceilings," that is, larger wage gaps among higher-earning men and women.

Women are more likely to work in small firms, to work in the informal sector, and to be concentrated in lower-paid occupations and sectors. Within firms, women are more likely than men to be temporary workers. Such employment segregation affects a number of economic outcomes by gender, including earnings, returns to education and experience, social security coverage, and exposure to shocks. Substantial gender-based occupational and sectoral segregation is seen in all countries and does not decline with development. In fact, employment segregation tends to increase as economies become more diverse with development. Economic growth and, in particular, urbanization appear to make occupational and sectoral segregation by gender more pronounced, particularly during the early stages of economic structural change.

Similarly, female-led enterprises tend to be smaller and more precarious than male-led enterprises (figure O.9). The micro, small, and medium firm sectors are important segments of most East Asian and Pacific economies and contribute a substantial fraction of GDP. Female-led enterprises across the region, particularly in the informal sector, have lower profits, are less likely than male-led enterprises to be registered,

FIGURE O.8 **In urban China and Indonesia, gender wage gaps are largest among low wage earners**

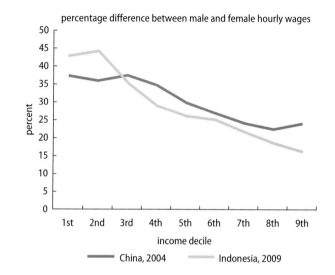

Sources: Chi and Li 2007; Sakellariou 2011.

FIGURE O.9 **Enterprises with female managers tend to be smaller**

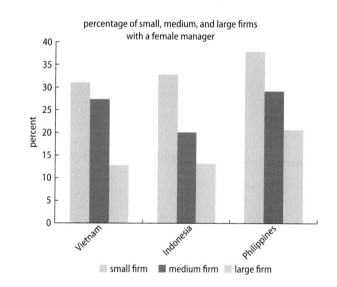

Source: World Bank staff estimates using Enterprise Surveys database for 2006–11.

have fewer employees and assets, and are more likely to be home based or to operate out of nonpermanent premises. Although female-owned and -managed enterprises are not inherently less productive, they tend to be smaller, less capitalized, and located in less remunerative sectors.

Women in the region still have less voice and influence than men. Women's household decision-making power in East Asia and the Pacific is relatively high, but levels of autonomy vary across the region. Women's autonomy in the household can be measured in several ways, including control of assets, freedom of physical mobility, and voice in decision making. By several of these measures, including control over large household purchases and visiting family and relatives, women in East Asian countries appear to have relatively high autonomy compared with women in other developing regions (World Bank 2011c). Women in the Pacific have relatively less control over their own earnings, however. Over 15 percent of women in the Marshall Islands, 15 percent in Samoa, and 13 percent in Tuvalu report that their husbands control their cash earnings (figure O.10). Moreover, 58 percent of partnered women in the Solomon Islands and 69 percent of partnered women in Vanuatu report that they have experienced some sort of controlling behavior by their partners. This includes preventing them from seeing family, wanting to know where they are at all times, forbidding contact with other men, and controlling their access to health care (SPC and NSO 2009; VWC 2011).

Women's voice and influence in the public domain—as measured by representation in national and local political assemblies—remains low. The share of female parliamentarians in East Asian and Pacific countries is just below the global average, at approximately 18 percent in 2011. Despite economic growth and development in the region, this figure has barely changed since 1990. Although the share of women in national assemblies varies considerably across the region, it is systematically lower in the Pacific than in East Asia (figure O.11). Indeed, in no country in the Pacific does the share of parliamentarians who are female exceed 10 percent, and four countries—the Federated States of Micronesia, Nauru, Palau, and the Solomon Islands—have no female parliamentarians.

The prevalence of gender-based violence is high in the region, and particularly so in the Pacific where the prevalence of domestic violence is among the highest in the world. As can be seen in figure O.12, 68 percent of ever-married women 15–49 years of age in Kiribati, 64 percent in the Solomon Islands, and 60 percent in Vanuatu have experienced physical or sexual violence at the hands of an intimate partner (SPC, Ministry of Internal and Social Affairs, and Statistics Division 2010; SPC and NSO 2009; VWC 2011). Although no nationally representative data exist for Papua New Guinea, studies conducted at the subnational level suggest that domestic violence is just as prevalent (Ganster-Breidler 2010; Lewis, Maruia, and Walker 2008). This violence is a linchpin to a bigger story; violence against women represents the extreme deprivation of voice and freedom among

FIGURE O.10   Who decides how wives' cash earnings are used varies widely across the region

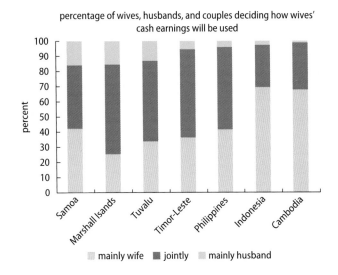

percentage of wives, husbands, and couples deciding how wives' cash earnings will be used

☐ mainly wife ■ jointly ▨ mainly husband

Sources: Demographic and Health Surveys, various years.

FIGURE O.11   **Women's representation in parliament is low, especially in the Pacific**

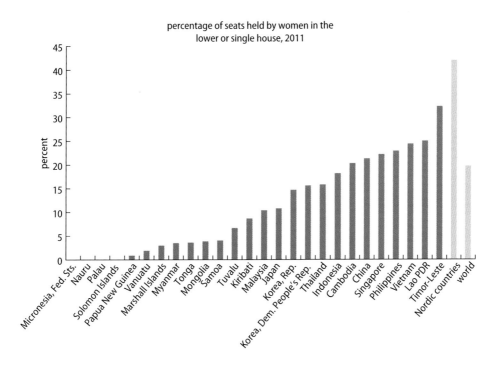

Source: PARLINE database (Inter-Parliamentary Union).

women and, as such, is often associated with a lack of agency in other dimensions.

Awareness is increasing that men and boys—not just women and girls—face gender-specific risks. Some countries in the region have started to experience a reverse gender gap in education; girls' secondary school enrollment now exceeds that of boys in China, Fiji, Malaysia, Mongolia, the Philippines, Samoa, and Thailand. Reverse gender gaps at the tertiary level are sometimes even starker: in Thailand, 122.4 females were enrolled for every 100 males in 2008. In addition, men across the region experience higher levels of morbidity and premature mortality related to substance abuse. The prevalence of smoking and drinking among males in East Asian and Pacific countries is much higher than the prevalence among females.

FIGURE O.12   **Violence against women is high in the region**

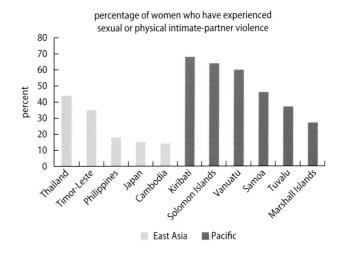

Sources: Demographic and Health Surveys, various years, and government surveys.
Note: Data for Thailand are for Bangkok and Nakhonsawan only, and data for Japan are for Yokohama only.

Why do many gender inequalities persist?

Low household incomes, weak service delivery, and traditional norms can impede gender equality in education and health

Where gender gaps in education are still observed, low income coupled with high costs of education can limit household demand for schooling. Traditional gender norms and practices also strongly influence household schooling decisions. Participants of focus group discussions in a qualitative research exercise in Papua New Guinea report, for example, that parents value boys' education over girls' education. The reason is that males will carry the family name and become household heads whereas females are expected to submit to their husbands and be caregivers and homemakers.

Weak systems of service delivery also constrain progress in education—overall as well as for girls. In Cambodia, Lao PDR, and Papua New Guinea, for example, school enrollments are low overall, and gender gaps persist. Low enrollment rates in Papua New Guinea also reflect limited physical access to schools and high dropout rates. Poor and sparse school infrastructure, poor teacher attitudes and attendance, lack of teachers in remote areas, and negative pupil behavior all contribute to low overall enrollments. Long distances to schools have been observed to make school attendance costly in both Cambodia and Lao PDR, particularly for girls, because long travel distances raise safety

FIGURE O.13 Women are concentrated in certain fields of study, such as education and medicine, but are underrepresented in law and engineering

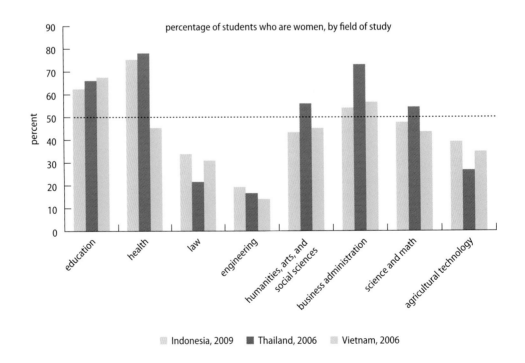

Source: Sakellariou 2011.

concerns among parents. The lack of toilets at many schools makes attendance more difficult for girls than boys.

Gender streaming in education largely reflects societal norms and expectations and has implications for gender inequalities in job placement and earnings. Substantial differences remain in the composition of education between men and women in the region (figure O.13). Economic returns or comparative advantage of females and males in different fields of study do not appear to explain education streaming. Social norms about appropriate work for women and men, role models in the labor market, and gender stereotyping in school curricula play important roles. In East Asia and the Pacific, teaching materials more frequently portray males than females in active and leadership roles. Women are often depicted as secretaries, assistants, nurses, and teachers whereas men are portrayed as doctors, politicians, or police officers. Gender streaming in education ultimately affects the type of work that women and men do and, importantly, affects their respective abilities to take advantage of existing and emerging economic opportunities.

Poor service delivery and cultural norms about birthing practices contribute in large part to high maternal mortality in several East Asian and Pacific countries. Poor access to quality obstetric health services, particularly in remote rural areas, places women at higher risk for maternal death. Rural areas tend to be less well served by the health system, and rural residents have much lower access to birth deliveries attended by trained staff than do urban residents. Evidence shows that poor health infrastructure and long distances to the nearest health center are both important barriers to reducing maternal mortality in Cambodia and Lao PDR. In addition, culture and tradition play an important role in the choice of health practices, such as the location of childbirth, the use of skilled birth attendants, and sterilization practices. Preferences can vary from birth deliveries at home to

deliveries in the forest, including beliefs that women do not need prenatal care or delivery supported by skilled attendants. These factors take a heavy toll on women during pregnancy and pose higher risks of mortality related to birth complications.

Strong son preference, intensified by declining fertility and the availability of prenatal sex-identification technology, underlies the observed skewed sex ratios at birth in a few East Asian countries (China, Vietnam, and to a lesser extent now, Korea). Parents' choices to keep and care for boys over girls can depend on social norms and values, different economic opportunities by gender, and the benefits parents expect from a son compared to a daughter, including material support in old age. Although many societies have some mild degree of preference for sons, the interplay of culture, state, and political processes can generate extreme patrilineality and highly skewed child sex ratios. In addition, the manifestation of son preference is influenced by public policies—for example, China's one-child policy and Vietnam's two-child policy—and the spread of prenatal sex-determination technology since the early 1980s. For these reasons, economic growth and development alone do not necessarily reduce son preference and sex ratios at birth in East Asia.

Gender norms about masculinity play a strong role in influencing the excessive tobacco and alcohol consumption observed among men in many parts of the region. Smoking and drinking are commonly viewed as masculine behaviors. Men and boys feel substantial pressure to accept gender stereotypes that they should be strong and tough. In contrast, social disapproval of women who smoke keeps the smoking prevalence among women very low in Vietnam, for example. Survey data indicate that the primary reason why most Vietnamese women do not use tobacco is the expectation that "women shouldn't smoke." Survey respondents consider this factor more influential in affecting smoking behaviors than health concerns.

Norms regarding women's household roles and disparities in productive resources constrain economic opportunity

Gender norms related to the allocation of time to household work affect women's opportunities in the labor market because they are expected to take primary responsibility for home and family in addition to any market role. Responsibility in the household fundamentally affects all outcomes in the market sphere—from where women work and what they do to how much they earn. Women work longer total hours than men and devote significantly more time to domestic and caregiving activities (figure O.14), particularly in households with small children. Many women temporarily leave the labor force when they must care for small children or the elderly. Trade-offs between household and market work can be particularly stark in rural areas, where women can spend long hours on domestic chores because of poor infrastructure and a lack of alternative childcare options. Indeed, differences in the types of work that women and men do, along with higher rates of female presence in the informal sector, are in part caused by

women's greater need for workplace flexibility to facilitate management of their dual household and market roles.

Female-headed households in the region tend to have less access to land because of the interaction of complex legal, social, and economic factors. In the majority of countries in East Asia, statutory law does not differentiate property inheritance by gender. However, parallel statutory and customary legal systems in a number of East Asian and Pacific Island countries mean that women are not treated equally to men in the implementation of the law. Gender inequalities persist also in access to other productive inputs and support services—from livestock holdings to agricultural extension services. Although evidence on access to credit is mixed across the region, female entrepreneurs in several countries, including Timor-Leste and Tonga, report greater difficulty than their male counterparts in accessing credit. Such disparities in access to productive resources continue to impede gender equality in access to economic opportunity.

A substantial share of the gaps in productivity and profits between female- and male-led firms can be accounted for by gender-based segregation of enterprises by sector, firm size, and firm characteristics. This "sorting" of firms is found among both formal and informal enterprises and reflects both gender norms regarding time allocation to household and market work and differential access to productive inputs. In Indonesia, for example, the food, retail, and garment manufacturing sectors—where female entrepreneurs are most likely to locate—are among the least capital-intensive and productive sectors (figure O.15). By contrast, the transport sector—where male entrepreneurs are most likely to locate—has higher productivity and capital intensity.

Broader constraints to business development, such as cumbersome registration procedures, affect both female- and male-led enterprises (figure O.16). The most important issues vary by country, but, within any given country, both male and female entrepreneurs often identify similar challenges—competition, difficulty in accessing finance,

FIGURE O.14   In Lao PDR, women—particularly those with young children—must balance household and market work

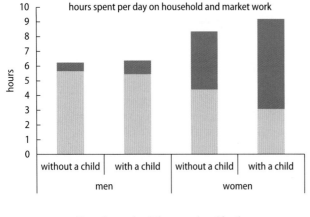

Source: World Bank staff estimates using Lao Socio-economic Survey, 2008.

and lack of electricity—and in comparable magnitudes. Evidence suggests that such constraints may be more onerous among small and informal firms than among larger firms, however, so to the extent that female-led firms are smaller and more likely to be informal, they are likely to be more adversely affected.

Gender inequalities in endowments and economic opportunity limit women's agency…

A woman's agency is affected in fundamental ways by her endowments and access to economic opportunities. Gender inequalities in educational attainment, economic assets, and own earnings can hinder women's abilities to influence their circumstances in the home, to enter and participate effectively in politics, or to leave bad or dangerous household situations. In Indonesia, for example, women with little or no education are less likely to participate in decisions involving their own health care, to make household purchases, or to engage regularly in social activities than women with at least a secondary education (BPS and ORC Macro 2003). In China and Cambodia, women with less education are less likely to enter politics—for reasons of norms or statute—than women with higher levels of education (Maffii and Hong 2010; Wang and Dai 2010). Moreover, worldwide evidence suggests that a woman's ownership and control of her own assets and income is associated with a decreased risk of intimate-partner violence (Agarwal and Panda 2007; ICRW 2006; Pronyk et al. 2006; Swaminathan, Walker, and Rugadya 2008). Women's income can also positively affect their accumulation of assets, which in turn positively affects their ability to leave an abusive partner, to cope with shocks, and to invest and expand their earnings and economic opportunities (World Bank 2011c).

… as do traditional norms regarding women's roles …

Social norms and practices can limit women's voice and influence in the home or in

FIGURE O.15   In Indonesia, female-led enterprises are clustered in lower-productivity and less capital-intensive industries

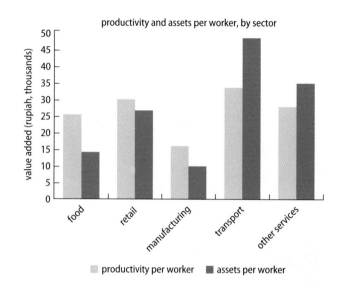

Source: World Bank staff estimates using Indonesia Family Life Survey 2007/2008.
Note: The graph shows productivity and assets per worker in five industries for firms with fewer than five workers. Productivity is measured by value added.

FIGURE O.16   Male- and female-led firms report similar constraints in Indonesia

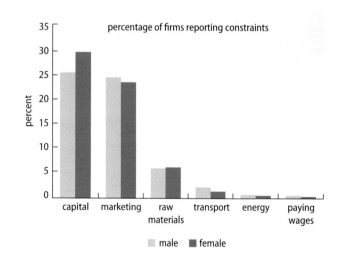

Source: World Bank staff estimates using Survey of Cottage and Small-Scale Firms (SUSI) 2002.

society. As previously noted, traditional norms about women's roles within the home constrain their economic opportunities and thus their decision-making power within the home. Traditional gender norms

FIGURE O.17   Men and, in some cases, women believe that men make better political leaders than women

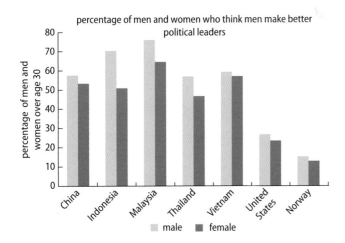

percentage of men and women who think men make better political leaders

Source: World Values Survey database, 2005–2009 data.
Note: Data for Indonesia, Malaysia, the United States, and Vietnam are for 2006; data for China and Thailand are for 2007; data for Norway are for 2008.

and social expectations also shape people's views about women's roles in the public sphere. Surveys conducted in several East Asian countries indicate, for example, that a majority of men—and sometimes a majority of women—think that men make better political leaders than women (figure O.17). Similarly, in parts of rural China, many people still think of women as less capable (*disuzhi*), and local norms dictate that they should confine their activities to the domestic settings (Wang and Dai 2010).

… and complex legal environments and, often, weak access to justice

The legal setting, along with people's access to justice, establishes the underlying environment in which women (and men) can exercise agency in the home and in society. Whether women and men are equally supported under the law, and whether their rights are protected in practice thus critically affect their voice and influence in society. Laws and access to justice also create the environment in which women and men can (or cannot) access resources and economic opportunity and accumulate assets, which also affects their agency. In several countries in the

region, the legal environment is affected by not only statutory but also customary law.

Plural legal environments, where both statutory and customary laws are practiced, can create important challenges to promoting gender equality in voice and influence. Statutory laws, customary (and sometimes religious) laws and practices can affect women's voice and influence in different ways when they bestow different rights by gender. Moreover, in practice, the interaction between statute and custom can mean that women's legal status varies substantially across ethnic (and religious) groups, even within a single country. This interaction can affect women's rights in marriage and divorce, reproductive health, education, asset ownership, inheritance, and freedom of mobility, among other things, which in turn can fundamentally affect the extent of women's agency.

In some countries, inadequate legal protection, weak implementation and enforcement, and social tolerance enable gender-based violence. Although more than three-quarters of countries in East Asia have strengthened legislation on domestic abuse in recent years (including, for example, Indonesia, Korea, Lao PDR, Thailand, and Vietnam), in the Pacific, more than 60 percent of countries still lack sufficient legislation on domestic violence (UNDP 2010). Even when countries have appropriate legislation in place, women remain unprotected by the legal system because the laws remain largely unenforced. A recent study found, for example, that officers in the Fiji Police Force Sexual Offences Unit often have unwelcoming attitudes when dealing with female victims (UNFPA 2008). The same is true in some areas in Cambodia, where many local officials still believe that a husband can threaten his wife despite the laws in place (UNDP Cambodia and VBNK 2010).

Emerging opportunities and risks in an increasingly integrated world

Several emerging trends in the region will present both new opportunities and new risks to achieving gender equality. East Asia and the Pacific are at the forefront of several global trends: increasing global economic integration, rising availability and use of ICTs, increased domestic and cross-border migration, rapid urbanization, and population aging. Because these trends have gender dimensions, they will affect the evolution of gender equality in the region. In many ways, these trends will bring with them new opportunities for gender equality. For example, increased economic integration, greater access to ICTs, and increased migration will likely all contribute to increased income earning opportunities for women. Along with new opportunities, however, these emerging trends will bring new risks.

Increasing global integration will likely continue to be an important source of nonagricultural employment growth for women, who are already highly represented in export-oriented sectors (figure O.18). And greater employment and earnings in export-orientated industries can contribute to greater female independence and autonomy in decision making. At the same time, increasing global integration can increase economic risk and uncertainty, as shocks are quickly transmitted across integrated markets. A number of studies find that while shocks do not necessarily have more adverse impacts on women than men, they do have gender-differentiated effects on outcomes as diverse as employment, earnings, labor force participation, education, health, and nutrition (see, for example, Bruni et al. 2011; Rodgers and Menon forthcoming).

Advances in ICTs are opening up opportunities for both men and women throughout the region. New and emerging technologies, if accessible, can help increase women's welfare through a number of channels by opening new economic opportunities, empowering women by breaking down information barriers, facilitating engagement of women in isolated communities in distance learning, and enabling them to take collective action. In Malaysia, for example, female entrepreneurs have created self-help cyber communities to

FIGURE O.18 **The share of female workers in export-oriented firms is relatively high**

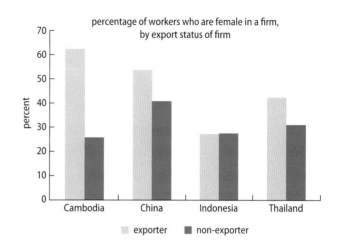

Source: World Bank staff estimates using Enterprise Surveys database for 2002–06.
Note: Share of female workers = female workers/total workers.

network and share information about starting and running a business. Limited evidence suggests that women in the region may still have lower access to information technologies than men, however. Although data from China show rapid growth in access to ICT services, Chinese women are still less likely to use the Internet or to subscribe to mobile phone services (figure O.19). Widening gender gaps in access to ICTs raise the risk of rising disparities in economic opportunity and voice going forward.

High economic growth and increased economic integration over the past three decades have spurred significant migration across the region. Women constitute nearly half of all migrants in East Asia and the Pacific and are increasingly migrating in search of better economic opportunity. Female migrants dominate a number of occupations and sectors, including labor-intensive manufacturing and export-oriented industries, and domestic work. Migration can provide women with increased economic opportunity, give them the chance to improve their knowledge and skills, and increase their agency through raising their contributions to family income.

FIGURE O.19   **China has seen remarkable growth in Internet use since 2000, but women's use trails men's**

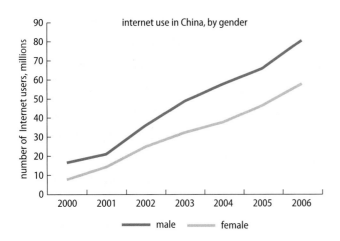

internet use in China, by gender

Source: CNNIC (China Internet Network Information Center) Internet Statistics.

At the same time, migration brings with it important gender-specific risks. For example, many female migrants work as domestic workers, an occupation with particularly weak worker protections in most countries. Female migrants are also disproportionately susceptible to human trafficking.

Many East Asian and Pacific countries are experiencing unprecedented levels of urbanization as migrants move to urban areas in search of economic opportunity. Between 2000 and 2015, Indonesia, China, and Cambodia are predicted to see an increase of the population residing in urban areas by 17, 13, and 9 percentage points, respectively (UN 2010). Urbanization affects all aspects of life, from the nature of employment to the availability of services to one's ability to rely on extended family and community networks for support. These changes almost certainly have gender-specific impacts. While urban areas can open up a wider range of economic opportunities for both men and women, women's ability to take advantage of new opportunities is likely to depend more fundamentally on the nature and availability of urban services—for example, whether transportation systems facilitate their safe travel to job sites or affordable child care can

compensate for the loss of extended family networks.

Finally, the high-income economies in East Asia are experiencing rapid population aging. Most emerging countries in the region have also begun this process; dependency ratios are already increasing in many middle-income countries in East Asia and the Pacific. Old-age dependency is expected to increase even more quickly in the coming decades (figure O.20). Population aging is likely to have gender-differentiated effects at all age levels. Gender differences in time devoted to caring for the elderly imply that in the absence of institutionalized care services, women are likely to bear the brunt of the increased demand for elder care (Dwyer and Coward 1992; Ofstedal, Knodel, and Chayovan 1999). In addition, while women tend to live longer than men, gender differences in education and labor force participation imply that women are less likely to be vested in formal pension systems and may have fewer assets to ensure a basic level of well-being in old age.

At present, these emerging trends have gender dimensions that are not commonly accounted for by policy makers. Nonetheless, these trends will generate a distinct set of challenges for promoting gender equality going forward. An important role for public policy, therefore, will be to support women (and men) in taking advantage of emerging opportunities while protecting them against the emerging risks.

Toward gender equality in East Asia and the Pacific: Directions for policy

The collection of evidence points to four priority areas where public policy can contribute to greater gender equality and more effective development in East Asian and Pacific countries:

- First, promoting gender equality in human development remains a priority where gender gaps in education are large or health outcomes are poor; closing gaps in human development, where they persist, is likely to yield high returns.

- Second, taking active measures to close gender gaps in economic opportunity is often warranted on both equity and efficiency grounds. Which policy levers will yield the highest returns depends on the structure of the country's economy and which specific constraints are most binding.
- Third, taking measures to strengthen women's agency—and to protect them from violence—is also called for across the region; strengthening women's voice and influence will contribute to the quality of development decision making and thus to development more broadly.
- Fourth, public policy has a critical role in fostering new opportunities and managing emerging risks associated with increasing global economic integration, the rising role of ICTs, increasing migration, rapid urbanization, and population aging.

The following sections examine policy approaches to promoting gender equality in East Asia and the Pacific in these four priority areas, drawing on recent experience from the region and beyond.

Promoting gender equality in human development

Closing persistent gender gaps in human development

In countries with unequal gender outcomes in education and health, the priority remains improving these outcomes. In East Asia and the Pacific, gender gaps in human development at the national level tend to persist where overall outcomes are low. In such cases, public action to strengthen countries' education and health systems will be called for to improve gender (as well as overall) outcomes. For countries with more localized gender disparities, for example, among specific ethnic groups or in remote, rural regions, more targeted interventions may be warranted. The exact constraints

FIGURE O.20   The old-age dependency ratio is increasing in most East Asian countries

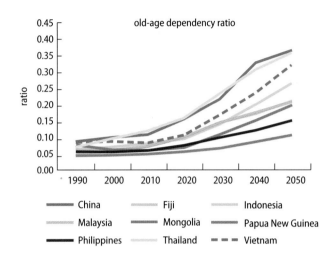

Sources: Data for 1990–2010: World Bank 2010; data for 2020–50: HNPStats Population Projections database.
Note: The old age dependency ratio is defined as the ratio of the elderly (ages 65 and above) to the working age population (ages 15–64).

vary by country context, but the evidence makes clear that both demand-side and supply-side factors are responsible for these poor human capital outcomes.

Policies can thus have an impact by improving service delivery (for example, through infrastructure, staffing, incentives, and use of ICTs) and implementing demand-side interventions (for instance, through cash transfers to poor households, information campaigns, and improved accountability). For example, Indonesia's school construction program in the 1970s significantly increased education attainment and future earnings (Duflo 2000). In Cambodia, a scholarship program targeted at girls and a related program targeted at boys and girls from low-income households led to an increase in school enrollment of at least 20 percentage points (Filmer and Schady 2008, 2009). Evidence indicates that to reduce maternal mortality, interventions that ensure basic infrastructure and improve accountability for service delivery are important. Approaches to providing services that take into account traditional norms and practices also show promise. For example, Malaysia has adopted programs that provide guidance that traditional birth attendants on hygiene practices, diagnosis of complicated cases, and information on the importance of prenatal care.

Reducing gender streaming in education
For East Asian and Pacific countries where gender equality in access to human capital is no longer the dominant concern, addressing education quality—specifically, gender streaming in education—will have high returns. Although concerted efforts in both education and the labor market will be needed to break gender "silos" in education and, consequently, in the economy, significant steps can be taken within the education system. One important step involves reform of school curricula to address the transmission of gender stereotypes through the education system. Cambodia, Lao PDR, and Vietnam have reviewed curricula and revised learning materials or encouraged better practice without gender stereotyping. Other active interventions may be warranted, including information campaigns, financial or nonfinancial incentives, and efforts to create gender-friendly environments in higher education. Providing information on wages and career paths in these programs before individuals choose their courses may help reduce gender streaming in training. In the United States, for example, the Science Connections program offered monthly science workshops for girls plus a summer science weekend for families to increase girls' knowledge about and interest in nontraditional careers in science. Scholarships that support women (and men) in entering nontraditional fields may provide another avenue for breaking down gender silos in education.

Promoting balanced sex ratios at birth
In the few countries with "missing girls" at birth, rooted in the prevalence of son preference, active measures are needed to address the issue. Even where laws against sex-selective abortion have been enacted, strong incentives to select the preferred gender still induce

expectant parents to bypass the law, and enforcement of such laws is difficult. Existing evidence suggests a more promising approach is to adopt policies that aim to enhance family perceptions of the value of daughters. While general policies to promote economic development may play a role, Korea's recent experience suggests that introducing interventions to influence norms and facilitate the spread of new social values may also be important, rather than relying on efforts to raise female education and labor force participation alone. Information campaigns, financial incentives, and improved social security for the elderly can all contribute to changing societal preferences and behaviors. China has been adopting several of these types of programs. For example, the National Population and Family Planning Commission scaled up the Chaohu pilot through the national Care for Girls campaign in 24 counties with severe gender imbalance. This campaign went beyond advocacy and media publicity alone; direct financial incentives for parents to raise daughters have also been introduced. Preliminary evidence suggests that these programs have had some impact on reducing imbalances in the sex ratio at birth.

Addressing male-specific gender issues
Paying attention to male- as well as female-specific gender issues is appropriate for reasons of basic welfare as well as for development effectiveness. In this context, initial signs of reverse gender gaps in education in several countries should be monitored closely. While the long-term implications of male disadvantage in education are still to be understood, depending on the underlying causes, it could have both economic and social consequences. Moreover, excessive tobacco and alcohol consumption among males in many parts of the region deserves policy attention; the social costs are usually higher than private costs because of the negative effects of these behaviors on other members of the society. Possible measures to tackle this challenge include providing information about the health risks of excessive tobacco and alcohol consumption, enacting

or increasing taxes on tobacco and alcohol, imposing regulatory measures on advertising, and restricting smoking in public sites. The Thai Health Promotion Foundation, for example, uses alcohol excise tax revenues to support the operation of an alcohol control center and a research center on alcohol consumption, to support advertising campaigns to reduce alcohol-related traffic accidents, and to promote abstinence and increase knowledge about the links between alcohol use and domestic violence.

Taking active measures to close gender gaps in economic opportunity

Mitigating trade-offs between women's household and market roles

Women often face stark time trade-offs between household and market work, particularly in rural areas. In such contexts, programs targeted at reducing women's time on household work—for example, through investment in infrastructure—are likely to increase women's ability to engage in market-based income-earning opportunities. In Lao PDR, for example, evidence indicates that having access to electricity extends the hours available for both productive and leisure activities, particularly for women and girls (World Bank 2011a).

Policies that support women in balancing their caregiving and market roles are also important in strengthening their access to economic opportunity. Access to affordable and accessible child care can be critical in this regard. Community child care centers, particularly those targeted at low-income neighborhoods, have been found to increase maternal employment in a number of Latin American countries. The importance of affordable child care, particularly as urban areas expand, can be seen from recent experience in the region. In Mongolia and China, reductions in subsidized child care in the 1990s and 2000s have significantly and negatively affected female labor force participation in urban areas.

Parental and paternity leave can promote greater parity between the sexes by facilitating a more equitable division of child-rearing responsibilities and allowing women to have the same opportunities as men for advancing their careers in the formal sector. Within the region, only Cambodia, Indonesia, and the Philippines currently have provisions for paternity leave. While the principles behind paternity (and parental) leave are appealing, evidence from the OECD on the take-up of paternity leave is mixed, suggesting that providing paternity leave alone is not sufficient to change the current gender division of child-rearing responsibilities within households; rather, such leave policies need to be combined with other approaches to breaking down gender norms regarding household caregiving.

Breaking down gender silos in the labor market

A key element of breaking down gender silos in the labor market involves supporting young women and men to invest in skills on the basis of their productivity rather than on the basis of gender norms and perceptions regarding "appropriate" occupations. Beyond efforts to reduce gender streaming in education, programs that help both women and men understand employment options outside of gender silos will likely improve the allocation of talent toward jobs in ways that improve both equality of economic opportunity and productivity. In Kenya, for example, a micro and small enterprise voucher program, called Jua Kali, provided its female beneficiaries information about wages in a range of occupations. Preliminary evaluation of the program suggests that 5 percent of women who received the information switched to more lucrative (often "male") jobs as compared to those who did not receive the information (Hicks et al. 2011).

Breaking down social norms and perceptions about gender roles in the workplace is an area where the public sector can lead by example, particularly with respect to enabling women as leaders and managers. The public sector is in a unique position to establish good practice in this regard by encouraging women's professional advancement, either through direct measures such as targets or quotas or through specialized training programs. In this context, the government of Malaysia has put in place a system of quotas for female managers in the public sector. In Mexico, the government initiated a system of grants to firms to address gender-related employment issues in the workplace, including fostering greater female participation in management.

Eliminating resource constraints on female-led farms and enterprises

Despite progress, women continue to have less access to a range of productive resources than do men as a function of their gender rather than because of their innate productive capabilities. Public policies thus have an important role to play in promoting gender equality in the control of productive inputs—whether land, agricultural extension, technology, or financial capital. Improving women's access to productive assets can play an important role in raising enterprise productivity in both the farm and nonfarm sectors. Following are some examples.

- Several countries in the region have made headway in recent years in increasing ownership and control of land. In response to concerns about persistent gender inequalities in land, several countries—including Indonesia, Lao PDR, and Vietnam—have recently adopted gender-sensitive reforms in land titling. Since the 2004 Land Law in Vietnam, all new land tenure certificates must include the names of both spouses. Qualitative assessment of the reform's effects in three provinces suggests that joint titling improves opportunities for women to access loans, empowers women in case of disputes, and leads to more mutual decision making (World Bank 2008). Because the reasons for women's lower access to land differ across the region—from unfavorable legal frameworks to cultural norms and practices that deem land to be a "male asset"—effective policies to increase female land holdings need to account for context-specific constraints.

- Gender inequalities in access to information and training, extension services, and other productive inputs constrain the productivity of female-led enterprises, both within and outside agriculture. In Papua New Guinea, where training and extension services are more likely to reach male than female farmers, a recently launched agriculture training program includes several components focused on closing the gap between rural women's economic needs and the inputs and services provided. Similarly, in Cambodia, the Cambodia-Australia Agricultural Extension Project has addressed two key constraints that have limited female farmers' access to agricultural support: the project has increased by 80 percent the number of female extension workers and undertakes special initiatives that account for female farmers' relatively lower levels of literacy.

- While evidence on access to finance in East Asian and Pacific countries is mixed, women do face particular challenges in accessing credit, especially given their poorer access to land, an important source of collateral. Beyond financial constraints, training programs that improve business skills may be implemented to address gender differences in entrepreneurial capital. In this context, an impact assessment of a women's entrepreneurship training program in Aceh, Indonesia, found that business planning and management training helped promote greater confidence among women trainees, create or strengthen social networks, and identify ways to improve the business environment for female entrepreneurs (ILO 2008).

As in the case of education and health, broad systemic weaknesses—whether in the form of cumbersome registration procedures, weak systems of financial intermediation, or lack of electricity—affect both female- and male-led enterprises. Evidence suggests that such constraints may be more onerous among small and informal firms than among larger firms and, therefore, may constrain female-led enterprises disproportionately. As a result, interventions that focus on improving the overall investment climate and particularly on promoting small business development will be important. Addressing systemic as well as gender-specific constraints will thus be critical to promoting gender equality in economic opportunity.

Creating an enabling environment for gender equality in employment

Public policy can strengthen the enabling environment for gender equality in formal employment. An important element of this is to ensure that women and men face a level legal playing field with respect to jobs and sectors. Labor regulations that result in asymmetries in the employability and costs of hiring male and female workers can be found across the region. Ostensibly protective legislation, in the form of restrictions on women working at night, working overtime, and working in so-called dangerous sectors, serves in practice to inhibit women's economic participation. Priority should be given to reducing labor market restrictions that limit women's employment options. Where the original concerns motivating these policies are still valid—for example, health and safety issues—measures should be taken to ensure that these concerns are addressed more directly through workplace safety codes, provision of safe and reliable transport infrastructure, and so on.

Active labor market policies are another means of overcoming gender differences in access for formal employment. For example, wage subsidies may induce employers to hire female workers whom they may not have otherwise considered, due to lack of information about their workplace productivity. This intervention thus provides the opportunity to reduce gender stereotypes by enabling employers to observe women's skills directly, and it can facilitate women gaining valuable labor market experience. Skills training programs may also enable women and men to move into professions outside of gender silos, particularly when paired with apprenticeship opportunities. Although evidence on the effect of active labor market policies in East Asia and the Pacific is limited, studies from Latin America and the Middle East suggest that well-designed active labor market policies can help improve women's employment outcomes.

Affirmative action policies have also been used to overcome gender-specific barriers to employment, whether caused by implicit or overt discrimination in hiring and promotion. The literature reflects some debate regarding the benefits and costs of affirmative action, but the collection of evidence (largely from high-income countries) suggests that carefully designed policies can help break down barriers to female employment with little or no adverse effects on firm productivity (World Bank 2011c). Affirmative action in hiring and promotion in the public sector can have important demonstration effects. In 2004, the government of Malaysia introduced a quota for the public sector of 30 percent female representation across all decision-making levels, including positions such as department heads or secretary-general (ASEAN 2008). Whereas in 2006, women held 24.6 percent of top positions in the public sector, by 2010, the figure had risen to 32.0 percent. More recently, the Malaysian government set a target for 30 percent of corporate board positions to be held by women by 2016.

Taking measures to strengthen women's voice and influence

Measures to increase women's endowments and economic opportunity, such as those described previously, contribute to strengthening their voice within the household and in society. Educated women in good health, with assets and income, are better able to act on their preferences and influence outcomes that affect themselves and others in society. In addition, several other policy approaches can directly promote women's agency and reduce gender-based violence.

Supporting initiatives to transform gender norms and practices

While gender norms may be persistent, they are far from static. Individual experiences as well as large-scale political and economic processes are capable of bringing about dramatic, and often rapid, social change. In East Asia, increasing economic integration and rising access to ICTs are not only transforming the economic landscape but also increasing flows of information in ways that may serve to transform gender norms in the region. Similarly, migration and rapid urbanization across the region are bringing with them the possibility of newly defined roles for men and women, as women and men alike are exposed to new ideas and production modalities.

The education system can be a vital source to change social norms that perpetuate gender inequality. The integration of gender equality principles into the school and professional curricula can address the value system of children early on and, over time, transform social norms (Utomo et al. 2009). Evidence of the positive effects of changing the curriculum is available for adults in Thailand where gender sensitivity was integrated into the curriculum in the Chulalangkorn medical school. Evaluation of the program showed that respondents were more aware of gender issues and tended to apply gender concepts and concerns in their work and personal lives (WHO GWH 2007).

Provision of information through television programming can also play a critical role in changing social norms, especially with respect to fertility and gender-based violence. Evidence shows that people can be prompted to rethink gender roles in society when they are exposed to new information and experiences that challenge existing norms. In Brazil, despite strong traditional norms in favor of having many children, increased exposure to the opposite behavior by popular women in soap operas led to a measurable decline in fertility (Chong and La Ferrara 2009; La Ferrara, Chong, and Duryea 2008). In India, increased exposure to television contributed to decreased acceptance of wife beating, lower fertility rates, and noticeable shifts away from son preference (Jensen and Oster 2008).

Strengthening the legal and institutional environment

Nearly all countries in the region have acceded to CEDAW, signaling commitment to adhering to internationally agreed-upon norms regarding gender equality.[7] An important pending agenda is to ensure that domestic legislation and the institutions of implementation and enforcement are aligned with countries' commitments. Where comprehensive legal reform is not possible, governments should identify priorities for action. For example, in contexts where women's agency within the home remains weak, a case exists for reforms to focus on rights in marriage and divorce, inheritance and maintenance laws, and protection of women from gender-based violence, which can strengthen the enabling environment for greater voice and influence in the household. As part of the process of monitoring progress toward gender equality, governments should undertake regular assessments to ensure that legal gaps are being filled and that relevant laws are being enforced. In countries where multiple systems of law coexist, assessing customary practices to ascertain whether they limit women's agency and then developing strategies to address these limiting factors will be important.

Strengthening the capacities of institutions to enforce the law and increasing knowledge of the law are also critical. Both financial and human investments need to be made to ensure that public sector personnel, such as judges and police, have the knowledge and capacity to actively enforce laws intended to protect women and to promote gender equality. Concerns have been raised that police forces in several countries in the region, including Indonesia, Malaysia, and Thailand, have been too passive in investigating trafficking and enforcing anti-trafficking laws (U.S. Department of State 2011). The Cambodian Women's Crisis Center began a community program that increases the awareness of violence against women and of the legal rights of women, including the law on domestic violence and protection that was implemented in 2005. The program promotes initiatives to strengthen enforcement of the law by providing information and training to local authorities and developing community support networks.

Increasing women's access to justice

Financial costs and travel distances often are de facto barriers to women accessing justice, even when the appropriate laws and institutions are in place. In that context, developing and implementing innovative methods to improve access to the judicial system can help women exercise their agency in the courts when needed. The use of mobile courts, for example, such as those in rural areas of China and Indonesia, provides a solution to the problem of accessibility and security for women who wish to exercise their rights in the legal system but who are unable to travel to the court. Technology, such as telephone hotlines and websites, can be used to undertake basic legal transactions. For women with few economic resources, waiving or subsidizing the costs of legal aid can help reduce financial barriers to accessing the judicial system.

Enabling women's participation in politics and policy making

Active measures to promote women's participation in policy making can be effective in increasing female representation in local and national assemblies in many contexts. A range of affirmative action mechanisms have been used in developing countries. Quotas have been used in several countries, sometimes in the form of constitutional changes to reserve a specified number of posts for women and sometimes through legislative and political party quotas. Quotas can be informal (and voluntary) or mandated formally at the subnational or national level (Dahlerup 2006). The idea is to provide temporary measures to break down barriers to the entry of women into politics. The suitability and impact of different forms of quotas or targets differ depending on the specific context. Data suggest that

these measures can be effective in increasing female representation in elected bodies. They can also help transform people's views about the efficacy of female political leaders (Beaman et al. 2009). At the same time, electoral quotas do impose constraints on the democratic process. It is thus important to keep this—as well as the expected benefits of increasing female participation in politics—in mind when determining when and how to enact such measures.

Pursuing a multipronged approach to reducing gender-based violence

Reducing gender-based violence requires action on a number of fronts: efforts to increase women's voice within the household; enactment and enforcement of appropriate legislation and strengthening of women's access to justice; provision of adequate support services for victims of violence; and use of the media to provide information on women's rights, to increase social awareness, and to shift social norms with respect to violence.

Countries that take a strong stance on gender-based violence legislation and enforcement can make positive strides against such violence in short periods of time. Cambodia saw a significant decrease in the incidence of domestic violence between 2000 and 2005, largely attributed to strong efforts by the country's Ministry of Women's Affairs, which introduced draft domestic violence legislation in 2001. Four years later, in October 2005, the National Assembly adopted the legislation. The new law criminalized acts of domestic violence, provided for the protection of victims, and enabled neighbors or local organizations to intervene if they witnessed domestic violence. As a complement to the law, women's organizations and other nongovernmental organizations carried out information campaigns to disseminate information on people's basic rights and responsibilities under the law.

Governments also need to provide adequate support services for victims when violence does occur. This support can include a range of services, from police and judiciary to health and social services. In Malaysia, the government established integrated one-stop crisis centers in hospitals that provide easy access to medical care and social services, and facilitate reporting of violence to specially trained police officers (World Bank 2011c).

Creating space for women's collective agency

While promoting women's individual agency is important, supporting women's collective agency can be an effective way to empower women to advocate for and promote effective public action toward gender equality. Experience from around the region highlights the potential of enabling women's collective agency. For example, during the debate in Cambodia leading to the 2005 Law on Prevention of Domestic Violence and Protection of Victims, the Cambodia Committee of Women, a coalition of 32 nongovernmental organizations, persistently lobbied the government and the Ministry of Women's Affairs to secure the legislation's passage. Similarly, in Fiji, the Fiji Women's Crisis Center campaigned successfully for the implementation of a nationally representative quantitative survey on violence against women; the results of this survey are scheduled to be released in 2012. Partnerships with the private sector, including women's business associations, can provide an important space for women to interact, learn, and advocate for gender equality. A recent initiative to increase women's participation in the private sector's dialogue with government in Cambodia, through the Government Private Sector Forum, has induced policy makers to undertake new initiatives addressing the needs of female entrepreneurs.

Fostering new opportunities, managing emerging risks

A new and important challenge for policy makers will be to help foster emerging opportunities and, in particular, to manage emerging risks associated with increasing economic integration, increasing access to ICTs, migration, rapid urbanization, and *population aging.* Many of the emerging opportunities can be fostered through the types of efforts to promote greater gender equality in endowments, economic opportunity, and agency discussed in this report. For example, where gender gaps in ICT use are growing, ensuring that women have access to these new technologies may require active measures similar to those discussed earlier to ensure equal access to other types of productive resources. Managing emerging risks, however, may require additional policy approaches, such as the following.

Greater economic integration will bring with it increased exposure to employment shocks that will have gender-differentiated effects. Adequately addressing the risks associated with economic integration will require designing social protection programs that take into account the different risks faced by female and male workers. Building on the lessons from recent economic crises, several developing countries, including some from East Asia and the Pacific, have begun to recognize the gender dimensions of risk and shocks in the design of programs. In Indonesia, for example, a conditional cash transfer program, Keluarga Harapan, targets households with members who are particularly vulnerable during times of crisis, such as pregnant and lactating women.

As female migration across the region increases economic opportunities for women, it creates new concerns about female migrants' welfare. Protecting female migrants from exploitative situations, including sex work and human trafficking, will also require a gender-aware approach. Greater protection through better laws, enforcement, and monitoring; improved information flows; and safety nets will better address the vulnerabilities specific to women traveling abroad. Specific areas for action include improving legal protections for female migrants, strengthening the monitoring and credibility of recruitment agencies, and developing and providing welfare and support services to assist female migrants. Governments in both sending and receiving countries will need to actively address the issue of human trafficking through prevention, protection, and prosecution. Gender-awareness training for people involved in the migration process will improve their ability to identify and assist abused female migrants, including those trafficked or at risk of being trafficked.

Growing urbanization in the East Asia and Pacific region has presented women not only with increased economic opportunities but also with particular challenges, such as limited access to child care and higher security risks in urban areas. Thus, policy makers need to ensure that child care, education, infrastructure, transportation, and water and sanitation policies take into account women's specific social and cultural needs. Rigorous laws and policies to protect women in urban areas from the risk of violence and exploitation are also warranted.

Rapid population aging in the region is likely to have important gender-differentiated effects, among other things, because older women may increasingly find themselves living as widows. Along with risks from urbanization and the breakdown of extended family support networks, these women are likely to find themselves at increased economic risk, having accumulated relatively few assets and mostly lacking access to formal social security. In this context, designing old-age income security programs that can protect women from destitution in old age will have an increasingly important role. In addition, policy makers should consider ways to strengthen care for the elderly to ensure that women do not bear an undue burden of caregiving as the region's population ages.

Filling knowledge gaps

Finally, while much has been learned from recent global and regional evidence on gender equality and development, much remains to be understood empirically to help inform effective public action. Continuing to close data and analytical gaps will thus be important to better understand policy priorities, the effects of specific interventions, and the costs and benefits of different policy options. To fill knowledge gaps, additional gender-disaggregated data need to be collected. Moreover, additional empirical analysis, both on long-standing gender issues and the gender implications of emerging trends in the region, will enable policy makers to better promote both gender equality and more effective development.

Notes

1. Sen (1999) defines freedoms and "unfreedoms" in five categories: (a) political freedoms, (b) economic facilities, (c) social opportunities, (d) transparency guarantees, and (e) protective security.

2. Adopted in 1979 by the United Nations General Assembly, CEDAW is often referred to as the international bill of rights for women. The convention defines what constitutes discrimination against women and provides an agenda for national action to end such discrimination. To date, it has been ratified by 187 countries worldwide (http://www.un.org/womenwatch/daw/cedaw/).

3. It is important to interpret these studies with caution, given the difficulty in establishing a causal relationship between gender equality in education and growth in cross-country studies.

4. Evidence from Africa and Latin America, for example, suggests that ensuring equal access to productive assets and technologies could significantly raise agricultural production and household income (Goldstein and Udry 2008; Quisumbing 1995; Udry 1996).

5. Globally, poverty and gender often interact to compound gender inequalities (World Bank 2011c). In East Asia and the Pacific, poverty does not appear to be as important a contributor to gender disadvantage in education as elsewhere, however. Survey data from several countries in the region indicate that gender gaps in enrollment do not vary substantially or systematically across income quintiles.

6. For data on land holdings, by gender, in China, see de Brauw et al. (2011). Data on other countries are based on World Bank staff calculations, using household survey data.

7. As of the end of 2011, only six countries in the world had not ratified CEDAW. Two of those countries are in the Pacific: Palau and Tonga (CEDAW, http://www.un.org/womenwatch/daw/cedaw/).

References

Agarwal, Bina. 2010a. "Does Women's Proportional Strength Affect Their Participation? Governing Local Forests in South Asia." World Development 38 (1): 98–112.

———. 2010b. *Gender and Green Governance: The Political Economy of Women's Presence Within and Beyond Community Forestry.* New York: Oxford University Press.

Agarwal, Bina, and Pradeep Panda. 2007. "Toward Freedom from Domestic Violence: The Neglected Obvious." *Journal of Human Development* 8 (3): 359–88.

ASEAN (Association of Southeast Asian Nations). 2008. "ASEAN Continues to Empower Women". http://www.aseansec.org/Bulletin-Feb-08.htm#Article-2.

Baulch, Bob, Truong Thi Kim Chuyen, Dominique Haughton, and Jonathan Haughton. 2002. "Ethnic Minority Development in Vietnam: A Socioeconomic Perspective." Policy Research Working Paper 2836, World Bank, Washington DC.

Beaman, Lori, Raghabendra Chattopadhyay, Esther Duflo, Rohini Pande, and Petia Topalova. 2009. "Powerful Women: Does Exposure Reduce Bias?" *Quarterly Journal of Economics* 124 (4): 1497–540.

Beegle, Kathleen, Elizabeth Frankenberg, and Duncan Thomas. 2001. "Bargaining Power within Couples and Use of Prenatal and Delivery Care in Indonesia." *Studies in Family Planning* 32 (2): 130–46.

BPS (Badan Pusat Statistik–Statistics Indonesia) and ORC Macro. 2003. *Indonesia Demographic and Health Survey 2002–2003.* Calverton, MD: BPS and ORC Macro.

Bruni, Lucilla, Andrew D. Mason, Laura Pabon, and Carrie Turk. 2011. "Gender Impacts of the Global Financial Crisis in Cambodia." World Bank, Washington, DC.

Chattopadhyay, Raghabendra, and Esther Duflo. 2004. "Women as Policy Makers: Evidence from a Randomized Policy Experiment in India." *Econometrica* 72 (5): 1409–43

Chi, Wei, and Bo Li. 2007. "Glass Ceiling or Sticky Floor? Examining the Gender Pay Gap across the Wage Distribution in Urban China, 1987–2004." MPRA Paper 3544, University Library of Munich, Germany.

Chong, Alberto, and Eliana La Ferrara. 2009. "Television and Divorce: Evidence from Brazilian Novelas." *Journal of the European Economic Association* 7 (2–3): 458–68.

CNNIC (China Internet Network Information Center) Internet Statistics. Beijing, China. http://www1.cnnic.net.cn/en/index/0O/index.htm.

Constable, Nicole. 2009. "Migrant Workers and the Many States of Protest in Hong Kong." *Critical Asian Studies* 41 (1): 143–64.

Cuberes, David, and Marc Teignier-Baqué. 2011. "Gender Inequality and Economic Growth." Background paper for the *World Development Report 2012*, World Bank, Washington, DC.

Dahlerup, Drude. 2006. "Introduction." In *Women, Quotas and Politics*, edited by Drude Dahlerup, 3–31. London and New York: Routledge.

de Brauw, Alan, Jikun Huang, Linxiu Zhang, and Scott Rozelle. 2011. "The Feminization of Agriculture with Chinese Characteristics." Background paper for the *World Development Report 2012*, World Bank, Washington, DC.

Do, Quy-Toan, Andrei Levchenko, and Claudio Raddatz. 2011. "Engendering Trade." Policy Research Working Paper 5777, World Bank, Washington, DC.

Duflo, Esther. 2000. "Child Health and Household Resources in South Africa: Evidence from the Old Age Pension Program." *American Economic Review* 90 (2): 393–98.

———. 2003. "Grandmothers and Granddaughters: Old-Age Pensions and Intrahousehold Allocation in South Africa." *World Bank Economic Review* 17 (1): 1–25.

Dwyer, Jeffrey, and Raymond Coward. 1992. "Gender and Family Care of the Elderly: Research Gaps and Opportunities." In *Gender, Families, and Elder Care*, edited by Jeffrey W. Dwyer and Raymond T. Coward, 151–62. London: Sage.

Edlund, Lena, and Rohini Pande. 2001. "Why Have Women Become Left-Wing? The Political Gender Gap and the Decline in Marriage." *Quarterly Journal of Economics* 117 (3): 917–61.

Enterprise Surveys database. World Bank/International Finance Corporation, Washington, DC. http://www.enterprisesurveys.org/.

FAO (Food and Agriculture Organization)/Sida Partnership Cooperation. 2010. *National Gender Profile of Agricultural Households, 2010: Report Based on the Lao Expenditure and Consumption Surveys, National Agricultural Census and Population Census.* Vientiane, Lao PDR: FAO.

Fehringer, Jessica, and Michelle J. Hindin. 2009. "Like Parent, Like Child: Intergenerational Transmission of Partner Violence in Cebu, the

Philippines." *Journal of Adolescent Health* 44 (4): 363–71.

Filmer, Deon, and Norbert Schady. 2008. "Getting Girls into School: Evidence from a Scholarship Program in Cambodia." *Economic Development and Cultural Change* 56 (3): 581–617.

———. 2009. "School Enrollment, Selection and Test Scores." Policy Research Working Paper Series 4998, World Bank, Washington, DC.

Ganster-Breidler, Margit. 2010. "Gender-Based Violence and the Impact on Women's Health and Well-Being in Papua New Guinea." *DWU Research Journal* 13.

Goldstein, Markus, and Christopher Udry. 2008. "The Profits of Power: Land Rights and Agricultural Investment in Ghana." *Journal of Political Economy* 116 (6): 981–1022.

Hicks, Joan, Hamory Hicks, Michael Kremer, Isaac Mbiti, and Edward Miguel. "Vocational Education Voucher Delivery and Labor Market Returns: A Randomized Evaluation Among Kenyan Youth," Report for Spanish Impact Evaluation Fund (SIEF) Phase II, World Bank, Washington, DC.

HNPStats Population Projections database. World Bank, Washington, DC. http://go.worldbank.org/H4UN4D5KI0

ICRW (International Center for Research on Women). 2006. *Property Ownership and Inheritance Rights of Women for Social Protection: The South Asia Experience.* Washington, DC: ICRW.

ILO (International Labour Organization). 2008. *Women's Entrepreneurship Development Aceh: Gender and Entrepreneurship Together (GET Ahead) Training Implementation: Impact Assessment.* Jakarta, Indonesia: ILO Jakarta Office.

———. Key Indicators of the Labour Market (KILM) database. International Labour Organization, Geneva, Switzerland. http://kilm.ilo.org/kilmnet/.

Indonesia Family Life Survey. 2007/2008. RAND Family Life Surveys, IFLS-4. http://www.rand.org/labor/FLS/IFLS/ifls4.html.

Jensen, Robert, and Emily Oster. 2008. "The Power of TV: Cable Television and Women's Status in India." University of Chicago, IL. http://home.uchicago.edu/eoster/tvwomen.pdf.

Johnson, Kiersten, Sovanratnak Sao, and Darith Hor. 2000. *Cambodia 2000 Demographic and Health Survey: Key Findings.* Calverton, MD: ORC Macro.

Judd, Ellen R. 2010. "Family Strategies: Fluidities of Gender, Community and Mobility in Rural West China." *China Quarterly* 204: 921–38.

Klasen, Stephan. 2002. "Low Schooling for Girls, Slower Growth for All? Cross-Country Evidence on the Effect of Gender Inequality in Education on Economic Development." *World Bank Economic Review* 16 (3): 345–73.

Knowles, Stephen, A. K. Lorgelly, and Dorian Owen. 2002. "Are Educational Gender Gaps a Brake on Economic Development? Some Cross-Country Empirical Evidence." *Oxford Economic Papers* 54 (1): 118–49.

La Ferrara, Eliana, Alberto Chong, and Suzanne Duryea. 2008. "Soap Operas and Fertility: Evidence from Brazil." BREAD Working Paper 172, Bureau for Research and Economic Analysis of Development, Duke University, Durham, NC.

Lewis, Ione, Bessie Maruia, and Sharon Walker. 2008. "Violence against Women in Papua New Guinea." *Journal of Family Studies* 14: 183–97.

Lott, John R., and Lawrence W. Kenny. 1999. "Did Women's Suffrage Change the Size and Scope of Government?" *Journal of Political Economy* 107 (6): 1163–98.

Maffii, Margherita, and Sineath Hong. 2010. "Political Participation of Indigenous Women in Cambodia." *Asien* 114–15 (April): 16–32.

Morrison, Andrew, Mary Ellsberg, and Sarah Bott. 2007. "Addressing Gender-Based Violence: A Critical Review of Interventions." *World Bank Observer* 22 (1): 25–51.

Naudeau, Sophie, Naoko Kataoka, Alexandria Valerio, Michelle Neuman, and Leslie Elder. 2011. *Investing in Young Children: An Early Childhood Development Guide for Policy Dialogue and Project Preparation.* Washington, DC: World Bank.

National Center for Injury Prevention and Control. 2003. *Costs of Intimate Partner Violence against Women in the United States.* Atlanta, GA: Centers for Disease Control and Prevention.

NIPH (National Institute of Public Health [Cambodia]), NIS (National Institute of Statistics [Cambodia]) and ORC Macro. 2006. *Cambodia Demographic and Health Survey 2005.* Phnom Penh, Cambodia, and Calverton, Maryland, USA: NIPH, NIS, and ORC Macro.

NSD (National Statistics Directorate [Timor-Leste]), Ministry of Finance [Timor-Leste], and ICF Macro. 2010. *Timor-Leste Demographic and Health Survey 2009–10.* Dili, Timor-Leste: NSD and ICF Macro.

Ofstedal, Mary Beth, John E. Knodel, and Napaporn Chayovan. 1999. "Intergenerational Support and Gender: A Comparison of Four Asian Countries." *Southeast Asian Journal of Social Sciences* 27 (2): 21–41.

PARLINE database on national parliaments. Inter-Parliamentary Union, Geneva. http://www.ipu.org/parline-e/parlinesearch.asp and http://www.ipu.org/wmn-e/world.htm.

PovcalNet (online poverty analysis tool). Development Research Group, World Bank, Washington, DC. http://iresearch.worldbank.org/PovcalNet/index.htm.

Pronyk, Paul M., James R. Hargreaves, Julia C. Kim, Linda A. Morison, Godfrey Phetla, Charlotte Watts, Joanna Busza, and John D. H. Porter. 2006. "Effect of a Structural Intervention for the Prevention of Intimate-partner Violence and HIV in Rural South Africa: A Cluster Randomized Trial." *Lancet* 2368 (9551): 1973–83.

Qian, Nancy. 2008. "Missing Women and the Price of Tea in China: The Effect of Sex-Specific Earnings on Sex Imbalance." *Quarterly Journal of Economics* 123 (3): 1251–85.

Quisumbing, Agnes. 1995. "Gender Differences in Agricultural Productivity: A Survey of Empirical Evidence." FCND Discussion Paper No. 5, Food Consumption and Nutrition Division, International Food Policy Research Institute, Washington, DC.

Rodgers, Yana, and Nidhiya Menon. Forthcoming. "Impact of the 2008–2009 Twin Economic Crises on the Philippine Labor Market." *World Development.*

Sakellariou, Chris. 2011. "Determinants of the Gender Wage Gap and Female Labor Force Participation in EAP." Paper commissioned for *Toward Gender Equality in East Asia and the Pacific: A Companion to the* World Development Report, Washington, DC: World Bank.

Sen, Amartya. 1999. *Development as Freedom.* Oxford, U.K.: Oxford University Press.

SPC (Secretariat of the Pacific Community), Ministry of Internal and Social Affairs [Republic of Kiribati], and Statistics Division [Republic of Kiribati]. 2010. *Kiribati Family Health and Support Study: A Study on Violence against Women and Children.* Nouméa, New Caledonia: SPC.

SPC (Secretariat of the Pacific Community) and NSO (National Statistics Office) [Solomon Islands]. 2009. *Solomon Islands Family Health and Safety Study: A Study on Violence against Women and Children.* Nouméa, New Caledonia: SPC.

SUSI (Survei Usaha Terintegrasi). 2002. *Integrated Survey of Cottage and Small-Scale Firms, Indonesia.* BPS. http://dds.bps.go.id.

Swaminathan, Hema, Cherryl Walker, and Margaret A. Rugadya, eds. 2008. *Women's Property Rights, HIV and AIDS, and Domestic Violence: Research Findings from Two Rural Districts in South Africa and Uganda.* Cape Town: HSRC Press.

Thomas, Duncan. 1995. "Like Father, Like Son, Like Mother, Like Daughter, Parental Resources and Child Height." Papers 95-01, RAND Reprint Series, Rand Corporation, Santa Monica, CA.

Udry, Christopher. 1996. "Gender, Agricultural Production, and the Theory of the Household." *Journal of Political Economy* 104 (5): 1010–14.

UN (United Nations Department of Economic and Social Affairs, Population Division). 2010. *World Urbanization Prospects: The 2009 Revision.* New York: United Nations. http://esa.un.org/unpd/wup/index.html.

UNDP (United Nations Development Programme). 2010. *Asia-Pacific Human Development Report. Power, Voice and Rights: A Turning Point for Gender Equality in Asia and the Pacific.* Colombo, Sri Lanka: Macmillan Publishers India Ltd. for UNDP.

UNDP (United Nations Development Program) Cambodia and VBNK. 2010. *Talking About Domestic Violence: A Handbook for Village Facilitators.* Phnom Penh: UNDP Cambodia and the Agencia Española de Cooperación para el Desarrollo.

UNFPA (United Nations Population Fund). 2008. An Assessment of the State of Violence against Women in Fiji. Suva, Fiji.

USCDC (United States Centers for Disease Control and Prevention). 2003. "Costs of Intimate Partner Violence against Women in the United States." USCDC, National Center for Injury Prevention and Control, Atlanta, Georgia.

U.S. Department of State 2011. *Trafficking in Persons Report 2011.* Washington, DC: U.S. Department of State.

Utomo, Iwu, Peter McDonald, Terence Hull, Ida Rosyidah, Tati Hattimah, Nurul Idrus, Saparinah Sadli, and Jamhari Makruj. 2009. "Gender Depiction in Indonesian School Text Books: Progress or Deterioration." *Australian*

Demographic and Social Research Institute, Australian National University.

VWC (Vanuatu Women's Centre) in partnership with the Vanuatu National Statistics Office. 2011. *Vanuatu National Survey on Women's Lives and Family Relationships.* Port Vila, Vanuatu: VWC.

Wang, Zhengxu, and Weina Dai. 2010. "Women's Participation in Rural China's Self-Governance: Institutional, Socioeconomic, and Cultural Factors in a Jiangsu County." Discussion Paper 69, China Policy Institute, University of Nottingham, U.K.

WHO GWH (World Health Organization, Department of Gender, Women and Health). 2007. *Integrating Gender into the Curricula for Health Professionals, Meeting Report, 4–6 December 2006.* Geneva: WHO.

World Bank. 2008. *Vietnam: Analysis of the Impact of Land Tenure Certificates with Both Names of Wife and Husband: Final Report.* Washington, DC: World Bank.

———. 2010. *World Development Indicators 2010.* Washington, DC: World Bank.

———. 2011a. "Lao PDR Country Gender Assessment." World Bank, Washington, DC.

———. 2011b. "Vietnam Country Gender Assessment." World Bank, Washington, DC.

———. 2011c. *World Development Report 2012: Gender Equality and Development.* Washington, DC: World Bank.

World Values Survey database. World Values Survey Association, Stockholm, Sweden. http://www.worldvaluessurvey.org/.

The State of Gender Equality in East Asia and the Pacific | 1

In recent decades, women across the globe have made advances toward gender equality. Literacy rates for young women and girls are higher than ever before, and gender gaps in primary education have closed in almost all countries across the world. In the past three decades, over half a billion women have joined the world's labor force (World Bank 2011b).

Strides toward gender equality in East Asia and the Pacific have been similarly noteworthy. Most countries in the region have either reached or surpassed gender parity in education enrollments. Health outcomes for both women and men have improved significantly. Female labor force participation rates in the region are relatively high. Yet despite considerable progress in this economically dynamic region, gender disparities persist in a number of important areas—particularly in access to economic opportunity and in voice and influence in society. For policy makers in East Asian and Pacific countries, closing these gender gaps represents an important challenge to achieving more inclusive and effective development.

The East Asia and Pacific region's significant economic growth, structural transformation, and poverty reduction in the past few decades have been associated with reduced gender inequalities in several dimensions. The region grew at 7 percent on average between 2000 and 2008 (figure 1.1), the structure of the region's economies has shifted away from agriculture toward manufacturing and services, and extreme poverty has fallen dramatically. Indeed, the share of the region's population living on less than US$1.25 a day has declined by more than 50 percent since 1990—from the highest poverty head count rate in the world to among the lowest (figure 1.2). Growth, structural transformation, and poverty reduction have been accompanied by considerable progress toward gender equality in several key areas, particularly education and health. Many countries in the region have experienced closing gender gaps in school enrollments and declining maternal and child mortality rates.

But growth and development have not been enough to attain gender equality in all its dimensions. Women still have less access than men to a range of productive assets and services, including land, financial capital, agricultural extension services, and new information technologies. Substantial employment segregation, or sorting, by gender still remains an issue. Women are less likely than

FIGURE 1.1   The East Asia and Pacific region has experienced rapid economic growth

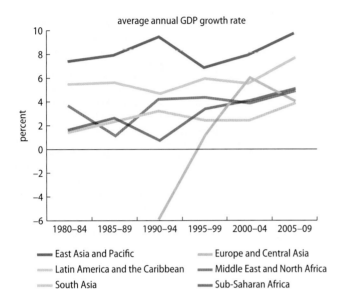

Source: World Bank 2010b.

FIGURE 1.2   Poverty reduction in the East Asia and Pacific region has been impressive

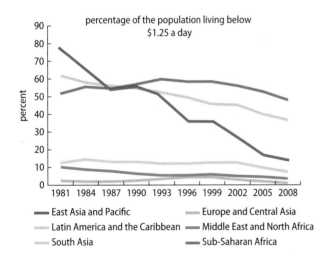

Source: PovcalNet.

in East Asian and Pacific countries still have a weaker voice and less influence than men, whether in household decision making, in the private sector, in civil society, or in politics. Moreover, women across the region remain vulnerable to gender-based violence, often at the hand of an intimate partner.

The East Asia and Pacific region is vast and diverse, with large differences in economic and social progress, including toward gender equality. Achievements in growth and development have not been uniform across the region. Although average annual gross domestic product (GDP) growth during the first decade of the 2000s neared 10 percent in China, it was close to zero in several small Pacific Island countries. By the end of the decade, levels of GDP per capita among the low- and middle-income countries of the region also varied widely, from US$623 in Timor-Leste to US$8,373 in Malaysia (WDI database). Nor has progress toward gender equality been uniform. Despite widespread progress toward gender equality in schooling, a few (mostly low-income) countries continue to face challenges in closing gender gaps in basic education. In spite of broad improvements in health outcomes, China—and, to a lesser extent, Vietnam—face significant imbalances in the ratio of boys to girls at birth, a function of prenatal sex selection stemming from the strong preference for sons in those societies. As a result, the region has more than a million "missing" girls at birth per year.

A number of Pacific Island countries face particular challenges with respect to promoting women's voice, influence, and empowerment—in both the private and public spheres. For example, although gender-based violence remains pervasive in the region, its prevalence in the Pacific is among the highest in the world. Data from Solomon Islands, Kiribati, and Vanuatu indicate that more than 60 percent of adult women have experienced physical or sexual violence during their lifetime, often at the hands of an intimate partner (SPC, Ministry of Internal and Social Affairs, and Statistics Division 2010; SPC and NSO 2009; VWC 2011).

men to work in formal sector jobs and more likely to work in poorly remunerated occupations and enterprises. And despite the closing of education gaps, women continue to be paid less than men for similar work. Women

Moreover, female representation in politics in the Pacific is among the lowest in the world. Although female political participation is relatively low worldwide—only 19.6 percent of the world's parliamentarians were women in December 2011—four of the eight countries in the world with no female parliamentarians were located in the Pacific (PARLINE database, 2011 data).

A growing body of literature also suggests that patterns of growth—not just levels—can affect gender equality by affecting incentives to invest in and to create opportunities for women and girls. Cross-country evidence indicates that gender gaps—in education and labor force participation, for example—tend to be smaller in countries that export more in relatively female labor–intensive sectors (Do, Levchenko, and Raddatz 2011). Recent studies have found that natural resource extraction, namely oil, reduces and discourages female labor force participation, which in turn reduces their political influence (Ross 2008; World Bank 2012a). Such distinctions in economic structure are relevant to the East Asia and Pacific context. East and Southeast Asian economies rely heavily on export-oriented manufacturing growth, whereas economies of the Pacific Islands are dominated by natural resource extraction, tourism, and remittances. These differences in economic incentives generated by distinct patterns of development may help to explain, at least in part, differences in progress toward gender equality in East Asian and Pacific countries, particularly with respect to voice and influence in society.

Several emerging trends in East Asian and Pacific countries will present both new opportunities and new risks to achieving gender equality. Much of the region is characterized by high levels of openness and *economic integration* with the rest of the world. Much of the region is also characterized by *migration* and *rapid urbanization* as the center of economic activities has moved from rural to urban areas. Increasing numbers of the region's citizens cross national boundaries in search of better economic opportunities. Swift declines

in fertility and mortality are dramatically changing the demographic profile of the region, and many countries will face rapid *population aging* in the coming years. The region also experiences rising adoption of new *information and communication technologies* (ICTs), which are breaking down information barriers, opening up new economic opportunities, and enabling collective action in many contexts.

These trends are likely to have important gender dimensions, generating both new opportunities and new risks for gender equality. For instance, while increased economic integration has contributed to higher demand for female labor in several East Asian countries, it has also increased workers' vulnerability to external shocks, with different effects on females and males. Similarly, while increased migration has opened up new economic opportunities in the region, it has also been accompanied by new risks—among the most severe is the risk of human trafficking.

This report clarifies empirically the relationship between gender equality and development and outlines an agenda for public action to promote gender equality in East Asian and Pacific countries. Written as a companion to the *World Development Report 2012: Gender Equality and Development* (World Bank 2011b), the report makes several distinct contributions to policy makers' understanding of gender, development, and public policy in the region. First, the analysis focuses on those issues and policy challenges that are particularly relevant to East Asian and Pacific countries. For example, compared with other developing regions, female access to basic education is no longer a first-order issue in many parts of the region. At the same time, gender stereotyping and gender "streaming" in education still represent critical challenges. These factors affect women's and men's aspirations and behaviors and contribute to persistent employment segregation. Second, the report examines the gender implications of several key emerging trends in the region: increased global economic integration, migration, urbanization, rapid population aging, and rising use of ICTs.

These trends have important gender dimensions that are not commonly accounted for by policy makers but that will generate a distinct set of challenges for promoting gender equality going forward. Third, the report accounts for intraregional diversity in a way that is not possible in a global report. Particular emphasis is placed, where possible, on the challenges faced by countries in the Pacific as distinct from those in East Asia. Finally, the report has undertaken extensive empirical analysis of gender equality using a newly created database of household surveys for the region. In doing so, the report has contributed significantly to the development of indicators and evidence on gender, development, and public policy that were not available previously.

This chapter examines the state of gender equality in the East Asia and Pacific region, highlighting both recent progress and pending challenges. Following the *World Development Report 2012*, the discussion focuses on gender equality in three domains: (a) endowments, (b) economic opportunity, and (c) agency (see box 1.1). *Endowments* are defined here as human capital and other productive assets that allow individuals to live healthy and productive lives. To analyze gender equality in endowments, the discussion focuses on education and health as well as other productive assets, such as land. *Economic opportunity* pertains to an individual's ability to fully and freely participate in and receive returns from their work in the economy. The report focuses on a range of economic indicators, including labor force participation, earnings, and employment segregation, whether in the labor market or in self-employment. *Agency* is defined as the ability of women and men to express themselves (exercise voice) in accordance with their preferences and to take actions on their own behalf to influence their surroundings. Since people exercise agency in all aspects of life, the report focuses on multiple dimensions: agency within a household and in several aspects of the public domain, including civil society, the private sector, and politics. The report also focuses on safety and security

as a dimension of agency, defining violence against women as the extreme deprivation of agency.

Why does gender equality matter for development?

Gender equality matters intrinsically. Nobel prize–winning economist Amartya Sen transformed the discourse on development when he argued that development is not only about raising people's incomes or reducing poverty, but rather involves a process of expanding freedoms equally for *all* people (Sen 1999).[1] Viewed from this perspective, gender equality is intrinsically valued. The near-universal ratification and adoption of the Convention on the Elimination of All Forms of Discrimination against Women (CEDAW)[2]—and the subsequent commitment of the international community to Millennium Development Goals 3 and 5—underscores a near-global consensus that gender equality and women's empowerment are development objectives in their own right.

Gender equality also matters for development. A growing body of empirical literature from around the world demonstrates that promoting gender equality is also good development policy, or as stated in the *World Development Report 2012* (2011b, 3), "Gender equality … is smart economics." Indeed, the literature shows that greater gender equality in endowments, access to economic opportunities, and agency can (a) contribute to higher productivity, income growth, and poverty reduction; (b) improve the opportunities and outcomes of the next generation; and (c) enhance development decision making. This section explores the evidence on these three pathways, in turn.

Gender equality can contribute to higher productivity and income growth

For households and economies to function at their full potential, resources, skills, and talent should be put to their most productive use. If societies allocate resources on

BOX 1.1   Defining and measuring gender equality

Gender refers to the social, behavioral, and cultural attributes, expectations, and norms that distinguish men and women. Gender equality refers to the extent to which men's and women's opportunities and outcomes are constrained—or enhanced—solely on the basis of their gender. This book focuses on gender equality in three domains: (a) *endowments*—human and productive capital; (b) *economic opportunity*—participation and returns in the economy; and (c) *agency*—the voice and influence of women in multiple dimensions in the private and public domains.

Gender equality can be conceptualized in two ways: in terms of equality of opportunities and equality of outcomes. Equality of opportunities measures inequalities that arise from circumstances beyond the control of individuals. Equality

of outcomes measures equality of results (World Bank 2011b). Both concepts can be useful, depending on the domain.

In some domains, such as in health and education, where gender equality in outcomes may be inherently valued, it is reasonable to focus on equality of outcomes. In contrast, equality of opportunities may be the more relevant conception of gender equality in the economic sphere, where people's preferences may lead to different outcomes, even if their opportunities are equal. Despite these distinctions, it is often difficult to distinguish opportunities from outcomes empirically. As such, though the book will rely on both conceptions of gender equality in its analysis, data limitations often necessitate that the evidence focuses on measuring outcomes.

the basis of one's gender, as opposed to one's skills and abilities, this comes at a cost. Indeed, the economic costs of gender inequalities—whether caused by the persistence of traditional norms or by overt discrimination—can be considerable. A recent study commissioned for the *World Development Report 2012* found that in the East Asia and Pacific region, output per worker could be 7 to 18 percent higher across a range of countries if female entrepreneurs and workers were to work in the same sectors, types of jobs, and activities as men, and have the same access to productive resources (Cuberes and Teignier-Baqué 2011).

Evidence suggests that misallocation of female skills and talent commonly begins before women enter the labor force, when families and societies underinvest in girls' schooling. A number of cross-country studies have found a robust inverse relationship between the size of the gender gap in education and GDP growth, controlling for average education levels and other factors associated with economic growth (see, for example, Klasen 2002; Knowles, Lorgelly, and Owen 2002).[3] Moreover, to the extent that young

women (or men) choose fields of study on the basis of their gender rather than their abilities, this too will exact costs not only on individuals' employment and earnings, but also on a country's economic productivity more broadly.

Gender inequalities in access to productive assets also have costs in terms of productivity and income. Microeconomic studies from a number of countries across developing regions show that female farmers and entrepreneurs are inherently no less productive than male farmers and entrepreneurs; rather, they tend to have less access to productive inputs.[4] A recent study by the Food and Agriculture Organization of the United Nations estimates that equalizing access to productive resources between female and male farmers could increase agricultural output in developing countries by 2.5 to 4.0 percent (FAO/Sida Partnership Cooperation 2010).

A number of studies show that gender-based violence also imposes significant costs on the economies of developing countries through lower worker productivity and incomes, lower human capital investments, and weaker accumulation of social capital

FIGURE 1.3 Girls' secondary school enrollments have converged to those of boys

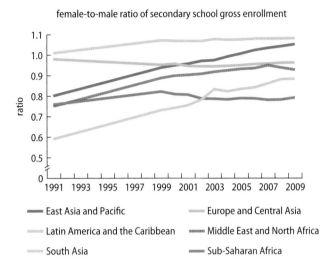

Source: World Bank 2011c.

FIGURE 1.4 Tertiary school enrollments of females have converged to those of males in East Asia and the Pacific

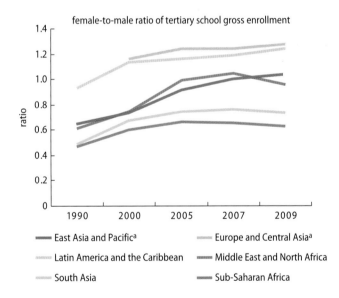

Source: UNESCO Institute for Statistics (UIS) Data Centre, 2009, 2011 data.
a. East Asia and Pacific includes developed countries. In this figure, Mongolia is included in Europe and Central Asia, not in East Asia and the Pacific.

(Morrison, Ellsberg, and Bott 2007). In addition to indirect costs, gender-based violence has large direct economic costs on society. A study in the United States found that the direct health care costs of intimate partner

violence against adult women were more than $4 billion USD in 1995 (USCDC 2003). Reducing gender-based violence would thus have significant positive effects on the region's economies by reducing health care costs and increasing investments in women's human capital, female worker productivity, and women's accumulation of social capital.

As the global economy becomes more integrated, the productivity effects associated with greater gender equality are likely to be increasingly important to East Asian and Pacific countries. A recent study on the relationship between gender and trade suggest that gender inequalities have become financially detrimental for countries in a world of open trade (Do, Levchenko, and Raddatz 2011). To participate effectively in an increasingly competitive world, countries will need to harness their resources efficiently by improving opportunities for all and allocating labor based on skill rather than by gender. Gender inequality, whether in endowments, economic opportunities, or in agency (voice), reduces a country's ability to compete in this increasingly globalized economic environment (World Bank 2011b).

Promoting gender equality is also an investment in the next generation

A large body of cross-country and country-specific literature shows that healthier, better educated mothers have healthier, better educated children, which can be expected to positively affect children's future productivity and economic prospects. The effects begin even before childbirth. In Timor-Leste, highly educated mothers and those in the wealthiest households are more likely to have their babies delivered by skilled birth attendants than less educated mothers and those from poorer households (NSD, Ministry of Finance, and ICF Macro 2010). Similarly, Demographic and Health Survey data show that Cambodian women with little education are relatively less likely to receive prenatal care and assistance from trained health personnel during birth deliveries than women with higher education (Johnson, Sao, and Hor 2000). A mother's health and nutrition status is also found to strongly affect children's physical health as well as cognitive and noncognitive abilities, which can have long-lasting developmental and societal consequences (Nadeau et al. 2011).

Higher labor force participation, greater income earned, and more assets held by women have also been shown to have positive effects on the next generation. In Indonesia, for example, women with a higher share of household assets before marriage tend to use more prenatal care and are more likely to have their births attended by skilled health care providers (Beegle, Frankenberg, and Thomas 2001). Similarly, in China, increasing adult female income by 10 percent of the average household income raised the fraction of surviving girls by 1 percentage point and increased years of schooling for both boys and girls. In contrast, a similar increase in male income reduced survival rates and educational attainment for girls with no impact on boys (Qian 2008). Studies from across developing and developed regions (for example, from places as diverse as Brazil, Ghana, South Africa, and the United States) show that income in the hands of women positively affects their female children's health (Duflo 2003; Thomas 1995); commonly, the marginal effects of income and assets in the hands of mothers are larger than the effects of similar income and assets in the hands of fathers.

Reductions in gender-based violence through greater female agency can also have important intergenerational benefits. Several studies show that experiencing domestic violence between parents as a child contributes to a higher risk of both women experiencing domestic violence as adults and of men perpetrating violence against their spouses (Fehringer and Hindin 2009). In Timor-Leste, 56.4 percent of women who were victims of spousal violence had a father who beat their mother (NSD, Ministry of Finance, and ICF Macro 2010). In Cambodia, women who reported that their mothers experienced domestic violence were more likely to experience physical and psychological domestic violence as well (NIPH, NIS, and ORC

FIGURE 1.5 **Gender gaps in secondary school enrollment vary substantially across countries**

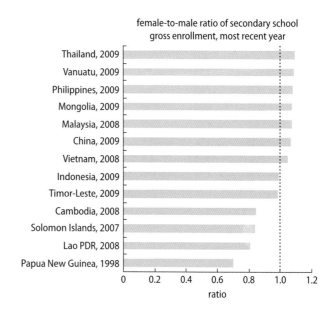

Source: World Bank 2011c.

FIGURE 1.6 **Gender gaps in education have reversed in several countries, particularly at the tertiary level**

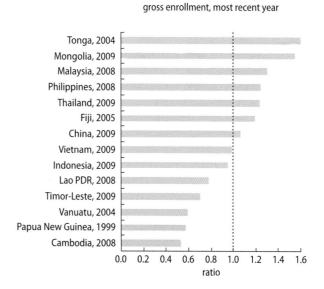

Source: UIS Data Centre, 2009, 2011 data.

FIGURE 1.7   **Minority populations in Vietnam often experience lower educational enrollments**

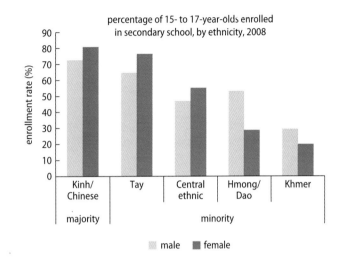

Source: World Bank 2011a.

FIGURE 1.8   **In Indonesia, gender gaps in enrollment do not vary substantially by household wealth**

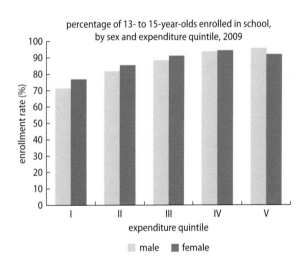

Source: World Bank estimates using Indonesia National Socioeconomic Survey 2009.

Macro 2006). Efforts that increase women's safety and security and that reduce domestic violence can thus lead to lower intergenerational transmission of violence within families.[5]

Strengthening women's voice can enhance the quality of development decision making

Several studies have shown that women and men have different policy preferences (Edlund and Pande 2001; Lott and Kenny 1999). Capturing these gender-based differences in perspective can lead to not only more representative but also better decision making. Evidence from India indicates that private firms can benefit from greater gender equality among the ranks of senior management. Other evidence from South Asia suggests the same is true with respect to development policy making. As an example, a study of women elected to local government in India found that female leadership positively affected the provision of public goods at the local level in ways that better reflected both women's and men's preferences (Chattopadhyay and Duflo 2004). Similarly, studies from rural India and Nepal found that when women who were previously excluded from decisions about local natural resource management had greater voice and influence, local conservation outcomes improved significantly (Agarwal 2010a, 2010b).

Women's collective agency can also be transformative, both for individuals and for society as a whole. For example, for a group of ethnic minority women in rural China, information sharing among them has helped empower them and raise their social standing in the Han-majority communities into which they married (Judd 2010). Migrant domestic workers in Hong Kong SAR, China, have been engaged in civic action focused on local migrant workers' rights as well as international human rights over the last 15 years (Constable 2009). These efforts have contributed to the enactment of laws that now provide migrant domestic workers in Hong Kong SAR, China, with some of the most comprehensive legal protections in the world.

Recent progress, pending challenges

Over the past few decades, many East Asian and Pacific countries have experienced considerable progress toward gender equality, at least in some dimensions. In others, gender disparities have been more persistent. This section reviews recent progress and pending challenges in achieving gender equality in *endowments, economic opportunity,* and *agency* in the region. It emphasizes where growth and development have contributed to advances in promoting gender equality and where this has not been sufficient. The section also highlights the considerable diversity of experience within the region as well as within countries. The basic gender profile developed here provides the foundation for the deeper analysis of gender, development, and public policy presented in subsequent chapters.

Endowments: Human and productive capital

Economic growth and poverty reduction in the region have been associated with rapid increases in female enrollment and convergence in the rates of school enrollment, across both genders and at all levels of education. In 2010, the region had the highest female-to-male enrollment ratio of all developing regions at the primary level. At the secondary enrollment level, only Latin America and the Caribbean had a higher ratio (figure 1.3). Although the female-to-male enrollment ratio in the East Asia and Pacific region is still below 1 at the tertiary level, it has been rising consistently over the past two decades (figure 1.4).

However, both overall enrollment rates and female-to-male enrollment ratios vary substantially across countries. Countries such as Cambodia, the Lao People's Democratic Republic, and Papua New Guinea still have relatively low enrollment levels and low female-to-male enrollment ratios, particularly at the secondary school level. Despite convergence in enrollment among the young, substantial gaps still remain in the educational endowments of adult populations. At the same time, the gender gap in education has reversed in several countries; girls' secondary enrollment rates now exceed those of boys in

FIGURE 1.9 Fertility rates have declined across the world

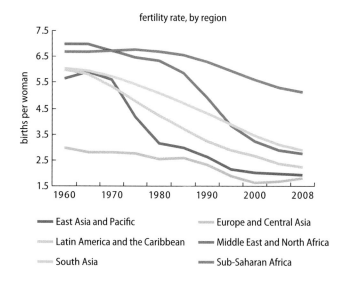

Source: World Bank 2010b.

FIGURE 1.10 Under-five mortality rates have declined sharply for both boys and girls

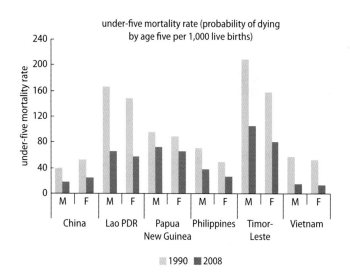

Source: World Health Organization (WHO) Global Health Observatory Data Repository.
Note: M = male, F = female. The under-five mortality rate is defined as the probability of death derived from a life table and expressed as the rate per 1,000 live births.

countries such as China, Malaysia, Mongolia, the Philippines, Thailand, Vanuatu, and Vietnam (figure 1.5). Reverse gender gaps at the tertiary level are sometimes even starker (figure 1.6).

FIGURE 1.11   Maternal mortality rates have declined across the world

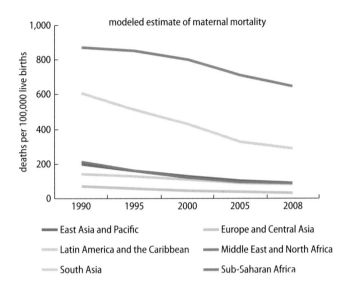

Source: World Bank 2010b.

FIGURE 1.12   Maternal mortality rate has declined in most countries in the region

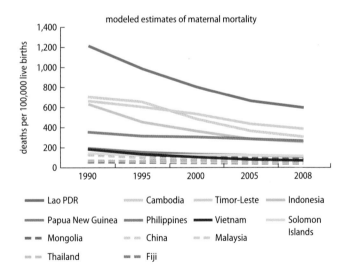

Source: World Bank 2010b.

Within countries, interactions between gender and other socioeconomic characteristics tend to exacerbate disparities in education. For example, economically disadvantaged and minority populations often experience lower educational enrollments. In Vietnam, school participation among 15- to 17-year-olds is substantially higher among the Kinh and Hoa (Chinese) majorities than among many of the 52 ethnic minority populations. Among the more economically disadvantaged and less well integrated Hmong, Dao, and Khmer minorities, far fewer girls attend school than boys (figure 1.7, and Baulch et al. 2002).

Globally, poverty and gender often interact to compound gender inequalities. Data suggest that, in East Asia and the Pacific, poverty is not as important a contributor to gender disadvantage in education as elsewhere in the world, however. Household survey data from several countries indicate that gender gaps in enrollment do not vary substantially or systematically across income quintiles. Indonesia, for example, actually shows a slight female advantage in enrollments among 13- to 15-year-olds from poorer households, but a slight female disadvantage exists among the wealthiest households (figure 1.8). Overall, data from the region suggest that gender gaps in enrollment tend to be smaller than enrollment gaps across income or wealth groups.

Although gender disparities in enrollment have closed, education streaming persists. Males and females differ in the types of education obtained. Data from Indonesia, Thailand, and Vietnam indicate that the fields of engineering and law are dominated by males, whereas the fields of education, health, and business administration are dominated by females.[6] This "gender streaming" in education contributes to persistent inequalities between women and men in access to economic opportunities.

Growth and development in the region during the past half century have also been associated with substantial improvements in key health indicators. Fertility rates have sharply declined, both in the region as well as across the world (figure 1.9). Under-five mortality rates have more than halved since 1990 for both boys and girls (figure 1.10). Noteworthy gains have been made in birth attendance by health professionals. In 2006, 87 percent of births were attended by physicians, compared to 47 percent in 1992. Gains of this magnitude were not witnessed in any other region. In addition, the East Asia and Pacific region has also seen declines in the maternal mortality rate, from approximately 200 deaths per 100,000 births in 1990 to 100 in 2008 (figure 1.11). The region has experienced consistent increases in both male and female life expectancy at birth since 1960. Female life expectancy in the region has increased from 48 to 74, and male life expectancy has increased from 45 to 70.

As with education, progress in health has been uneven across the region. Maternal mortality remains high in lower-income countries and in parts of the Pacific, especially in Cambodia, Lao PDR, Papua New Guinea, and Timor-Leste (figure 1.12). In Lao PDR, for example, maternal mortality rates were approximately 580 deaths per 100,000 births[7] in 2008, among the highest in the world. Indonesia's maternal mortality rate remains high compared to other countries in the region at similar levels of development. Women in remote rural areas commonly have limited access to health care, which significantly raises the risks associated with pregnancy and childbirth. Although Vietnam has, on average, experienced noteworthy declines in maternal mortality over the past decade, progress has been much slower in remote and ethnic minority regions (World Bank 2011a).

Moreover, more than a million girls and women per year are "missing" in East Asia. Despite growth and development, the problem of missing girls remains. The term *missing women* was first coined by Sen (1992) to refer to the phenomenon that many low-income countries have far fewer women than men, relative to what is observed in developed countries. Sen argued that this imbalance in sex ratios reflected severe forms of gender bias in affected societies. At birth, biological differences between males and females imply that approximately 105 boys are born for every 100 girls. Nonetheless, China, Vietnam, and, until recently, the Republic of Korea have experienced substantial deviations from the biological norm. Moreover, the trend over time in China has been alarming (figure 1.13). The

number of missing girls at birth per year in China, calculated by comparing the sex ratio at birth in China to those in high-income countries, increased from 890,000 in 1990 to 1,092,000 in 2008 (World Bank 2011b).

Men face gender-specific health risks as well. For example, men are more likely to experience higher morbidity and premature mortality related to substance abuse, war and conflict, and violence. Cambodia experienced considerable declines in the male population during the Khmer Rouge regime of Pol Pot; so did Vietnam during its war era. Moreover, differences between men and women in the incidence of tobacco use are higher in East Asia and the Pacific than in other regions of the world; the gender differential in heavy episodic alcohol consumption is also particularly stark. Male abuse of tobacco and alcohol in the region has important effects on men's health and mortality rates, which in turn can impose significant costs on economic productivity and growth.

Gender disparities still exist in access to and control of productive resources. Gender disparities in access to and control of land or farm inputs are pervasive around the world and remain issues in the region, despite significant growth and development. Women remain less likely to own land (or hold formal land titles) than men. Moreover, data from Cambodia, China, Indonesia, Lao PDR, Mongolia, Timor-Leste, and Vietnam indicate that when women—or, specifically, female-headed households—do own land, they typically have smaller holdings.[8] A recent study of women's land holdings in post-tsunami Aceh similarly found that women's land holdings were considerably lower than men's (World Bank 2010).[9] Female-headed households in the region also tend to have poorer access to other productive inputs and support services, including livestock holdings and access to agricultural extension services.

Despite improvements in women's access to microcredit, important challenges remain in accessing enterprise finance. Women also have traditionally had less access to capital than men. This disparity has been compounded by their poorer access to land, an

FIGURE 1.13 **East Asia has a highly skewed male-to-female ratio at birth**

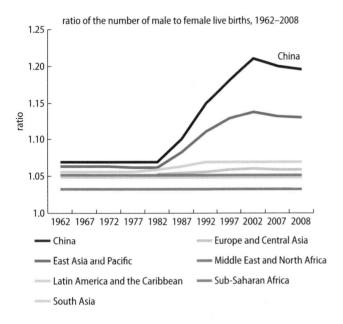

Source: HNP Stats (Health Nutrition Population Statistics) database, 2010 data.

FIGURE 1.14 **The East Asia and Pacific region has high female labor force participation rates**

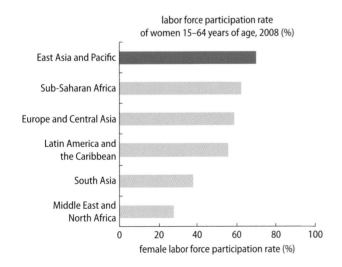

Source: World Bank 2010b.

important source of collateral. In response to gender disparities in access to credit, the microfinance movement has focused on increasing women's access to capital across the world. Of the 106.6 million poorest clients worldwide who have been reached by microcredit initiatives by the end of 2007, 83.2 percent were women (Daley-Harris 2009). Evidence on access to credit among female and male entrepreneurs tells a more nuanced story. Among micro- and small firms in Indonesia, both female-and male-run enterprises cite access to finance as their most significant business constraint, with the share of female-run firms reporting this constraint only slightly higher than the share of male-run firms (30 versus 25 percent, respectively). Among small and medium enterprises in nine East Asian and Pacific countries, only in Timor-Leste and Tonga do a greater share of female-lcd enterprises report access to credit as a significantly more important constraint than reported by their male counterparts.[10]

Economic opportunity: Participation and returns in the economy

The East Asia and Pacific region has the highest average female labor force participation rate and highest ratio of female-to-male labor force participation in the developing world. In 2008, 70.1 percent of females were participating in labor market activities (figure 1.14) and the gender gap in labor force participation was approximately 15 percentage points. In comparison, the average female labor force participation rate in Latin America and the Caribbean was 55 percent in 2008, and in Europe and Central Asia it was 58 percent. Gender gaps in labor force participation were 27 and 16 percentage points, respectively.

Both the levels of female participation and their trends over time vary substantially in the region. In Cambodia, China, Lao PDR, Thailand, and Vietnam, for example, female labor force participation was over 75 percent in 1980. Between 1980 and 2008, these countries witnessed declines in female participation of 5 to 10 percentage points despite strong economic growth. Except for China, these declines were mirrored in similar drops in male participation, leaving the ratio of female-to-male participation rates unchanged. Participation rates in Korea, Indonesia, and Malaysia were significantly lower than the regional average: in 1980, only 45 percent of working-age females participated in the labor market. Female participation rates have increased over time in Korea and Indonesia, while they have remained stagnant in Malaysia despite strong economic growth. Female labor force participation varies substantially in the Pacific, ranging from over 75 percent in Vanuatu to 40 percent in Fiji in 2008 (figure 1.15). Although time series data for these countries are limited, female participation rates appear to have risen over time in countries with lower initial rates. In Fiji, female participation rates rose from 25 percent in 1980 to 40 percent in 2008, and in Tonga they rose from 45 percent to 57 percent over the same period.

Structural transformation in the region's economies has changed the type of work conducted. The region has seen a significant decline in the fraction of the workforce in the agricultural sector over the past half century, from approximately 60 percent in 1991 to just over 40 percent in 2008 (figure 1.16). Between 1960 and 2000, agriculture's share of total employment declined from over 80 percent to under 50 percent in Thailand, while in Indonesia the share declined from 70 percent to approximately 45 percent (Butzer, Mundlak, and Larson 2003). Nonetheless, agriculture remains important: in 2008, it was still the largest sector of employment in the region (figure 1.16).

Women's labor market responses to structural transformation have, in part, reflected country-specific patterns of development. Thailand, for example, moved from a heavy concentration of workers in agriculture in 1980 to a rising employment share in the industrial and service sectors (figure 1.19). The early 1990s saw a substantial movement of females away from agriculture and out of the workforce. Similar patterns were

FIGURE 1.15   Female labor force participation varies substantially across countries

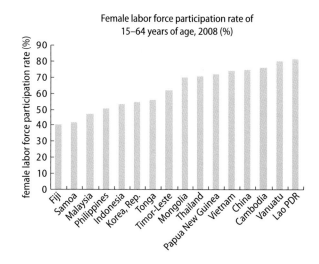

Source: World Bank 2010b.

FIGURE 1.16   The fraction of the workforce employed in agriculture has declined in the East Asia and Pacific region

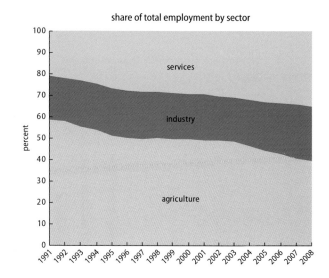

Source: World Bank 2010b.

seen in the United States during the early stages of the transition away from agriculture (Goldin 1995; Mammen and Paxson 2000). In Indonesia, by contrast, female labor force participation increased by 9 percentage points between 1980 and 2007,

FIGURE 1.17 **The evolution of sectoral composition by gender varies across countries**

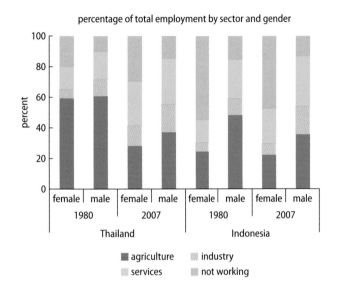

percentage of total employment by sector and gender

- agriculture
- services
- industry
- not working

Source: World Bank 2010b.

FIGURE 1.18 **Women are more likely than men to be temporary workers**

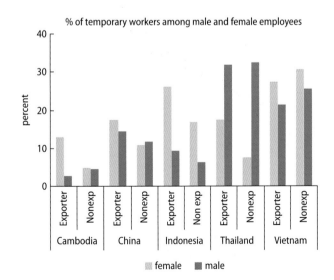

% of temporary workers among male and female employees

- female
- male

Source: World Bank estimates using Enterprise Surveys database, 2002–06 data.
Note: Nonexp = nonexporter. The percentage of temporary workers is calculated as a ratio of female temporary workers to female total workers. The analysis includes only manufacturing firms.

with the bulk of the increase coming from the service sector (figure 1.17). Participation in industry and agriculture was fairly similar in 1980 and 2007.

Labor market "sorting"—or employment segregation—along gender lines is pervasive, by industry, occupation, formality, and flexibility of employment. Such sorting affects a number of economic outcomes, by gender, including earnings, social security coverage, the intensity of work conducted, returns to education and experience, and exposure to shocks. In many countries in the region, women are more likely than men to conduct unpaid family labor in agriculture and in the informal sector (Asia Foundation et al. 2006; Asian Development Bank and World Bank 2005; World Bank 2011a). In addition, women are more likely to be found in some occupations—such as teaching and nursing—and are less likely to be found in others, such as mining. Within the manufacturing sector, women are more likely to be found in industries such as textiles and food processing, and are also found in large and export-oriented firms. Within firms, women are more likely than men to be temporary workers (figure 1.18).

Female- and male-led enterprises also tend to have distinct characteristics. Female-led enterprises across the region tend to be smaller than male-led enterprises (figure 1.19). They are more likely to operate in the informal sector (less likely to be registered) and to be home based or operate out of nonpermanent premises. In general, female-led enterprises across the region have fewer employees and assets and command lower profits. Although female-owned and -managed enterprises are not inherently less productive, they tend to be less capitalized and located in less-remunerative sectors. For instance, in Indonesia, female-led enterprises are more likely to locate in the food and garment manufacturing sectors. These sectors are among the least capital intensive and least productive; male-led enterprises are more likely to locate in sectors such as transportation and other services, which are among the most capital intensive, most productive sectors.

Women in East Asian and Pacific countries continue to undertake the majority of household work, in addition to market work, a function of longstanding gender norms regarding the division of labor within households. In many contexts, this tradition constrains women's economic opportunities, affecting their labor force participation, choice of sectors and occupations, time worked, and, ultimately, earnings. Global evidence indicates that women tend to work longer hours than men once both market and nonmarket work are taken into account (Ilahi 2000; World Bank 2001, 2011b). The composition of work also varies substantially by gender. Men devote relatively more time to market work, while women devote more time to domestic activities. Evidence from East Asia is consistent with global patterns. Recent data from Lao PDR indicate that both gender differences in hours worked and in the division of labor between market and nonmarket work are exacerbated once families have children (figure 1.20).

Cross-country evidence on wages indicates that in the low- and middle-income countries of East Asia, women earn between 70 and 80 percent of what men earn for similar work (figure 1.21). These gender wage gaps are partially attributable to differences in education, experience, and industrial choice across men and women. Differences in education endowments, experience, and industrial and occupational segregation explain up to 30 percent of the gender wage gap in East Asian countries (Sakellariou 2011).

While gender wage gaps have evolved over time, they have not always narrowed with growth and development. In Vietnam, the process of economic transition from a centrally planned to a market-based economy has been associated with a sharp reduction in the gender pay gap among salaried employees. The average gender pay gap halved between 1993 and 2008, with the majority of the contraction evident by 1998 (Pham and Reilly 2007; Sakellariou 2011). Pay gaps still persist between men and women, however, with women earning on average 75 percent of the male wage in 2009 (Pierre 2012). By

FIGURE 1.19   **Enterprises with female management tend to be smaller**

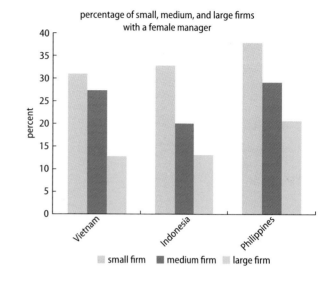

Source: World Bank estimates using Enterprise Surveys database, 2006–11 data.

FIGURE 1.20   **In Lao PDR, women—particularly those with young children—must balance household work commitments with market work**

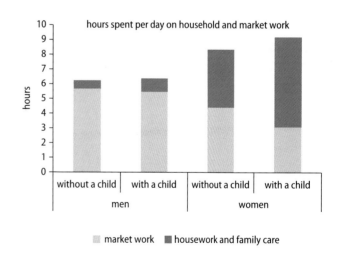

Source: World Bank estimates using the Lao Expenditure and Consumption Survey (LECS) (LSB Lao PDR 2008).

contrast, in the Philippines, the wage gap widened between 2000 and 2009, a change that has been partly attributed to growing differences between men and women in terms of their returns to education and other characteristics (Sakellariou 2011). In Indonesia, the average wage gap and the gap by age cohorts widened between 1976 and 1999 (Dhanani and Islam 2004), although more

recent evidence suggests that the gender gap declined between 1997 and 2009 (Sakellariou 2011). This complex relationship between growth, development, and the gender wage gap reflects a number of factors at the country level, including changes in the structure of the economy and in labor demand; the gender composition of the labor force; changes in the relative education, skill, and experience among male and female workers; and labor market institutions and policies, each of which affect how remuneration is evolving by gender.

FIGURE 1.21   **Women in East Asia still earn less than men**

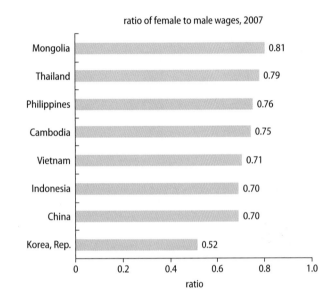

ratio of female to male wages, 2007

Country	Ratio
Mongolia	0.81
Thailand	0.79
Philippines	0.76
Cambodia	0.75
Vietnam	0.71
Indonesia	0.70
China	0.70
Korea, Rep.	0.52

Source: World Bank 2010b.

Agency: Women's voice and influence

East Asian and Pacific countries have experienced progress and pending challenges in achieving gender equality in *agency*—women's voice and influence—as with endowments and economic opportunity discussed above. Agency refers to the ability of women and men to take action on their own behalf, in accordance with their preferences, and to influence outcomes that affect them in both private and public domains (box 1.2). The ability to influence

---

BOX 1.2   **Defining and measuring agency: Women's voice, influence, and participation**

All individuals in a society have ideas and preferences on how to use scarce resources and live their lives. The ability of women and men to express and act on their preferences is affected by—and also affects—their ownership and control over endowments and their access to economic opportunities (Kabeer 1999). This ability to act on one's preferences and influence outcomes is referred to as *agency*. Changes in individuals' characteristics in a household, community, or society will affect the strength of their voices relative to others, and may also influence their preferences. This influence, in turn, will affect choices made at the household, community, and societal levels. For example, an increase in the education of a woman will affect investments in the education and health of her children through its effect on both her relative bargaining power

within the household (the extent to which her voice is heard) and, potentially, her preferences.

The concept of agency has multiple dimensions that have been measured empirically. The form of agency that is most frequently measured is the decision-making power of men and women (Mason 2005; McElroy 1990). Agency may be more explicitly measured by examining women's mobility in the public domain, their participation in public action, and the incidence of gender-based violence (Kabeer 1999). Finally, gender differences in bargaining power within a household have been assessed by examining the extent to which people's choices change when factors affecting their bargaining power, such as education, relative earnings, or asset holdings, also change (Duflo 2003; Quisumbing and Maluccio 2003; Thomas 1990, 1995).

one's life by making choices and taking action is also a key dimension of well-being in and of itself (Sen 1999). This report focuses on agency in three domains: (a) agency in the household and individual decisions, examined through household decision making, control of resources, and reproductive decisions; (b) agency in the public sphere, examined through gender-based participation and representation in the private sector, civil society, politics, and public institutions; and (c) safety and security as an expression of agency, examined through the prevalence of gender-based violence. Although the East Asia and Pacific region has made progress in several domains of agency, progress has been uneven across countries, and many challenges still remain.

Women in East Asia experience relatively high levels of agency at the household level compared with other developing regions. Cross-country data suggest that women across the wealth distribution in East Asia have a greater say in decisions regarding large household purchases and also experience greater freedom to visit family and relatives without husbands' permission than do women in other developing regions. Women in East Asia also have as great as or greater control over their own earnings compared with women in other developing regions (figure 1.22). Reductions in fertility rates and unwanted fertility, defined as gaps between actual and desired fertility, observed in most East Asian and Pacific countries suggest that women may have gained greater control over their reproductive decisions.

Women's voice in the public domain remains weak despite economic development. Women have relatively low levels of representation in political assemblies, whether at the national or local levels. For example, women make up only one-fifth of national parliamentarians worldwide. The share of female parliamentarians in East Asian and Pacific countries is just below the global average, at approximately 18 percent in 2011 (figure 1.23). Female representation in the region has not increased over time with economic

FIGURE 1.22 Women in East Asia and the Pacific have more control over earnings and household decisions across all wealth quintiles than women from other developing regions

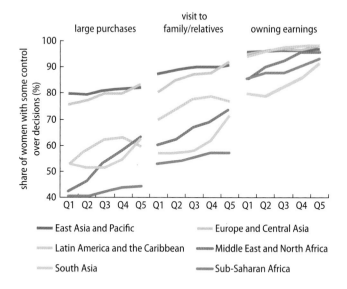

Source: *World Development Report 2012* team estimates based on Demographic and Health Surveys of 40 countries, 2003–09.

FIGURE 1.23 Female representation in parliament in East Asian and Pacific countries has hardly changed since the 1990s

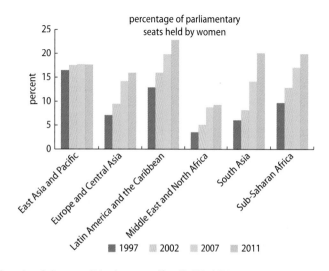

Source: Inter-Parliamentary Union data extracted from World Bank 2011c.

FIGURE 1.24   Women's representation in parliament is low, especially in the Pacific

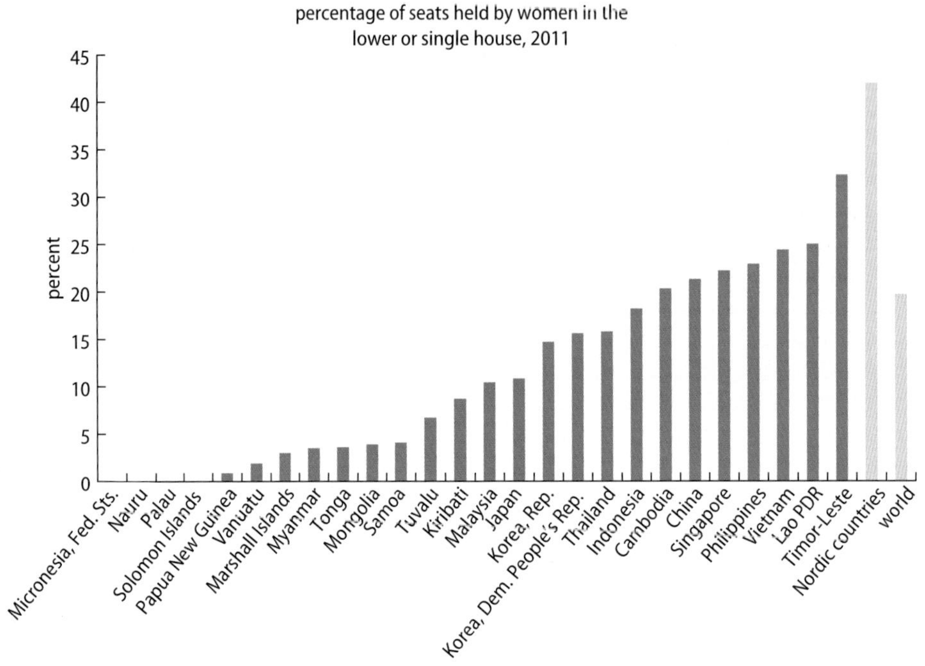

percentage of seats held by women in the
lower or single house, 2011

FIGURE 1.25   Men, and in some cases women, believe that men make better political leaders than women

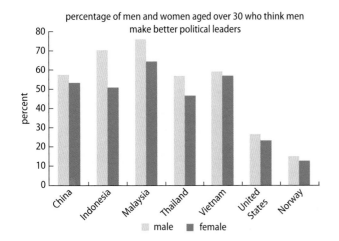

percentage of men and women aged over 30 who think men
make better political leaders

■ male   ■ female

growth and development; the overall share of female parliamentarians in East Asian and Pacific countries has barely changed since the late 1990s. This trend stands in contrast to other developing regions, where levels of female political representation have tended to increase, at least since 2000.[11]

The share of women in national parliaments varies tremendously across the region. In December 2011, the highest levels of female representation were found in Timor-Leste (32.3 percent), Lao PDR (25.2 percent), Vietnam (24.4 percent), the Philippines (22.9 percent), and China (21.3 percent) (figure 1.24). Female parliamentary representation in the Pacific is systematically lower. Four countries in the Pacific—the Federated States of Micronesia, Nauru, Palau, and the Solomon Islands—do not have female representation in parliament (PARLINE database, 2011 data). Countries that have enacted temporary special measures to

THE STATE OF GENDER EQUALITY IN EAST ASIA AND THE PACIFIC

promote female participation in political leadership have achieved higher levels of female representation in parliament, with an average of 27 percent in 2009, compared with 14 percent for those countries without temporary special measures.

Barriers to political representation in the region are likely to reflect perceptions held by both men and women that female politicians make less competent political leaders than their male counterparts (figure 1.25). Evidence from India suggests that people's perceptions of women as political leaders improve with exposure to elected female officials (Beaman et al. 2009). However, these perceptions may take time to evolve, reinforcing the case for concerted action to support increased political participation on the part of women. Ensuring that women are represented, whether at the local or national level, is a first step to ensuring that their views and preferences are expressed in public policies.

Women from East Asian and Pacific countries have experienced improvements in their voice and ability to actively participate in civil society and grassroots movements. Civil rights groups tackle a variety of issues to improve gender equality in their community, country, and region. PEKKA, a program in Indonesia, was created to address the needs of widows of conflict in Aceh and now provides training for village paralegals that focuses on domestic violence and family law. The program also holds district forums to bring together judges, prosecutors, police, nongovernmental organizations, and government officials to raise awareness of gender issues (PEKKA 2012; World Bank 2011b). As another example, the Fiji Women's Crisis Center lobbied for a nationally representative quantitative survey on violence against women, to be released in 2012.

Women's representation in top management and participation in ownership is high relative to other developing regions and the world average. In the East Asia and Pacific region, women are represented among the owners in over 50 percent of small, medium, and large firms, higher than in any other developing region (figure 1.26). Female participation in management in the region and all other regions is lower than participation in ownership, however, indicating that women have a more limited voice in running a firm and making decisions, despite being represented in the ownership of the firm.

The incidence of physical, sexual, psychological, and emotional violence against women is high throughout the region, particularly in the Pacific, where the prevalence of domestic violence is among the highest in the world (figure 1.27). Human trafficking is also a growing concern throughout the region. An increase in female migration in the past decade has brought about increased economic opportunities as well as increased risk of being trafficked. The ILO estimates that Asia and the Pacific account for over half of all trafficked victims worldwide, with an estimated 1.36 million victims, most of whom are women and girls (ILO 2008).

Gender-based violence and human trafficking of women and girls are often enabled by the lack of and weak enforcement of relevant legislation. Although the factors that enable gender-based violence are multiple and complex, the phenomenon is exacerbated by a lack of adequate legal protections in many countries, most notably in the Pacific islands. A recent United Nations Development Programme study (UNDP 2010) found that although more than three-quarters of the countries in East Asia have strengthened legislation on domestic abuse in recent years, more than 60 percent of countries in the Pacific still lacked relevant legislation on domestic violence. Police forces in several countries in the region—including Indonesia, Malaysia, and Thailand—have been criticized for limited commitment in investigating trafficking and enforcing antitrafficking laws (U.S. Department of State 2011).

Fostering new opportunities, managing emerging risks

As the region continues to develop, several emerging trends will present both new opportunities and new risks to achieving gender equality. As noted earlier, East Asian

FIGURE 1.26   More women hold top management positions in East Asia and the Pacific than in other developing regions

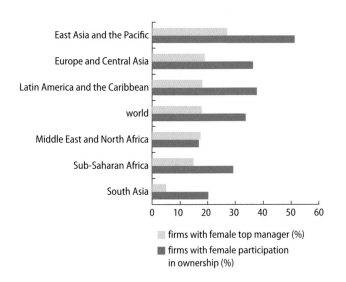

Source: Enterprise Surveys database, World Bank.

FIGURE 1.27   Violence against women is high in East Asia and the Pacific

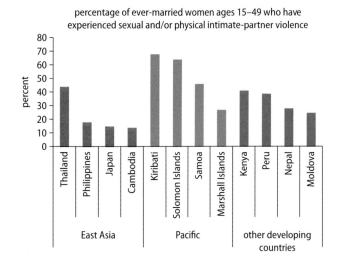

Source: Demographic and health surveys, various countries and years, and government surveys.

and Pacific countries are at the forefront of several global trends: (a) increasing *global economic integration*, (b) increasing domestic and cross-border *migration*, (c) rapid *urbanization*, (d) *population aging*, and (e) rising availability and use of *information and communication technologies (ICTs)*. Though not commonly accounted for by policy makers, these trends are likely to have important gender dimensions and, thus, affect the evolution of gender equality in the region. They will create new challenges for public policy, in terms of fostering new opportunities and managing new risks to gender equality.

Greater global economic integration brings with it substantial economic opportunities for women, but also potentially greater exposure to economic shocks. For example, evidence indicates that greater global economic integration can contribute to greater access to economic opportunities for women and reduce gender wage gaps by raising demand for female workers in export-oriented enterprises (Boserup 1970; Schultz 2006). At the same time, increasing global integration also increases risk and uncertainty, as shocks are more quickly transmitted across integrated markets. A number of studies find that although shocks do not necessarily have more adverse effects on women than men, they do have gender-differentiated effects on outcomes as diverse as employment, earnings, labor force participation, education investment, health, and nutrition (see, for example, Bruni et al. 2011; Rodgers and Menon forthcoming).

Opportunities gained by migration are balanced by new risks. High economic growth and increased economic integration in the region over the past three decades has spurred significant migration across the region. Women make up nearly half of all migrants in East Asia and the Pacific and in some countries represent the majority of new migrants. Migration can provide both women and men with access to new economic opportunities, which commonly differ by gender because of employment segregation in receiving areas' labor markets. At the same time, migration brings with it important gender-specific risks. For example, many female migrants find employment as domestic workers, a sector with particularly weak worker protection in most countries. Female migrants are also disproportionately susceptible to risks associated with human trafficking.

Unprecedented levels of urbanization are affecting all aspects of life: from the nature of employment to the availability of services to individuals' ability to rely on extended family and community networks. Many East Asian and Pacific countries are experiencing unprecedented levels of urbanization as migrants move to urban areas in search of economic opportunity. Between 2000 and 2015, Indonesia, China, and Cambodia are predicted to see an increase in the percentage of the population residing in urban areas by 17, 13, and 9 percentage points respectively (World Bank 2009). These changes almost certainly have gender-specific impacts. Although urban residence can open up a wide range of economic opportunities for both men and women, women's ability to take advantage of new opportunities is likely to depend fundamentally on the nature and availability of urban services—for example, whether transport systems facilitate safe travel of women to job sites or if affordable child care can compensate for the loss of extended family networks.

Aging populations will represent another challenge for women. The high-income economies in East Asia are experiencing rapid population aging.[12] Most emerging countries in the region have also begun this process; dependency ratios are already increasing in many middle-income countries in East Asia and the Pacific. Old-age dependency is expected to increase even more quickly in the coming decades (figure 1.28). Population aging is likely to have gender-differentiated impacts at all age levels. Gender differences in the time devoted to caring for the elderly imply that in the absence of institutionalized care services, women are likely to bear the brunt of the increased demand for elder care (Dwyer and Coward 1992; Ofstedal, Knodel, and Chayovan 1999). Moreover, while women tend to live longer than men, gender differences in education and labor force participation imply that women are less likely to be vested in formal pension systems and may have fewer assets to ensure a basic level of well-being in old age.

Advances in ICTs are revolutionizing the ways in which both men and women in the region are exposed to ideas, share knowledge, and networks. Existing evidence suggests, however, that women in the region still have lower access to information technologies than men. New and emerging technologies, if accessible, can help to empower women by opening new economic opportunities, breaking down information barriers, helping women in isolated communities engage in distance learning or commerce, or enabling women to take collective action. As in the case of other productive resources, such as land, machinery, or credit, growing gender gaps in access to ICTs could lead to widening gaps in access to economic opportunities as well as in voice and influence in society.

Gender equality in East Asia and the Pacific: A roadmap to the report

This chapter has provided a basic profile on the status of gender equality in East Asia

FIGURE 1.28 **The dependency ratio is increasing in most East Asian countries**

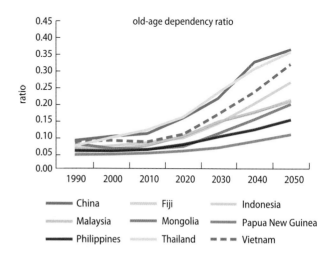

Sources: World Bank 2010b (1990–2010 data); HNP Stats—Population Projection (projections for 2020–50).
Note: Old-age dependency ratio is defined as the ratio of the population 65 years of age and older over working-age population (15–64 years of age).

and Pacific countries, taking into account several important factors that have characterized development in the region. Specifically, the region has experienced significant growth, poverty reduction, and economic structural transformation over the past several decades. The chapter has shown that rapid growth and development have been accompanied by reduced gender inequalities in several key dimensions, most notably in education and several key aspects of health. In East Asian countries, women's voice and influence, whether in the home or in the economy, are relatively strong. In many ways, women in the region are better positioned than ever before to participate in, contribute to, and benefit from their economies and societies.

At the same time, the evidence demonstrates clearly that economic growth and development alone are not enough to attain gender equality in all its dimensions. While gender outcomes in education and health have been responsive to growth, other areas have proved "stickier"; significant gender inequalities persist in a number of important areas despite development. Women still have less access than men to a range of productive assets and services. There remains substantial employment segregation, by gender. And despite closing of education gaps, women continue to earn less than men. Moreover, women in the region still have weaker voice and influence than men, whether in household decision making, in the private sector, in civil society, or in politics. And women across the region remain vulnerable to gender-based violence.

Progress toward gender equality has been uneven across and within countries. Despite widespread progress, a few, mostly low-income, countries have yet to close gender gaps in basic education. In China (and to a lesser extent Vietnam), significant imbalances occur in the sex ratio at birth, reflecting strong son preference in those societies. At the same time, a number of Pacific island countries face particular challenges with respect to promoting women's voice and influence. Specifically, women

in Pacific countries experience among the highest rates of violence against women in the world; they also have among the lowest levels of female representation in politics. Within countries, gender frequently interacts with other socioeconomic characteristics, such as ethnicity or geographic remoteness, resulting in specific subgroups of countries' populations facing a double disadvantage.

Because many aspects of gender inequality do not disappear automatically with growth and development, and because persistent gender inequalities impose high costs on women and girls and on societies more broadly, a case can be made for public policy to promote gender equality. In this context, an important contribution of this report will be to (a) clarify empirically the relationship between gender equality and development, (b) analyze the factors that contribute to or impede gender equality in its different dimensions, and (c) identify effective avenues for public action to promote gender equality and, thus, more effective development in East Asian and Pacific countries.

To achieve these objectives, the three chapters that follow focus on providing a deeper understanding of the factors affecting gender equality in endowments, economic opportunity, and agency. Specifically, chapter 2 examines in more depth the evidence on gender dimensions of human and physical capital accumulation. Chapter 3 analyzes access to economic opportunity, including the factors affecting female labor force participation, employment segregation across occupations and industries, and persistent gender gaps in wages and earnings. And chapter 4 focuses on factors that enhance or constrain women's voice and influence in society, both in the private and public domains. In carrying out their analyses, each of these chapters seeks to frame an agenda for effective public action moving forward.

The report also analyzes the gender dimensions of several important emerging trends in the region. Specifically, chapter 5 examines globalization and economic integration, increasing migration, rapid urbanization, population aging, and enhanced access to ICTs, while identifying both the emerging opportunities and the emerging risks to gender equality that these phenomena entail. Building on the in-depth analyses presented in chapters 2 through 5, chapter 6 then outlines directions for public policy to promote gender equality and, thus, more effective development in East Asian and Pacific countries. The report concludes by framing a forward-looking agenda for analysis and action—to continue to fill knowledge gaps and strengthen public policy responses to promoting gender equality in the region.

Notes

1. Sen (1999) defines freedoms and "unfreedoms" in five categories: (a) political freedoms, (b) economic facilities, (c) social opportunities, (d) transparency guarantees, and (e) protective security.
2. Adopted in 1979 by the United Nations General Assembly, CEDAW is often referred to as the international bill of rights for women. The convention defines what constitutes discrimination against women and provides an agenda for national action to end such discrimination. To date, it has been ratified by 187 countries worldwide (http://www.un.org/womenwatch/daw/cedaw/).
3. These studies must be interpreted with caution, given the difficulty of establishing a causal relationship between gender equality in education and growth in cross-country studies.
4. Evidence from Africa and Latin America, for example, suggests that ensuring equal access to productive assets and technologies could significantly raise agricultural production and household income (Goldstein and Udry 2008; Quisumbing 1995; Udry 1996).
5. A number of studies show that gender-based violence itself imposes significant costs on the economies of developing countries, for example, through lower worker productivity and incomes, lower human capital investments, and weaker accumulation of social capital (see Morrison, Ellsberg, and Bott 2007, for a review of key findings). For related evidence from the United States, see National Center for Injury Prevention and Control 2003.
6. Similarly, in Organisation for Economic Co-operation and Development (OECD) countries,

men are more likely to be found among mathematics and computer science graduates than women and have also been found to outperform women in mathematics (Schleicher 2008).
7. The maternal mortality ratio (MMR) in Lao PDR is a modeled estimate, to make it comparable to MMR estimates in other countries. Lao PDR's national estimate in 2005 was lower, at 405 deaths per 100,000 births (WDI database).
8. For data on land holdings by gender in China, see de Brauw et al. (2011). Data on other countries are based on World Bank staff calculations, using household survey data.
9. Evidence from other parts of Indonesia suggests that land ownership patterns, by gender, can differ in important ways, depending on local norms and customs. In the matrilineal region of West Sumatra, Indonesia, for example, at the time of marriage, husbands commonly own more forest land than their wives, and wives commonly own more paddy land (Quisumbing and Maluccio 2003).
10. Self-reported information on credit constraints, by gender, should be interpreted with caution. As discussed below, female-led enterprises are smaller, use less capital, and operate in difference sectors than male-led enterprises, making direct comparisons of self-reported credit constraints, by gender, difficult.
11. In the Europe and Central Asia region, female representation in national assemblies fell substantially following the dissolution of the Soviet Union, although levels increased again between 2000 and 2008.
12. For example, the share of the elderly (age 65 or above) in Hong Kong SAR, China; Japan; Korea; Singapore; and Taiwan, China, is above 10 percent and is expected to increase substantially in the next two decades.

References

Agarwal, Bina. 2010a. "Does Women's Proportional Strength Affect Their Participation? Governing Local Forests in South Asia." *World Development* 38 (1): 98–112.
———. 2010b. *Gender and Green Governance: The Political Economy of Women's Presence Within and Beyond Community Forestry.* New York: Oxford University Press.
Asia Foundation, ADB (Asian Development Bank), CIDA (Canadian International Development Agency), NDI (National Democratic Institute),

and World Bank. 2006. *Indonesia: Country Gender Assessment.* Manila, Philippines: Asia Foundation, ADB, CIDA, NDI, and World Bank.

Asian Development Bank and World Bank. 2005. "Mongolia: Country Gender Assessment." Country Planning Document, Asian Development Bank, Manila, Philippines.

AusAID. 2008. *Making Land Work: Vol. 1. Reconciling Customary Land and Development in the Pacific.* Canberra: Australian Agency for International Development.

Baulch, Bob, Truong Thi Kim Chuyen, Dominique Haughton, and Jonathan Haughton. 2002. "Ethnic Minority Development in Vietnam: A Socioeconomic Perspective." Policy Research Working Paper 2836, World Bank, Washington, DC.

Beaman, Lori, Raghabendra Chattopadhyay, Esther Duflo, Rohini Pande, and Petia Topalova. 2009. "Powerful Women: Does Exposure Reduce Bias?" *Quarterly Journal of Economics* 124 (4): 1497–540.

Beegle, Kathleen, Elizabeth Frankenberg, and Duncan Thomas. 2001. "Bargaining Power within Couples and Use of Prenatal and Delivery Care in Indonesia." *Studies in Family Planning* 32 (2): 130–46.

Boserup, Ester. 1970. *Woman's Role in Economic Development.* London: Earthscan.

Bruni, Lucilla, Andrew Mason, Laura Pabon, and Carrie Turk. 2011. "Gender Impacts of the Global Financial Crisis in Cambodia." World Bank, Washington, DC.

Butzer, Rita, Yair Mundlak, and Donald Larson. 2003. "Intersectoral Migration in Southeast Asia: Evidence from Indonesia, Thailand and the Philippines." Policy Research Working Paper 2949, World Bank, Washington, DC.

Chattopadhyay, Raghabendra, and Esther Duflo. 2004. "Women as Policy Makers: Evidence from a Randomized Policy Experiment in India." *Econometrica* 72 (5): 1409–43.

Constable, Nicole. 2009. "Migrant Workers and the Many States of Protest in Hong Kong." *Critical Asian Studies* 41 (1): 143–64.

Cuberes, David, and Marc Teignier-Baqué. 2011. "Gender Inequality and Economic Growth." Background paper for *World Development Report 2012*, World Bank, Washington, DC.

Daley-Harris, Sam. 2009. "State of the Microcredit Summit Campaign Report 2009." Microcredit Summit Campaign, Washington, DC.

de Brauw, Alan, Jikun Huang, Linxiu Zhang, and Scott Rozelle. 2011. "The Feminization

of Agriculture with Chinese Characteristics." Background paper for *World Development Report 2012*, World Bank, Washington, DC.

Dhanani, Shafiq, and Iyanatul Islam. 2004. *Indonesian Wage Structure and Trends, 1976– 2000.* International Labour Organization: Geneva.

Do, Quy-Toan, Andrei Levchenko, and Claudio Raddatz. 2011. "Engendering Trade." Policy Research Working Paper 5777, World Bank, Washington, DC.

Duflo, Esther. 2003. "Grandmothers and Granddaughters: Old-Age Pensions and Intrahousehold Allocation in South Africa." *World Bank Economic Review* 17 (1): 1–25.

Dwyer, Jeffrey, and Raymond Coward. 1992. "Gender and Family Care of the Elderly: Research Gaps and Opportunities." In *Gender, Families, and Elder Care*, edited by Jeffrey W. Dwyer and Raymond T. Coward, 151–62. London: Sage.

Edlund, Lena, and Rohini Pande. 2001. "Why Have Women Become Left-Wing? The Political Gender Gap and the Decline in Marriage." *Quarterly Journal of Economics* 117 (3): 917–61.

Enterprise Surveys (database). International Finance Corporation and the World Bank. http://www.enterprisesurveys.org/.

Esteve-Volart, Berta. 2004. "Gender Discrimination and Growth: Theory and Evidence from India." Development Economics Discussion Paper Series 42, Suntory and Toyota International Centres for Economics and Related Disciplines, London School of Economics and Political Science, London.

FAO (Food and Agriculture Organization)/Sida Partnership Cooperation. 2010. *National Gender Profile of Agricultural Households, 2010: Report Based on the Lao Expenditure and Consumption Surveys, National Agricultural Census and Population Census.* Vientiane, Lao PDR: FAO.

Fehringer, Jessica, and Michelle J. Hindin. 2009. "Like Parent, Like Child: Intergenerational Transmission of Partner Violence in Cebu, the Philippines." *Journal of Adolescent Health* 44 (4): 363–71.

Goldin, Claudia. 1995. "The U-Shaped Female Labor Force Function in Economic Development and Economic History." In *Investment in Women's Human Capital,* edited by T. Paul Schultz. Chicago: University of Chicago Press.

Goldstein, Markus, and Christopher Udry. 2008. "The Profits of Power: Land Rights and

Agricultural Investment in Ghana." *Journal of Political Economy* 116 (6): 981–1022.

General Statistics Office of Vietnam. 2009. *Result of the Survey on Household Living Standards 2008*. Hanoi: Statistical Publishing House. http://www.gso.gov.vn /default_en.aspx?tabid=515&idmid=5&Item ID=9647.

HNP Stats (Health Nutrition Population Statistics) database. World Bank, Washington, DC. http://data.worldbank.org/data-catalog/ health-nutrition-population-statistics.

Ilahi, Nadeem. 2000. "The Intra-household Allocation of Time and Tasks: What Have We Learnt from the Empirical Literature?" Working Paper 13, World Bank Development Research Group, World Bank, Washington, DC.

ILO (International Labour Organization). 2008. *Women's Entrepreneurship Development Aceh: Gender and Entrepreneurship Together (GET Ahead) Training Implementation. Impact Assessment*. Jakarta, Indonesia: ILO Jakarta Office.

Indonesia National Socioeconomic Survey (SUSENAS). 2009. Badan Pusat Statistik, Jakarta, Indonesia. http://dds.bps.go.id/eng/ aboutus.php?id_subyek=29&tabel=1&fl=3.

Johnson, Kiersten, Sovanratnak Sao, and Darith Hor. 2000. *Cambodia 2000 Demographic and Health Survey: Key Findings*. Calverton, MD: ORC Macro.

Judd, Ellen R. 2010. "Family Strategies: Fluidities of Gender, Community and Mobility in Rural West China." *China Quarterly* 204: 921–38.

Kabeer, Naila. 1999. "Resources, Agency, Achievements: Reflections on the Measurement of Women's Empowerment." *Development and Change* 30 (3): 435–64.

Klasen, Stephan. 2002. "Low Schooling for Girls, Slower Growth for All? Cross-Country Evidence on the Effect of Gender Inequality in Education on Economic Development." *World Bank Economic Review* 16 (3): 345–73.

Knowles, Stephen, A. K. Lorgelly, and Dorian Owen. 2002. "Are Educational Gender Gaps a Brake on Economic Development? Some Cross-Country Empirical Evidence." *Oxford Economic Papers* 54 (1): 118–49.

Lao Statistics Bureau. 2008. Lao Expenditure and Consumption Survey. Vientane, Lao PDR. http://www.nsc.gov.la/index.php?option=com_ content&view=article&id=50&Itemid=73& lang=.

Lott, John R., and Lawrence W. Kenny. 1999. "Did Women's Suffrage Change the Size and

Scope of Government?" *Journal of Political Economy* 107 (6): 1163–98.

Mammen, Kristin, and Christina Paxson. 2000. "Women's Work and Economic Development." *Journal of Economic Perspectives* 14 (4): 141–64.

Mason, Karen Oppenheim. 2005. "Measuring Women's Empowerment: Learning from Cross-National Research." In *Measuring Empowerment: Cross-Disciplinary Perspectives*, edited by D. Narayan, 89–102. Washington, DC: World Bank.

McElroy, Marjorie. 1990. "The Empirical Content of Nash-Bargained Household Behavior." *Journal of Human Resources* 25 (4): 559–83.

Morrison, Andrew, Mary Ellsberg, and Sarah Bott. 2007. "Addressing Gender-Based Violence: A Critical Review of Interventions." *World Bank Observer* 22 (1): 25–51.

Naudeau, Sophie, Naoko Kataoka, Alexandria Valerio, Michelle Neuman, and Leslie Elder. 2011. *Investing in Young Children: An Early Childhood Development Guide for Policy Dialogue and Project Preparation*. Washington, DC: World Bank.

National Center for Injury Prevention and Control. 2003. *Costs of Intimate Partner Violence against Women in the United States*. Atlanta, GA: Centers for Disease Control and Prevention.

NIPH (National Institute of Public Health [Cambodia]), NIS (National Institute of Statistics [Cambodia]), and ORC Macro. 2006. *Cambodia Demographic and Health Survey 2005*. Phnom Penh, Cambodia, and Calverton, Maryland, U.S.: National Institute of Public Health, National Institute of Statistics, and ORC Macro.

NSD (National Statistics Directorate [Timor-Leste]), Ministry of Finance [Timor-Leste], and ICF Macro. 2010. *Timor-Leste Demographic and Health Survey 2009–10*. Dili, Timor-Leste: NSD and ICF Macro.

Ofstedal, Mary Beth, John E. Knodel, and Napaporn Chayovan. 1999. "Intergenerational Support and Gender: A Comparison of Four Asian Countries." *Southeast Asian Journal of Social Sciences* 27 (2): 21–41.

PARLINE database on national parliaments. Inter-Parliamentary Union, Geneva. http:// www.ipu.org/parline-e/parlinesearch.asp and http://www.ipu.org/wmn-e/world.htm.

PEKKA (Perempuan Kepala Keluarga [Women Headed Household Empowerment Program]).

Accessed 2012. http://www.pekka.or.id/8/index.php.

Pham, Hung T., and Barry Reilly. 2007. "The Gender Pay Gap in Vietnam, 1993–2002: A Quantile Regression Approach." *Journal of Asian Economics* 18 (5): 775–808.

Pierre, Gaelle. 2012. "Recent Labor Market Performance in Vietnam through a Gender Lens." Policy Research Working Paper 6056, World Bank, Washington, DC.

PovcalNet: An Online Poverty Analysis Tool. World Bank, Washington, DC. http://iresearch.worldbank.org/PovcalNet/index.htm.

Qian, Nancy. 2008. "Missing Women and the Price of Tea in China: The Effect of Sex-Specific Earnings on Sex Imbalance." *Quarterly Journal of Economics* 123 (3): 1251–85.

Quisumbing, Agnes. 1995. "Gender Differences in Agricultural Productivity: A Survey of Empirical Evidence." FCND Discussion Paper No. 5, International Food Policy Research Institute, Washington, DC.

Quisumbing, Agnes R., and John A. Maluccio. 2003. "Resources at Marriage and Intra-household Allocation: Evidence from Bangladesh, Ethiopia, Indonesia, and South Africa." *Oxford Bulletin of Economics and Statistics* 65: 283–327.

Quota Project: Global Database of Quotas for Women. Institute for Democracy and Electoral Assistance and Stockholm University. "About Quotas." http://www.quotaproject.org/aboutQuotas.cfm.

Rand, John, and Finn Tarp. 2011. "Does Gender Influence the Provision of Fringe Benefits? Evidence from Vietnamese SMEs." *Feminist Economics* 17 (1): 59–87.

Rodgers, Yana, and Nidhiya Menon. Forthcoming. "Impact of the 2008–2009 Twin Economic Crises on the Philippine Labor Market." *World Development*.

Ross, Michael. 2008. "Oil, Islam and Women." *American Political Science Review* 102 (1): 107–23.

Rozelle, Scott, Xiao-yuan Dong, Linxiu Zhang, and Amelia Hughart. 2002. "Opportunities and Barriers in Reform China: Gender, Work, and Wages in the Rural Economy." *Pacific Economic Review* 7 (1).

Sakellariou, Chris. 2011. "Determinants of the Gender Wage Gap and Female Labor Force Participation in EAP." Paper commissioned for *Toward Gender Equality in East Asia and the Pacific: A Companion to the* World

Development Report. Washington, DC: World Bank.

Schleicher, Andreas. 2008. "Student Learning Outcomes in Mathematics from a Gender Perspective: What Does the International PISA Assessment Tell Us?" In *Girls' Education in the 21st Century*, edited by M. Tembon and L. Fort. Washington, DC: World Bank.

Schultz, Paul. 2006. "Does the Liberalization of Trade Advance Gender Equality in Schooling and Health?" Discussion Paper 935, Economic Growth Center, Yale University, New Haven, CT.

Sen, Amartya. 1992. "Missing Women." *British Medical Journal* 304: 587–88.

———. 1999. *Development as Freedom*. Oxford, U.K.: Oxford University Press.

SPC (Secretariat of the Pacific Community), Ministry of Internal and Social Affairs [Republic of Kiribati], and Statistics Division [Republic of Kiribati]. 2010. *Kiribati Family Health and Support Study: A Study on Violence against Women and Children*. Nouméa, New Caledonia: SPC.

SPC (Secretariat of the Pacific Community) and NSO (National Statistics Office) [Solomon Islands]. 2009. *Solomon Islands Family Health and Safety Study: A Study on Violence against Women and Children*. Nouméa, New Caledonia: SPC.

Thomas, Duncan. 1990. "Intra-household Resource Allocation: An Inferential Approach." *Journal of Human Resources* 25 (4): 635–64.

———. 1995. "Like Father, Like Son, Like Mother, Like Daughter, Parental Resources and Child Height." Papers 95-01, RAND Reprint Series. Santa Monica, CA: RAND Corporation.

Udry, Christopher. 1996. "Gender, Agricultural Production, and the Theory of the Household." *Journal of Political Economy* 104 (5): 1010–14.

UIS (UNESCO Institute for Statistics) Data Centre. Montreal, Canada. http://www.uis.unesco.org/Pages/default.aspx.

UNDP (United Nations Development Programme). 2010. *Asia-Pacific Human Development Report. Power, Voice and Rights: A Turning Point for Gender Equality in Asia and the Pacific*. Colombo, Sri Lanka: Macmillan Publishers India Ltd. for UNDP.

UNESCAP (United Nations Economic and Social Commission for Asia and the Pacific), Statistics Division. 2003. *Integrating Unpaid Work into*

National Economics. Bangkok, Thailand: UNESCAP.

UNESCO (United Nations Educational, Scientific and Cultural Organization) Institute for Statistics (UIS). 2009. *Global Education Digest*. Montreal, Quebec: UIS.

———. 2011. *Global Education Digest*. Montreal, Quebec: UIS.

USCDC (United States Centers for Disease Control and Prevention). 2003. "Costs of Intimate Partner Violence against Women in the United States." USCDC, National Center for Injury Prevention and Control, Atlanta, Georgia.

U.S. Department of State. 2011. *Trafficking in Persons Report 2011*. Washington, DC: U.S. Department of State.

VWC (Vanuatu Women's Centre) in partnership with the Vanuatu National Statistics Office. 2011. *Vanuatu National Survey on Women's Lives and Family Relationships*. Port Vila, Vanuatu: VWC.

WDI (World Development Indicators) database. World Bank, Washington, DC. http://data.worldbank.org/data-catalog/world-development-indicators.

WHO (World Health Organization) Global Health Observatory Data Repository. WHO, Geneva, Switzerland. http://apps.who.int/ghodata/.

World Bank. 2001. *Engendering Development: through Gender Equality in Rights, Resources, and Voice*. Washington, DC: World Bank.

———. 2009. "Gender Equality in East Asia: Progress and the Challenges of Economic Growth and Political Change, East Asia Update." World Bank, Washington, DC.

———. 2010a. "Indonesia—Reconstruction of Aceh Land Administration System Project." World Bank, Washington, DC.

———. 2010b. *World Development Indicators 2010*. Washington, DC: World Bank.

———. 2011a. "Vietnam Country Gender Assessment." World Bank, Washington, DC. http://documents.worldbank.org/curated/en/2011/11/15470188/vietnam-country-gender-assessment.

———. 2011b. *World Development Report 2012: Gender Equality and Development*. Washington, DC: World Bank.

———. 2011c. *World Development Indicators 2011*. Washington, DC: World Bank.

———. 2012a. *Capabilities, Opportunities and Participation: Gender Equality and Development in the Middle East and North Africa Region*. Washington, DC: World Bank.

World Values Survey. World Values Survey Association, Stockholm, Sweden. http://www.worldvaluessurvey.org/.

Gender and Endowments: Access to Human Capital and Productive Assets

2

Access to human capital and productive assets allows individuals to live healthy and productive lives. Imbalances between such opportunities for men and women are costly to individual welfare, to society, and to development. The East Asia and Pacific region has experienced remarkable economic growth and poverty reduction, combined with the spread of education and progress in health, from the latter half of the 20th century to the first decade of the 21st century. Therefore, East Asian and Pacific countries are well placed to promote strong improvements in access to human capital and productive assets for many men and women in those countries.

This chapter analyzes gender differences in endowments—defined here as the human and productive capital that enables opportunities to improve welfare—in the region and lays the foundations for discussing countries' policy priorities in chapter 6. It examines factors underlying those gender differences and identifies what drives progress toward gender equality or impedes it. The analysis relies on the framework of the interactions among households, markets, and institutions to understand gender outcomes in endowments. The chapter also acknowledges the links among the different types of endowments—specifically, education, health, and assets—and their ties with economic opportunities and agency.

Education and health are areas in which gender equality has generally been the most responsive to growth and development in the region.

- Girls' enrollment has recently caught up with that of boys, except in several countries and subpopulations experiencing slower progress in education overall.
- Health indicators such as infant and maternal mortality have also had impressive gains, except in several places with slower economic progress.
- However, the region has a large number of "missing girls," a persistent issue that appears not to be mitigated by growth and development.
- Gender equality in assets has been less responsive to development than that in education and health and is constrained by complex legal, social, and economic factors.

These messages, stemming from the analysis in this chapter, help shape thinking about policies to promote gender equality in endowments. The end of the chapter sets the stage for that purpose, and a policy discussion placing priorities in the endowment domain in the broader development context will follow in chapter 6.

Substantial progress toward gender equality in education

Closing gender gaps in education is beneficial.[1] Several cross-country analyses find a positive relationship between female education and growth in gross domestic product (GDP) (Klasen and Lamanna 2009; Knowles, Lorgelly, and Owen 2002). Looking at households within countries, an extensive literature has found clear evidence of correlation between mothers' education and children's education and health, particularly children's health and nutrition status (Schultz 1993; Thomas and Strauss 1992).[2] Women are usually the primary child rearers, and a mother with more education is likely to provide better child care. In East Asian and Pacific countries, analyses of national demographic and health surveys (DHS) show that Cambodian women with little education are less likely than educated women to receive antenatal care and assistance from trained health personnel during birth deliveries (Johnson, Sao, and Hor 2000). Similarly, in Timor-Leste, highly educated mothers are most likely to have their births delivered by skilled attendants (88 percent), as are mothers in the wealthiest households (69 percent) (NSD, Ministry of Finance, and ICF Macro 2010).

Gender equality in endowments can feed into development indirectly through links to gender equality in economic opportunities and agency. Promoting equal access to education through investing in girls' education can broaden girls' economic opportunities and raise their income. Better economic opportunities and higher income, in turn, have positive intergenerational effects: income in women's hands is likely to improve children's health (Duflo 2003; Fiszbein and Schady 2009; Thomas 1995). More education and more income also empower women and provide them with more bargaining power, voice, and representation, as discussed in chapter 4.

Recognizing the importance of educating girls, many East Asian and Pacific countries have made great strides toward equal enrollment between girls and boys, as highlighted in chapter 1. The discussion on education in this chapter will analyze these patterns in more depth to understand their dynamics and their determinants. These gains in enrollment have responded to changes in both supply-side and demand-side factors that enable better education outcomes. Where the education system as a whole is lagging, progress on the gender front has also been limited. However, even with equal enrollment, quality of education and choice of education streams still affect girls and boys differently and have strong implications for young people's school-to-work transition.

Closing gender gaps in enrollment

Most countries in the East Asia and Pacific region have seen narrowing gender gaps in school enrollment and completion over the past two decades. Girls' and boys' enrollment rates are now roughly on par at all levels, including tertiary education. The female-to-male enrollment ratio in secondary school has approached parity in most countries. Tertiary enrollment ratios between females and males are more dispersed, but mostly on an upward trend. The East Asia and Pacific region has performed better than any other region at increasing both enrollment levels and the female-to-male enrollment ratio. In 2010, the region had the highest primary school female-to-male ratio of enrollment among the developing regions and the second highest secondary school ratio.[3]

The narrowing of the gender gap in education since the 1990s is observed not only at the aggregate level, but also for the poor and nonpoor alike. Figure 2.1 shows the ratio of female-to-male enrollment rates in upper secondary schools for children in the poorest quintile. In most of the countries depicted, female and male enrollment rates have been converging among the poor. Similar patterns are also observed for primary and lower secondary education.

Another way to see the significant progress in education is through the smaller gender gaps in youth literacy compared to gaps in adult literacy. Figure 2.2 shows Indonesia and Cambodia as illustrative examples, but the observations are similar in many East Asian and Pacific countries. Younger generations are more likely to be literate. Although the gender gaps in adult literacy can be stark, such as in the case of Cambodia, youth literacy tends to be more equal across genders in both urban and rural areas, as well as across income quintiles. The gender gaps in literacy continue to close over time.

Given this trend, some countries are even starting to experience a reverse gender gap in education: girls' secondary enrollment exceeds that of boys in China, Fiji, Malaysia, Mongolia, the Philippines, Samoa, and Thailand. The reverse gender gaps at the tertiary level are often even starker: in Thailand, 122 females were enrolled for every 100 males in 2008. In Samoa, for example, boys underperform relative to girls in both enrollment and academic achievement. Secondary education enrollment rates have consistently been higher for girls, by a large margin that is widening over time. In tests taken in year 4 and year 6 (the first year of education starts at age 5), girls significantly outperform boys in all three tested subjects: English, Samoan, and numeracy. A significantly higher proportion of boys than girls have been at risk of not achieving minimum competencies in these years. In year 8, girls still outshine their male counterparts in all subjects, including science and mathematics, but the gender gap is smaller than it was in the earlier years (Jha and Kelleher 2006). Box 2.1 discusses the reverse gender gap in education in further detail.

Persistent gender gaps in some countries and subpopulations

Despite progress in narrowing the enrollment gaps between genders, several countries and disadvantaged populations within countries still experience more visible gender disparities than elsewhere in the region.[4] Girls' enrollment has not caught up with that of boys in

FIGURE 2.1   Enrollment for both genders has been converging even among the poorest populations

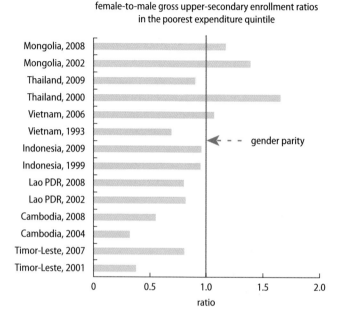

Sources: World Bank estimates using country household income and expenditure surveys: Cambodia Socio-Economic Survey (CSES) (NIS Cambodia), 2004, 2008 data; Indonesia National Socioeconomic Survey (SUSENAS) (BPS Indonesia), 1999, 2009 data; Lao Expenditure and Consumption Survey (LECS) (LSB Lao PDR), 2002, 2008 data; Living Standards Measurement Survey (LSMS) (NSO Mongolia), 2002, 2007–08 data; Thailand Socio-Economic Survey (SES) (NSO Thailand), 2000, 2009 data; Timor-Leste Demographic and Health Survey (DHS) (NSD Timor-Leste), 2001, 2007 data; Vietnam Household Living Standards Surveys (VHLSS) (GSO Vietnam), 1993, 2006 data.

FIGURE 2.2   Gender gaps in youth literacy are smaller than gender gaps in adult literacy

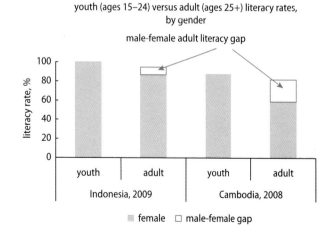

Source: World Bank estimates using CSES (NIS Cambodia), 2008 data; SUSENAS (BPS Indonesia), 2009 data.

BOX 2.1   Various parts of the world experience reverse gender gaps in education

Although the phenomenon of reverse gender gaps in education is relatively new in the East Asia and Pacific region, it has long been documented in parts of Europe and Central Asia, Latin America and the Caribbean, and Organisation for Economic Co-operation and Development (OECD) countries. The range of possible reasons varies, including returns in the labor market, norms and gender identity, and the school environment itself. Interactions between households, markets, and institutions influence households' decisions about education investments; therefore, explanations for gender gaps in education are also context-specific. Given the differential payoffs in the labor market, by which men tend to earn more (discussed further in chapter 3), men have an incentive to drop out earlier to join the labor force. Social norms perceiving men as the breadwinner and stressing masculinity values may reinforce this incentive. In many cases, such as Mongolia or the Philippines, male underachievement in education is most stark among the poor. The figure below shows the biggest enrollment disadvantage for boys in the poorest quintile. A U.K. Department for International Development study in Botswana and Ghana also shows a similar relationship between economic disadvantages and boys' underperformance (Dunne and Leach 2005). The study identifies reasons for boys dropping out in the studied areas of Ghana as related to employment opportunities.

The school environment itself may perpetuate this set of norms and gendered identity. Gender

stereotyping and how masculinity values within society are reflected in the classroom have been argued to contribute to high dropout rates among boys in the Caribbean (Bailey and Bernard 2003; Davis 2002; Figueroa 2000). Marks (2001) shows that, by age 14, girls in the United Kingdom start to substantially outperform boys in English. Boys' lower performance in the United Kingdom has been attributed to the use of more "female-oriented" reading materials, with suggestions that the inclusion of more factual, male-oriented works could increase male performance. Other authors argue that teachers have low expectations of boys' behavior and academic effectiveness, which contributes to the levels of boys' underachievement in Latin America and the Caribbean (Davis 2002; Figueroa 2000; Martino and Berrill 2003). The experimental literature on test grading suggests that there can be gender bias, but it is context-specific. For example, Lavy (2008) finds that male high school students in Israel face discrimination in teachers' test grading, but Hanna and Linden (2009) do not find such bias in their study in India. The United Nations Children's Fund (UNICEF 2004) outlines the role that poverty has to play in boys' underachievement in the Caribbean and Latin America, where governments have become increasingly aware that boys and young men are more likely to be alienated from school if they come from poor socioeconomic backgrounds.

BOX FIGURE 2.1.1   The biggest enrollment disadvantage for boys in the Phillippines is among the poor

Gross enrollment rates in secondary schools, by expenditure quintile, gender, and region, Philippines, 2006

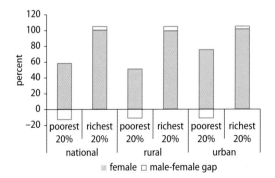

Source: World Bank estimates using Family Income and Expenditures Survey 2006 (NSCB Philippines 2006).

those places where the overall education level, regardless of gender, is also low.

Gender gaps in enrollment are still high in Papua New Guinea and low-income countries such as Cambodia and the Lao People's Democratic Republic. Papua New Guinea's education enrollment rates are among the lowest in the region, although they have been rising moderately in recent years. In 2007, primary education (elementary preparation through grade 8) gross enrollment rates were 73 percent for males and 66 percent for females (Papua New Guinea Department of Education 2009). Cambodia and Lao PDR have notable gender gaps in enrollment at the secondary level, and more extreme gaps at the tertiary level. The secondary education enrollment ratio is about 8 females to 10 males in Cambodia and Lao PDR. Household survey data from 2008 indicate that although the ratio in enrollment between girls and boys is about equal up to around age 14 in Cambodia, it diverges significantly above that age. In 2008, at the tertiary level, 40 percent of urban males in Cambodia were enrolled, whereas only about 20 percent of urban females were. Enrollment rates in these countries are relatively low regardless of gender. Cambodia's adult literacy rates are also among the lowest in the region, with the biggest gender gaps. About 60 percent of Lao women are literate, as opposed to 80 percent of Lao men and more than 90 percent of people in most other countries in the region.

In some cases, girls in disadvantaged populations have the lowest education outcomes when gender interacts with other forms of vulnerability, such as poverty and ethnicity. Household survey data indicate that girls in the poorest quintile in rural areas in Cambodia and Lao PDR have the lowest secondary school enrollment rates in these countries.[5] In Lao PDR, gender gaps in school enrollment can be particularly stark among the Hmong population. Girls in the Hmong and Mon-Khmer groups have a lower chance of being in school than boys of the same ethnicity, a disparity that has been slow to change over the past decade. They are only half as likely to be enrolled in lower-secondary

school as Lao-Tai (ethnic majority) boys and girls (figure 2.3). Even in Vietnam, where girls do not lag boys at the aggregate level, the Hmong and Dao populations have stark gender gaps in secondary school enrollment: about one girl to two boys is enrolled in secondary school (figure 2.4).

FIGURE 2.3 **Girls in some ethnic minority groups in Lao PDR lag even further in enrollment**

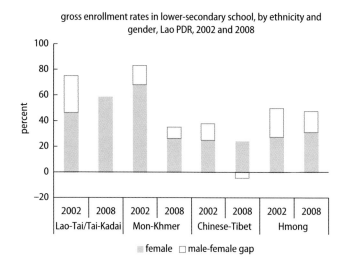

gross enrollment rates in lower-secondary school, by ethnicity and gender, Lao PDR, 2002 and 2008

Source: World Bank estimates using LECS (LSB Lao PDR), 2002, 2008 data.

FIGURE 2.4 **Girls in some ethnic minority groups in Vietnam lag even further in enrollment**

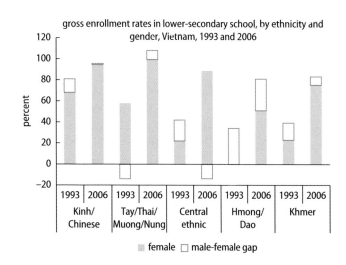

gross enrollment rates in lower-secondary school, by ethnicity and gender, Vietnam, 1993 and 2006

Source: World Bank estimates using VHLSS (GSO Vietnam), 1993, 2006 data.

Explaining progress and pending challenges in education

The observed gains in enrollment in East Asian and Pacific countries have been a result of changes in supply-side and demand-side factors that enable better education outcomes. Explanations for progress (or lack of progress) in reducing gender gaps in education are context-specific because interactions between households, markets, and institutions influence households' decisions regarding education investments. Responding to market returns, or payoff, to investment choices, household decisions reflect individual preferences, constraints, and the relative bargaining power of members. The returns to education, as determined in the labor market, play a role. Costs or prices in the form of direct costs (fees and uniforms), indirect costs (distance to schools), and opportunity costs (wages earned outside of school) also matter and can be shaped by markets and institutions. Households have preferences, which may be influenced by cultural norms. They face budget and possible credit constraints. With this framework in mind, the following discusses how changing demand-side constraints for households (for example, household income), institutional factors that affect the supply side (for example, reducing the cost/prices of education), and returns to educating girls each and together have led to more gender equality in school enrollment in most East Asian and Pacific countries.

First, factors affecting households' demand for education matter. Household survey data in East Asian and Pacific countries indicate that enrollment is always higher for children in richer families. Poor households in developing countries tend to face borrowing constraints, and, under limited budgets, they tend to invest more in sons than in daughters. Income gains are thus likely to raise school participation relatively more for girls than for boys, as empirically documented in various countries (World Bank 2001). Behrman and Knowles (1999) showed that, in Vietnam, the income elasticity of demand for education was 6 percent lower for boys than for girls in

terms of passing grades per year of schooling. Cash transfers that alleviate households' budget constraints have also been shown in many developing countries to increase children's school enrollment. In the East Asia and Pacific region, a program providing scholarships to poor girls in lower secondary schools in Cambodia had large effects on girls' enrollment, an increase of 20 percentage points. A related program for both boys and girls had similar positive impacts on enrollment and attendance for boys and girls (Filmer and Schady 2008, 2009). Economic development in the region, which brings more stable income to households, also helps protect girls' education. In times of income shocks, families with girls are more likely to reduce education expenditure, as shown, for example, in the case of Indonesia (Cameron and Worswick 2001).

In addition, norms and preferences also affect households' demand for schooling by gender. Changing norms in some contexts have led to changing girls' status relative to boys. Qualitative research through focus group discussions in six communities in Fiji suggests that parents value girls' education more now than in previous generations (Chattier 2011). In Indonesia today, compared to a few decades ago, women exhibit a stronger preference for fewer children and stronger emphasis on children's health and education. Evidence suggests that this preference change is associated with decreased preference for sons over daughters compared to the past (Niehof 2003). Indonesian parents now appear to intrinsically value daughters no less than sons (Kevane and Levine 2003).

Second, changes in formal institutions, such as better service delivery and easier access to schools, have improved the supply of education services and lowered the cost of education. Economic development in East Asian and Pacific countries has been associated with better infrastructure and service delivery, either through the public or private sector. The massive school construction program in Indonesia in the 1970s led to significant increases in education attainment and earnings, presumably through reducing costs in terms of distance to schools (Duflo 2000).

Although expanded service delivery might not specifically target girls, the benefits to them may be disproportionately high because distance and safety tend to be larger barriers for girls than for boys.

Third, in some cases, better employment opportunities and returns to education for females could have encouraged parents to educate girls. Cross-country evidence shows that trade liberalization, which has led to the expansion of nonagricultural jobs for women, is positively linked to greater human capital and more gender equality in human capital investment (Schultz 2006). The literature also shows that education investments do respond to expected returns as long as information on these returns is available (Jensen 2010; Oster and Millet 2010). Longer expected durations of receiving payoff from daughters, as a result of declining maternal mortality, can also induce parents to invest more in girls' education (Jayachandran and Lleras-Muney 2009). However, this component might have played a relatively small role in the East Asia and Pacific region, because the gender wage gaps and gender gaps in expected returns to education are still large, as discussed in chapter 3.[6]

What constrains progress? Where the education system as a whole is lagging, progress on the gender front has also been limited. Although the explanations can be very context-specific, factors related to both the supply and demand sides of education in certain places constrain progress in education outcomes in general as well as outcomes for girls.

Low household income coupled with high costs of education can limit the demand for schooling, but income is usually not the whole story. Families in Cambodia have raised concerns about bearing the high direct cost of education (Velasco 2001). However, in both Cambodia and Lao PDR, gender gaps in secondary school enrollment exist even in the top quintile, as shown in figure 2.5, even though gaps across quintiles are still much larger. Aside from income, norms also have a strong influence on household decisions. Participants of focus group discussions in a qualitative research exercise in Papua New Guinea report that males will carry the family name

FIGURE 2.5 **Even girls in wealthier households in Cambodia and Lao PDR lag behind boys**

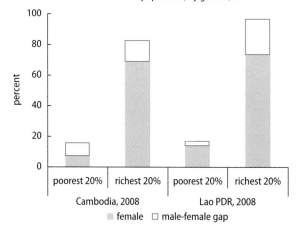

gross enrollment rates in upper-secondary schools of the poorest 20% and richest 20% of population, by gender, 2008

Source: World Bank estimates using CSES (NIS Cambodia), 2008 data; LECS (LSB Lao PDR), 2008 data.

and become household heads. Females are expected to submit to their husbands and be caregivers and homemakers; therefore, parents value boys' education over girls' (Tararia 2011). In Lao PDR, social norms about gender roles within the family may mean that girls face higher opportunity costs of schooling as a result of their socially defined value in home production. Poor rural girls spend the fewest hours in school but spend almost three hours a day fetching water, collecting firewood, and caring for other household members (King and van der Walle 2007). Families' perception of lower benefits from educating girls is also expected to impede girls' enrollment in Cambodia, particularly after puberty (Velasco 2001).

The supply side of service delivery also matters for the overall school enrollment as well as for girls' enrollment. Low enrollment rates in Papua New Guinea are the result of limited access and high dropout rates. Apart from demand-side factors such as affordability of school fees and low perceived value of education, elementary schools have not been established in many communities. Even urban areas may have inadequate capacity to admit all children wanting to enroll. Poor learning environments (often due to lack of education materials), poor school infrastructure, poor teacher attitude and attendance, lack of teachers in remote areas, and negative pupil behavior all contribute to poor enrollment overall (Papua New Guinea Department of Education 2009). In Cambodia and Lao PDR, distance to schools is an important barrier. The unequal provision of schools in Lao PDR makes schooling more costly for girls than for boys (King and van der Walle 2007). In Cambodia, given the long distance to schools, boys can live in *wats* (temples) while attending secondary school, but girls have no comparable system of accommodation. Anecdotal evidence suggests that the lack of toilets at school also makes it more difficult for girls than boys to attend (Velasco 2001).

Limitations of the education system in catering to the rural poor and ethnic minorities mean that the gender disadvantage can interact with these forms of vulnerability. In East Asian and Pacific countries, gender gaps within a subpopulation are usually smaller than the enrollment gaps across income quintiles or between major and minor ethnicities. In Vietnam, though the ethnicity gap in primary enrollment had almost closed between 1993 and 2006, enrollment at the secondary level among the Kinh/Chinese majorities is still substantially higher than that among ethnic minority populations. With the exception of the Hmong and Dao, the gender gaps within an ethnicity in Vietnam are smaller than the gaps across ethnicities (figure 2.4). In Cambodia, the gap in secondary school enrollment across quintiles also exceeds gender gaps within any given quintile. Similar patterns occur in many other East Asian and Pacific countries. These large enrollment gaps across subpopulations suggest that making general improvements in the education system to reach vulnerable subpopulations could contribute to closing the gender gap as well.

Gender streaming in education

Education investment does not end at enrollment. In middle-income countries in the East Asia and Pacific region, despite the closing of gaps in enrollment and attainment, male and female students still differ in their choice of education streams. Factors that influence gender streaming in education are important because such streaming affects the occupations and sectors that men and women engage in and the income that they subsequently earn. The persistence of these patterns implies a gender-differentiated school-to-work behavior and ultimately sustains gender inequalities in job placement and earnings. Earning gaps across genders largely reflect differences across occupations and sectors of employment. Chapter 3 discusses further the importance of education and labor market streaming to labor market outcomes.

Evidence from the East Asia and Pacific region shows clear patterns of specialization by field among men and women, implying differences in skill profiles when they enter the labor market. Figure 2.6 shows the fraction of females in each field of study in Indonesia, Thailand, and Vietnam. The fields of engineering and law are heavily dominated by males, whereas the fields of education, health, and business administration are dominated by females. These gender differences in the choice of field are larger than the gender differences in enrollment or completion rate in secondary or tertiary education in these countries. The positive relationship between being male and choosing science, engineering, or law is statistically significant in multinomial logit analysis that also accounts for characteristics such as parental education and area of residence (Sakellariou 2011). Education streaming can be slow to change over time and is less responsive to economic growth than enrollments, at least as evidence indicates for Indonesia from 1997 to 2009. Such gender streaming in education is not unique to East Asian and Pacific countries, but is evident in many Organisation for Economic Co-operation and Development (OECD) countries as well. Flabbi (2011) documents that in many OECD countries, relatively more women enter the social sciences and business, whereas more men enter the fields of engineering and architecture.

There could be multiple reasons for men and women's choice of education streams. The relative payoff of the different streams—the expected returns from labor markets—is likely to influence this decision. However, evidence shows that women do not necessarily take up fields with the highest premiums in the labor market wage. Table 2.1 shows the estimated returns to selected fields of study in Indonesia. The returns to studying engineering are high for females and are much higher than the returns to studying education. For example, among the female adults surveyed in 2009, the premium for having studied engineering (rather than religion) was 72 percent higher earnings, whereas the premium for having studied education was 17 percent. This phenomenon has been the case for over a decade, since the late 1990s. Still, females prefer education to engineering. As a note of caution, the data available here do not account for differences in unobservable characteristics, such as ability, among those choosing the different fields. They capture only monetary returns, whereas nonmonetary aspects—such as values, attitudes, and social expectations about women as mothers and homemakers or caregivers—also play an important role in influencing the decisions. Nonetheless, the statistics indicate what appear to be persistent patterns that also translate into patterns in the labor market, as discussed in chapter 3.

An alternative explanation for the gender pattern of subject choice might be that females and males have different comparative advantages in a particular field based on their academic performance. The available evidence from the East Asia and Pacific region does not support this hypothesis. Female students do not systematically perform worse than males in key subjects related to male-dominant fields of study. As shown in figure 2.7, results of the Trends in International Mathematics and Science Study (TIMSS) indicate no evident female disadvantage in math and science scores in Indonesia, Malaysia, the Philippines, and Thailand, unlike the findings for OECD countries that men tend to outperform

FIGURE 2.6 **Women are concentrated in certain fields of study, such as education and health, but are underrepresented in law and engineering**

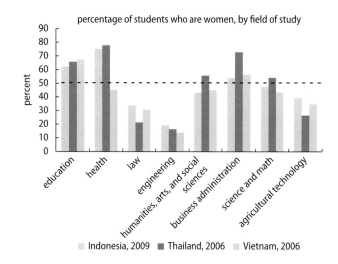

Source: Sakellariou 2011.

TABLE 2.1 **Labor market returns to studying engineering are high relative to studying education**

Indonesia: Estimated return to field of study within tertiary education, % change in wage

Field of study (compare to field: Religion)	1997		2006		2009	
	Males	Females	Males	Females	Males	Females
Law	19.7	20.9	32.3	28.4	47.7	23.4
Health	39.1	10.5	32.3	20.9	56.8	39.1
Engineering	58.4	78.6	44.8	68.2	60.0	71.6
Education	13.9	10.5	13.9	15.0	23.4	17.4

Source: Sakellariou 2011.
Note: The estimates were based on a log wage regression, accounting for characteristics such as experience and experience squared, marital status, and urban/rural residence. All coefficients were significant at the 5 percent level.

women in math and computer science (Schleicher 2008). Moreover, female students outperform their male counterparts in reading assessments, according to the Programme for International Student Assessment (PISA) test scores.

In addition, norms and expectations may influence preferences and, therefore, the choice of education streams. They can do so, for example, through shaping parental expectations, shaping role models in the labor market, or shaping how gender roles are portrayed in school curricula. Social norms

FIGURE 2.7   Girls tend to outperform boys in several subjects

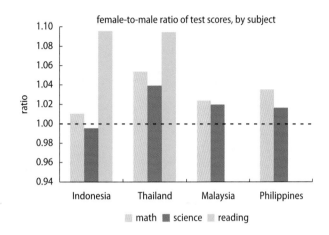

female-to-male ratio of test scores, by subject

■ math ■ science ■ reading

Source: World Bank staff calculations based on TIMSS and PISA data.
Note: Mathematics and science scales for eighth graders are from Trends in International Mathematics and Science Study (TIMSS). The most recent year of data is 2003 for the Philippines and 2007 for all other countries. The reading scale is from the Programme for International Student Assessment (PISA).

about the role of females as "homemakers" translate into expectations about appropriate jobs for men and women. In fact, women in the region work more hours and devote more time to caregiving and housework activities than do men. Given these expectations, female students may be inclined to choose a field of study that will lead to a job with sufficient time flexibility, such as teaching, to balance their home and labor market work. The absence of female role models in the labor market has also been identified as a contributing factor in influencing aspiration and career choice and in limiting women's access to nontraditional careers. These topics will be explored further in chapter 3. The rest of this section focuses on how gender roles are portrayed in school curricula, which pertains directly to the education system.

School curricula are important in shaping choices of what to study, because the way in which gender roles are portrayed in school curricula is believed to affect children's performance and aspirations. Curriculum analysis in developed and developing countries alike indicates that gender stereotyping in curricula as well as differences in teacher interactions with male and female students affect the probability of girls staying

in school. Gender stereotyping also appears to affect school performance of boys and girls in key subjects (science, mathematics, and computer science), ultimately contributing to varying occupational choices (Blumberg 2007; Eccles and Blumenfeld 1985; Shel 2007). This literature also explores the importance of the gender roles conveyed through teaching materials in shaping boys' and girls' aspirations. School, in particular primary school, is believed to be a key element of socialization for children and a site where key social values are transmitted. Teaching materials and teachers' feedback are the key instruments through which ideas about what are appropriate areas of study and acceptable professions for men and women are conveyed to children (Blumberg 2007; Nielsen and Davies 2010).

In East Asia and the Pacific, the results of a 2008 review of the Education for All (EFA) initiative indicated that teaching materials in the region included stereotypical portrayals of boys and girls (UNICEF 2009). The findings, in line with those of other regions, stress that boys appear more than girls in the learning materials, that they are portrayed as more active (and girls as more passive), and that they are shown more frequently in leadership roles.[7] The report also found a lack of female role models in teaching materials. Women in the school textbooks reviewed were portrayed as secretaries, assistants, nurses, and teachers more frequently than men, who often appeared as doctors, politicians, or police officers.

Qualitative research involving both the review of teaching materials and classroom observation, which allows for an analysis of teacher-student interaction, has highlighted how gender stereotyping frequently exists in curricula in a number of East Asian countries. Table 2.2 provides selected examples of the types of issues typically encountered in curriculum analysis. Male characters tend to be portrayed as dominant in the public sphere. Similar stereotyping is also observed in other countries throughout the region. The depiction of women in China's history and social sciences manuals illustrates this point. Guo

TABLE 2.2 **School curricula in a number of East Asian countries have gender stereotyping**

	China	Vietnam	Korea, Rep.
Overall visibility of male and female characters or authors in teaching materials	The proportion of male characters rises from 48 percent in books for four-year-olds to 61 percent in books for six-year-olds. Female characters appear most commonly in reading materials for very young children.	In grade 4 and grade 5 texts analyzed, most authors mentioned and quoted were men (74 of 85 and 77 of 84, respectively). In mathematics textbooks analyzed, female characters were found to appear in illustrations more often than men (some grade levels); however, they were associated with less challenging activities.	
Stereotypical portrayals of men and women in textbooks	Portrayal of male and female characters follows gender role stereotypes in both mathematics and social sciences manuals. An analysis of mathematics texts indicated that male characters were 74 percent of those in stimulating activities, and female characters represented 70 percent of those in passive roles. Men and boys are therefore typically portrayed as courageous, independent, and ambitious, in contrast to "passive, obedient, neat, cooperative girls."	Stereotypical depictions of men/boys and women/girls are present in a range of textbook illustrations and exercises. A detailed analysis of materials used in grades 1 to 5 highlights the following depictions of men and boys: (a) heroes/courageous; (b) strong/able to do complicated and physically challenging jobs; (c) knowledgeable/smart; (d) naughty; (e) creative; and (f) leaders. In contrast, women and girls are typical portrayed as (a) "nice and lovely"; (b) caring (as teachers, nurses); (c) clean and ordered; and (d) weak/emotional. Interestingly, men and boys were also more frequently associated with "forbidden" or dangerous activities.	Curricular materials presented traditional gender roles with women depicted doing housework versus office work for men, male characters leading activities and female characters assisting.
Student-teacher interaction	Teacher-pupil and peer interaction were observed to follow the same stereotypes, to girls' disadvantage.		Explanations and examples reflecting gender role stereotypes were used frequently, as was gender-discriminative language. Male students had more teacher-student interaction opportunities as well as more social contact with teachers. Teachers were found to discipline male students more severely than female students.

Sources: Jung and Chung 2005 (Korea); Ross and Shi 2003; Shi and Ross 2002 (China); UNESCO/Vietnam Ministry of Education and Training 2011 (Vietnam).

and Zhao's (2002) analysis of elementary language textbooks highlights that only about a fifth of the historical characters portrayed are female. When they are depicted, they also tend to be portrayed in stereotypical roles. For example, an influential female leader of the Chinese Communist Party is depicted twice: once mending Premier Zhou Enlai's clothes, and another time bringing an umbrella to a guard on a rainy day.

Improvements and remaining gender issues in health

Promoting better health is important for enhancing welfare. Most societies recognize that girls and boys should have equal access to the elements of a healthy life and that maternal health is important. Mothers' health and nutrition affect their children's physical health as well as cognitive abilities.

Research shows that delays in cognitive and overall development from the time a child is conceived up to age six, a sensitive period for brain formation, have long-lasting consequences that are difficult to compensate for later on in life (Naudeau et al. 2011). In addition, improved maternal health has been shown to enable women to reconcile work and motherhood, playing a role in raising married women's labor force participation in the United States (Albanesi and Olivetti 2009). Research has also shown that with declining maternal mortality—that is, longer life expectancy for women—parents can expect a longer duration of payoff from daughters. In Sri Lanka, for example, parents respond with more investment in girls' education (Jayachandran and Lleras-Muney 2009).

To explain progress or the lack thereof in gender equality in health in East Asian and Pacific countries, this section focuses on several key indicators, such as fertility, child and maternal mortality, and sex ratios at birth. The chapter argues that many, but not all, health outcomes for males and females in the East Asia and Pacific region have improved with development. Fertility and child and maternal mortality have had impressive gains, except in a few places with slower economic progress, such as Cambodia, Lao PDR, and Timor-Leste. However, the region still has a large number of missing girls, a persistent issue not easily mitigated by growth and development. The chapter also sheds light on two behavioral health issues associated with high adult mortality risks for men in all East Asian and Pacific countries: excessive smoking and drinking.

Declines in fertility and child and maternal mortality rates

Gender differences in health outcomes reflect biological differences as well as gender-differential behaviors, which are difficult to separate in what we observe. As a result of biological factors determining life expectancy, for example, women tend to live longer than

men (Eskes and Haanen 2006). Male mortality is naturally higher than female mortality for the first six months of life (Waldron 1998). Men and women are also susceptible to different diseases, such as different types of cancer. Aside from biological differences, health outcomes are also affected by differences in behaviors and health investments that could disadvantage one particular gender. Thus, many unobservable factors affect morbidity and mortality for men and women.

This chapter focuses on mortality risks throughout the life cycle. For early childhood and childbearing periods, most East Asian and Pacific countries have experienced impressive progress in narrowing the gender gaps in the infant mortality rate and reducing maternal mortality. However, as discussed later, the period before birth is a concern, with male-skewed sex ratios at birth in several parts of the region.

Many East Asian and Pacific countries have significantly improved several health outcomes during the past two decades. Fertility rates went down sharply. Infant and child mortality rates for both boys and girls have declined substantially since 1990, closing the gender gaps in the infant mortality rate. The maternal mortality rate (MMR) has also been declining, and, along with the female infant mortality rate in most of East Asia and the Pacific, is now low relative to the region's income level. Figures 2.8 and 2.9 illustrate this point: in the cross-country graphs of these health indicators and income measured in terms of purchasing power parity, most East Asian and Pacific countries lie below the downward-sloping curve representing this relationship.

As in the case of education, progress has not been uniform across the region. First, China differs from other East Asian and Pacific countries in having a high rate of female child mortality relative to that of males. Figure 2.10 graphs the ratio of female-to-male child mortality as well as the ratio for infant mortality against GDP per capita. Although in most countries, male infant and child mortality rates are slightly higher than female rates, consistent with biological

FIGURE 2.8   Maternal mortality is lower in higher-income countries

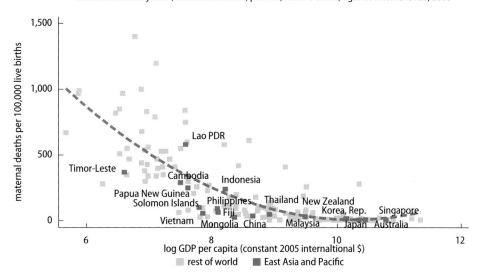

maternal mortality rate (modeled estimate, per 100,000 live births) against income levels, 2008

Source: World Bank estimates using World Development Indicators (WDI) database, 2011 data.

FIGURE 2.9   Female infant mortality is lower in higher-income countries

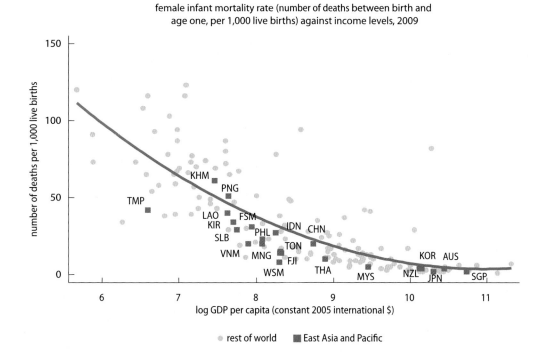

female infant mortality rate (number of deaths between birth and
age one, per 1,000 live births) against income levels, 2009

Source: World Bank estimates using WDI database, 2011 data.

factors, China experiences the opposite patterns. Females face higher mortality risks
during infancy than males, and even more so
before birth, which is related to the "missing
girls at birth" topic, discussed later.

Second, the maternal mortality rate is
still a serious concern in a number of places
in the region. As shown in figure 2.8, Lao
PDR experienced more than 500 maternal
deaths per 100,000 births in 2008, a rate

FIGURE 2.10   Most East Asia and Pacific region countries do not have female-skewed under-five mortality and infant mortality, except China

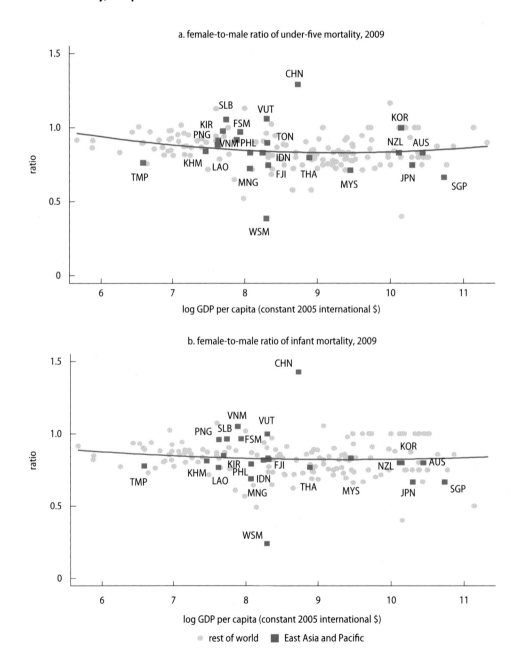

Source: World Bank estimates using WDI database, 2011 data.

much higher than other countries of similar income level. The maternal mortality rate remains high—above 240 maternal deaths per 100,000 births—in several other countries, such as Cambodia, Papua New Guinea, and Timor-Leste, despite progress over the past two decades. Indonesia's maternal mortality rate remains high compared to other countries in the region at similar levels of development. Even for Vietnam, a country that has successfully reduced this rate at the national level, maternal health outcomes still lag in rural areas and among ethnic minority groups (World Bank 2011b).

Growth, development, and improvements in health outcomes

Maternal health and child health outcomes are a result of many factors, including service delivery related to the public and private health systems, prices and the availability of insurance mechanisms, and the demand side of households' fertility and health-seeking behavior. Social norms regarding childbirth practices also have an important influence in many contexts.

The East Asia and Pacific region has experienced a substantial reduction in total fertility, and this decline has reduced the risk of maternal health complications and death. The share of women using modern contraception, and thus are presumably more able to control fertility, has been on the rise in many East Asian and Pacific countries. Government policies such as family planning programs and China's one-child policy were expected to control fertility. Schultz and Yi (1997) argued that institutional reforms, such as the replacement of the collective production team with the household responsibility system starting in 1979 in China, may have contributed to the decline in fertility in the long term for a variety of reasons. For example, the intensified market competition encouraged parents to educate children and focus on the "quality" rather than quantity of children. Increased mobility and migration for better economic opportunities were also linked to delayed childbearing. The declining

fertility in Thailand before 1980 has been attributed to government subsidies to public and private family planning systems and to the rapid increase in female education (Schultz 1997). Breierova and Duflo (2004) showed that higher education among females in Indonesia led them to have fewer children early on, and the increased education of mothers and fathers led to lower child mortality. As mentioned earlier, in Indonesia today, compared to a few decades ago, women exhibit a stronger preference for fewer children and for more per-child investment in health and education. This preference change is attributed to decreased preference for sons over daughters and to other social changes (Niehof 2003).

Growth and development in the East Asia and Pacific region have contributed to the region's progress in improving maternal and reducing child mortality through a combination of demand-side and supply-side factors. In fact, living in a high-income country is generally associated with lower risk of maternal death and of female infant death (as well as male infant death, not shown). Figures 2.8 and 2.9 show a negative relationship between a country's maternal mortality rate or female infant mortality rate with its GDP per capita level.

For the household, rising income in the region appears to have positive impacts on health outcomes. A rise in income may loosen the incentives to differentiate health investments across boys and girls. Evidence from a large data set of developing countries shows that, on average, a one-unit increase in log GDP per capita is associated with a decrease in mortality of between 18 and 44 infants per 1,000 births. This negative relationship holds true even when various factors—such as the mother's characteristics, weather shocks, conflicts, and the quality of institutions—are accounted for. Female infant mortality is more sensitive to changes in economic conditions than male mortality (Baird, Friedman, and Schady 2007).[8] In addition, households with higher income can afford more health services, such as the use of prenatal care and hospitals during births. Figure 2.11 shows that in every East Asian and Pacific country

FIGURE 2.11   **Women in wealthier households are more likely to have births assisted by trained medical staff**

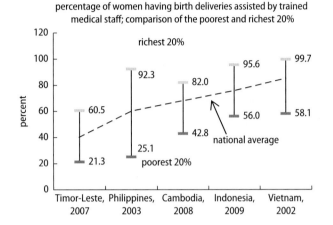

percentage of women having birth deliveries assisted by trained medical staff; comparison of the poorest and richest 20%

Sources: World Bank staff estimates using household income and expenditure surveys of various countries and years; World Bank Health, Nutrition, and Population Statistics (HNPStats) database.

examined, women living in richer households are more likely to have birth deliveries assisted by trained health professionals. In some cases, such as Cambodia, Indonesia, and the Philippines, the gap between the richest quintile and the bottom one can be fourfold. Over time, as the economy grows, the rate of professional birth attendance increases. Lastly, the availability of insurance mechanisms has also made health care more affordable. The expansion of health insurance coverage in Vietnam, from 25 percent of the population in 2004 to 40 percent in 2006, and serious efforts by the government to extend health insurance to the poor and ethnic minorities in recent years have had positive impacts on access to health services (World Bank 2011b).

Rising women's income, in particular, is likely to have contributed to this progress in health outcomes. Many countries in the East Asia and Pacific region have experienced recent increases in female labor force participation. Evidence shows that income in the hands of women positively affects children's health, particularly girls' (Thomas 1995 for Brazil, Ghana, and the United States; Duflo 2003 for South Africa). Higher female labor force participation has been shown to improve their bargaining power and to increase the

height of children in Mexico (Atkin 2010). In Indonesia, women with a higher share of household assets tend to use more prenatal and delivery care (Beegle, Frankenberg, and Thomas 2001).

Improvements in health technologies as well as improvements in the institutions supplying health care have led to lower costs of health services and better health outcomes. Medical progress with the medicalization and hospitalization of childbirth contributed to the substantial decline in maternal mortality in the United States in the first half of the 20th century (Albanesi and Olivetti 2009). These technologies, when adopted in developing countries in the region, were likely to have similar effects. In addition, the share of women using modern contraception has been on the rise in many East Asian and Pacific countries. Contraception use allows families to control fertility and avoid extremely short periods between pregnancies, which tend to pose higher health risks. Countries shown in figure 2.12 with a high rate of contraceptive use (for example, the high and increasing rate in Vietnam) are also those with a low maternal mortality rate.

The functioning of the health system, including infrastructure, medical facilities, and equipment and staffing, is also key to improving health outcomes. Experience in the United States illustrates the importance of public health investments: two-thirds of the decline in overall infant mortality and the entire decline in excess female infant mortality in the early 20th century were attributable to clean water and sanitation (Cutler and Miller 2005). Vietnam's remarkable achievements in bringing down child and maternal mortality have been attributed to a general strengthening of the health system (World Bank 2011b). The gains in the share of births attended by professionals from 2000 to 2008 were very impressive in the East Asia and Pacific region, larger than any other developing region. Bringing better services closer to women can change their patterns of use, and positively affects health outcomes. Frankenberg and Thomas (2001) analyzed the Indonesia Family Life Survey panel data using community-level fixed effects to measure the impacts of a major expansion in midwifery services between 1990 and 1998 on health and pregnancy outcomes for women of reproductive age. The authors showed that the addition of a village midwife to communities between 1993 and 1997 was associated with a significant increase in body mass index (BMI) for women of reproductive age, as well as an increase in birthweight of newborns. Frankenberg et al. (2009) further investigated the reasons behind this impact on outcomes. The presence of village midwives appears to have increased women's receipt of iron tablets and influenced their choice of childbirth practice away from reliance on traditional birth attendants toward delivery attended by skilled professionals.

Ongoing challenges: Poor service delivery and social norms

East Asian and Pacific countries with high maternal mortality rates—Cambodia, Indonesia, Lao PDR, Papua New Guinea, and Timor-Leste—are precisely those with low rates of contraceptive use and low rates of births delivered by professionals (figure 2.12 and figure 2.13). As shown in figure 2.12, Pacific Island countries tend to have low contraceptive prevalence compared to East Asia. In Timor-Leste, for example, only 20 percent of women ages 15–49 use contraception, leading to high fertility rates and very short periods between pregnancies. Less than 20 percent of the births in Timor-Leste are assisted by professionals. Delivery at home without professional help and without easy access to a functioning referral center poses high risks, particularly in case of complications. The absence of the factors that explain progress elsewhere is at play in these countries. Demand-side constraints from households, including social norms about pregnancy and birthing practices, and poor supply-side provision of care explain the poor health outcomes in these countries. The exact reasons can vary from context to context.

On the supply side, poor access to quality obstetric health services, particularly among

FIGURE 2.12 **Contraceptive prevalence varies across East Asian and Pacific countries**

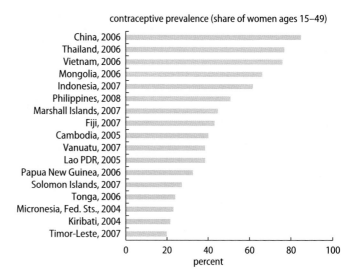

Source: WDI database, Gender Statistics.

FIGURE 2.13 **The percentage of births attended by skilled professionals varies across East Asian and Pacific countries**

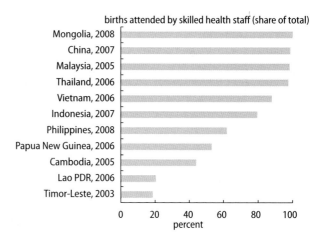

Source: WDI database, Gender Statistics.

rural and remote areas, places women at a high risk for maternal death. Rural areas tend to be less well served by the health system; figure 2.14 shows that rural residents have substantially less access to birth deliveries attended by trained staff than urban residents. Delivery in the home is of particular concern for poor, rural women because they

FIGURE 2.14   **Women in rural areas are less likely to have births assisted by trained medical staff**

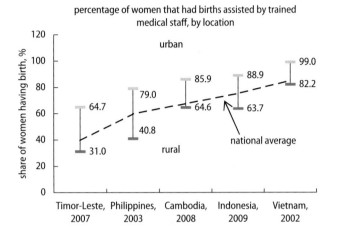

percentage of women that had births assisted by trained medical staff, by location

Source: World Bank estimates using household income and expenditure surveys of various countries and years, and the HNPStats database (see figure 2.1 country sources).

lack the basic sanitary conditions needed for a safe delivery. In Lao PDR, almost 90 percent of rural women deliver at home, compared to approximately a quarter of urban women. As a result, a large disparity in maternal mortality rates exists between urban and rural areas in Lao PDR: 170 versus 580 maternal deaths per 100,000 births, respectively (GRID 2005). In Timor-Leste, 59 percent of urban births are assisted by skilled providers, compared with 21 percent of births in rural areas. This rate is 69 percent in the capital city, Dili, but less than 10 percent in the Oecussi region (NSD, Ministry of Finance, and ICF Macro 2010). Moreover, the quality of prenatal care in Timor-Leste, is limited: in 2001–02, only 41 percent of those giving birth were protected against neonatal tetanus, a major cause of neonatal death (ADB 2005). Substantial disparities also exist across provinces in Indonesia: Jakarta has 97 percent of births attended by a skilled provider, whereas Maluku has only 33 percent (World Bank 2010).

Long distance to the nearest health center and the poor infrastructure available can both impose high costs of access. Although high in all rural areas, the rate of deliveries at home is highest among highland women

in Lao PDR for whom both distance and infrastructure are of concern (GRID 2005). Chine-Tibet women appear to be most at risk during childbirth since they are very likely to give birth outside of a hospital, and a large majority of Chine-Tibet villages (76 percent) lack safe water, let alone access to other sanitary measures (GRID 2005). In Cambodia, low quality health care, the poor state of rural roads, lack of transport, and poor access to a clean water supply have all been shown to impede progress in bringing down maternal mortality (UNIFEM, World Bank, ADB, UNDP, and DFID/UK 2004).

On the demand side, tight household budgets can constrain the use of health care services. Women from poor families may not be able to afford health costs, which are the major barrier to seeking health care in many developing countries. As shown earlier in figure 2.12, in every country examined, women living in poor households are less likely to have birth deliveries assisted by trained health professionals. In Indonesia, the community health insurance scheme, Jamkesmas, has had little effect on facility-based deliveries because some delivery costs such as transport and costs for family members were not covered (World Bank 2010). Even in a country with health insurance subsidies for the poor, only 60 percent of Vietnamese women in the poorest quintile had births attended by trained medical staff in 2002 (figure 2.11).

Women's access to reproductive health care could be constrained by norms. Culture and tradition play an important role in the choice of health practices, such as the location of childbirth, the use of birth attendants, and sterilization practices. For instance, following a traditional practice, a number of Mon-Khmer women in Lao PDR deliver neither in the home nor in a medical center, but rather in the forest (GRID 2005). In Cambodia, cultural beliefs that pregnancy and childbirth are part of the natural process lead families to perceive that women do not need prenatal care or delivery supported by skilled attendants. Thus, many women continue heavy physical labor and long work hours during pregnancy and immediately after childbirth

(UNIFEM, World Bank, ADB, UNDP, and DFID/UK 2004).

The evidence suggests that improving service delivery is key to reducing gender disparities and improving health outcomes. Given the central influence of social norms in birthing practice, service delivery could and should be strengthened by providing services in a culturally acceptable way. Policy implications and recommendations are discussed at the end of this chapter and in chapter 6.

Missing girls at birth

One concerning issue that persists in the East Asia and Pacific region despite tremendous growth and development is the phenomenon of missing girls, particularly at birth. The term "missing women" was first coined by Sen (1992) to refer to the phenomenon that many low-income countries have far fewer women than men, relative to what is observed in developed countries. At birth, the biological norm is approximately 105 boys born for every 100 girls. Yet, the male-female ratio at birth in East Asian and Pacific countries far exceeds that of other regions, mainly driven by China's ratio.

The United Nations Department of Economic and Social Affairs, Population Division, database provides estimates of cross-country sex ratios at birth over time. These estimates are based on projections from national census data and a fertility modeling exercise. Estimates from national statistical offices can sometimes stem from more recent data and, as a result, differ from the United Nations (UN) projections. According to the UN projections, over the 2005–10 period, 120 boys in China were born for every 100 girls. Outside of China, new concerns are emerging regarding Vietnam's rising sex ratio at birth. According to the Vietnam General Statistical Office's Annual Population Change Surveys, the sex ratio at birth increased regularly from 2004 and crossed the 110 threshold in 2005 (UNFPA 2009). Yet, trends in sex ratios in the East Asia and Pacific region are not all bad news. Korea has seen declining sex ratios at birth, from 110.2 in 1998 to

106.4 in 2008, now close to the biological ratio (figure 2.15).[9]

Sex ratios at higher order births (that is, second children or above) are usually worse than the average ratios. In China, although the sex ratio at birth in the 1980s was within the normal range for the first birth, it became unbalanced at higher orders—1.3 for the fourth or later child in 1989. The sex ratios for higher order births, conditional on earlier female births, were even more starkly skewed toward males (Zeng et al. 1993). Chung and Das Gupta (2007) used data from Korea's 2003 fertility survey to show that the sex ratio at birth after the first birth was 129 if all the previous births were girls, and it is 112 if at least one previous birth was a boy. This difference was even starker among those women who stated that having a son was imperative. More recently in Korea, even though the average sex ratio at birth is close to the normal range, the ratios for the third birth and for the fourth or higher births were still 116 and 124, respectively, in 2008 (figure 2.15). In Vietnam, the pattern is unusual in that the sex ratio for the first birth is already higher than that for the second birth and similar to that for the third or later births. However, according to Vietnam's 2006 population survey, the sex ratio at birth for third-order

FIGURE 2.15 Sex ratios at higher order births are still of concern, even though the overall sex ratio at birth has approached the normal range in the Republic of Korea

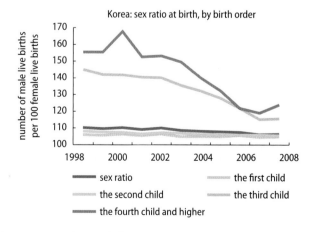

Source: Korea National Statistical Office. http://kostat.go.kr.

births decreases with the previous number of sons born between 2000 and 2006: 110.3 for no sons, 103.5 for one son, and 102.2 for two sons (UNFPA 2009).

As a consequence, in comparison to other developing regions, the number of missing girls at birth in the East Asia and Pacific region, particularly in China, dominates the excess mortality risks for females after birth. The *World Development Report 2012* uses the following methodology to calculate the number of missing girls at birth and excess female deaths in other parts of the life cycle. The number of missing girls at birth is calculated by comparing the sex ratio at birth in a particular country to the ratio in high-income countries (105.9 boys for 100 girls). Throughout the age distribution, excess female mortality is calculated by comparing the mortality risks of females relative to males in a particular age group in a country with the mortality risks in a reference group of high-income countries.[10] Table 2.3 shows the number of missing girls at birth and excess female deaths per year, calculated using this methodology. Excess female mortality in infancy and during the reproductive years have substantially decreased in China. An estimated 71,000 girls under age five were missing in China in 2008, consistent with figure 2.10, which shows China's high

female-to-male child mortality ratio. However, the most worrisome period of missing girls in China is at birth. The number of missing girls at birth increased from 890,000 in 1990 to 1,092,000 in 2008. Missing girls as a fraction of the total number of female births increased from 8.6 percent in 1990 to 13.3 percent in 2008.

The reason for the missing girls phenomenon has been attributed to son preference (Das Gupta 2005). Parents' preferential choices to keep and care for boys over girls can depend on social norms and values, different economic opportunities by gender, and what benefits parents expect from a son or a daughter. As an example of how economic opportunities influence parental choice over the gender of their child, Qian (2008) showed that the sex ratio at birth is responsive to returns in the labor markets for women in rural China. An increase in women's income relative to men's led to higher survival rates for girls. Another example of how parental choice responds to changing economic conditions is the rise of marriage migration—cross-border marriages between women from Southeast Asia and men from East Asia. Through a 2007 survey of three migrant-sending communities in southern Vietnam, Bélanger and Tran (2011) documented an enhanced status of emigrating daughters sending remittances

TABLE 2.3   **The East Asia and Pacific region, mainly driven by China, is characterized by its large number of missing girls at birth**

Missing girls at birth and excess female deaths throughout the age distribution (1,000s per year)

	At birth		Under 5		5–14		15–49		50–59		Total (under 60)	
	1990	*2008*	*1990*	*2008*	*1990*	*2008*	*1990*	*2008*	*1990*	*2008*	*1990*	*2008*
China	890	1,092	259	71	21	5	208	56	92	30	1,470	1,254
India	265	257	428	251	94	45	388	228	81	75	1,255	856
Sub-Saharan Africa	42	53	183	203	61	77	302	751	50	99	639	1,182
South Asia (excluding India)	0	1	99	72	32	20	176	161	37	51	346	305
East Asia and Pacific (excluding China)	3	4	14	7	14	9	137	113	48	46	216	179
Middle East and North Africa	5	6	13	7	4	1	43	24	15	15	80	52
East and Central Asia	7	14	3	1	0	0	12	4	4	3	27	23
Latin America and the Caribbean	0	0	11	5	3	1	20	10	17	17	51	33
Total	1,212	1,427	1,010	617	230	158	1,286	1,347	343	334	4,082	3,882

Source: World Bank 2011c.

Note: Estimates are based on data from World Health Organization (WHO) 2010 and United Nations Department of Economics and Social Affairs, Population Division (UN DESA 2009).

back home, and, consequently, a change in families' preference for having girls.

Most societies have some mild degree of preference for sons (Williamson 1976), but the manifestation of extreme sex ratios comes from rather extreme son preferences. The interplay of culture, the state, and political processes appears to generate extreme patrilineality and highly skewed child sex ratios such as in the case of China, northwest India, and Korea (Das Gupta 2009). Chung and Das Gupta (2007) argued that son preference in Korea is correlated with factors such as lower socioeconomic status, rural area residence, higher parental control in terms of arranged marriage and co-residence with the parents, and lower education of the woman.

In addition, the manifestation of son preference is also influenced by public policies and the availability of technology. China's one-child policy and Vietnam's two-child policy, though intended to reduce fertility, may have put additional pressure on the incentives to have a son and intensify the skewed sex ratios. In fact, Ebenstein (2010) showed evidence of a positive correlation between the fines imposed by China's one-child policy and the sex ratio. With development and the introduction of prenatal sex determination technology (ultrasound) in the early 1980s, male-to-female sex ratios became unnaturally very high in a few East Asian countries. Li and Zheng (2009) found a strong impact of the B-ultrasound technology on the sex ratio of second-order births for rural mothers in Fujian province, China, but no effect among first-born children. The recent increase in Vietnam's sex ratio at birth may be related to supply-side factors, that is, access to quality sex determination technology, rather than to an increasing preference for sons. Ultrasound technology first started to appear in major hospitals in Vietnam during the mid-1990s and was subsequently offered through the private sector (Bélanger et al. 2003), but the quality and availability of medical equipment have improved during the past 10 years. Thus, the proportion of mothers with prior knowledge of the sex of

their fetus rose from 60 percent in 2003 to 73 percent in 2007 (UNFPA 2009).

With prior knowledge of the sex of the fetus, families can discriminate through less prenatal investment or even through abortion. In China, as well as other countries with prevalent son preference, mothers are 5 percent more likely to acquire prenatal care and visit an antenatal clinic 10 percent more frequently when pregnant with a boy (Bharadwaj and Nelson 2010). Bélanger and Khuat (2009) examined the timing of abortion among 885 married women in an obstetric hospital in Hanoi, Vietnam, in 2003 to study sex-selective abortions, which generally happen during the second trimester of pregnancy. They found that women with more daughters and without a son were more likely to have a second-trimester than a first-trimester abortion. Their estimates suggest that 2 percent of all abortions by women with at least one prior child were intended to avoid a female birth.

Given the factors discussed earlier, the literature shows mixed evidence on whether development mitigates or worsens son preference and sex ratios at birth in Asia (Chung and Das Gupta 2007). Development can bring about substantial normative changes within the entire society together with improvements in individuals' socioeconomic situations, as argued in the case of Korea by Chung and Das Gupta (2007). However, cross-country evidence shows that modernization does not appear to bring down son preference. In South Asia, son preference is greater for women with more education and is increasing over time (Filmer, Friedman, and Schady 2008). Unbalanced sex ratios at birth could be worsened by economic development, as sometimes argued in the literature, since highly educated and wealthier women tend to have better access to technologies. Vietnam's 2006 population survey shows that the sex ratio at birth is high for women who have a graduate education (113), have highest grade of 10 and above (111), work in a foreign organization (117), and have previous knowledge of the baby's sex (111). Of women with a graduate degree, 87 percent knew the gender of their child, whereas no more than 28 percent of illiterate women had prior knowledge of the sex of their child. The sex ratio at birth increases with the level of education, rising from 103 for illiterate women to 113 for women with a graduate education (UNFPA 2009).

The collection of evidence has several implications for approaches to address the unbalanced sex ratios at birth, and China has taken active measures in this direction. General policies to promote economic development might play a role, but Korea's recent experience suggests that active measures to influence social norms and facilitate the spread of new values may be very important, in addition to relying on raising female education and labor force participation alone (box 2.2).

Risky behaviors in men

Men, as well as women, experience gender-specific health risks. Men tend to bear the burden of higher morbidity and premature mortality related to substance abuse, war and conflict, and violence. The latter tends to be

---

BOX 2.2   Recent improvement in the sex ratio at birth in the Republic of Korea

Since the 1970s, Korea has experienced significant industrialization and urbanization, coupled with increases in women's education and labor force participation. It is also the first Asian country to register a decline in the proportion of missing girls, from the most male-skewed sex ratios at birth in the mid-1990s to ratios within the normal range by 2008. As the sex ratios at birth are usually argued to be a manifestation of son preference, Chung and Das Gupta (2007) used data from fertility surveys to measure trends in son preference directly. They documented a continuous decline from 1985 to 2003 in the fraction of Korean women who reported that they must have a son, from almost 50 percent to less than 20 percent. Women with similar characteristics demonstrated lower son preference over time.

Both the process of development and public policies since the 1950s have influenced the factors underlying son preference in Korea. First, the impacts of development were expected to work in many ways, as argued by Chung and Das Gupta (2007): (a) higher earning prospects increased individuals' independence of family lineage; (b) retirement savings reduced financial dependence on children in old age; (c) urban life setting reduced the focus on traditional filial duty and promoted female-inclusive social networks; (d) females' greater economic and physical mobility enhanced the value of daughters; and (e) urban life, with assets associated with nonfarm activities and less pressure from customary laws, facilitated gender equity in inheritance. Through a decomposition exercise, the authors attributed the observed reduction in son preference

in larger part to changing social norms (changes in son preference within all education and urban/rural population groups) and in smaller part to increased urbanization and education (changes due to movements between education and urban/rural population groups). Another methodological approach using simulations of an economic model also implies the impact of development: as the Korean society becomes richer, households that initially selected boys will select girls because of increasing bride price and declining marginal benefits from unmarried sons (Edlund and Lee 2009).

Second, though the role of Korean public policies in this process is a mixed story, its experience suggests that interventions to influence social norms and facilitate the spread of new values may be very important, as opposed to reliance on raising female education and labor force participation alone. Active policies to promote rapid economic development in Korea played a role early on in breaking down previous norms of son preference as well as raising female education and labor force participation. And changing social norms contributed relatively more to the decrease in son preference. Reforms to policies that directly constrain women's status, such as the Family Law established in 1958, which stipulated male family headship and inheritance only through the male line, were slow to follow. With several women's movements demanding greater gender equity since the establishment of democracy in 1987, this law went through major reforms in 1990, but the system of male household headship was not officially abolished until 2005 (Chung and Das Gupta 2007).

context-specific rather than an issue that prevails in the region. However, two behavioral health issues—smoking and drinking—are more concerning among men than women in all East Asian and Pacific countries, as well as globally.[11] Figures 2.16 and 2.17 show the prevalence of smoking and drinking among males compared with the prevalence among females. All East Asian and Pacific countries lie above the 45-degree line, implying a much higher rate among males than females.

The gender difference in the incidence of tobacco use is higher in the East Asia and Pacific region than in other developing regions, even though male dominance in smoking and drinking is a global phenomenon. The data for China in 2006 show that tobacco use is 60 percent for men and 4 percent for women. Similar gender gaps exist throughout the region. Indonesia (62 percent versus 5 percent), Korea (53 percent versus 6 percent), Lao PDR (64 percent versus 15 percent), and Tonga (62 percent versus 15 percent) lead the region in the largest gender differentials in tobacco prevalence (figure 2.16).

Alcohol consumption can vary from occasional drinking to heavy episodic drinking, and the gender differential in the latter is particularly stark. Although women are less likely to report drinking at all in many countries in the East Asia and Pacific region, they do come close to men in places such as Japan and Mongolia (figure 2.17). In terms of heavy episodic drinking (binge drinking) and chronic heavy drinking, data show a large gender gap: overall in the region, men are more than twice as likely to be heavy episodic drinkers. According to the World Health Organization 2011 Global InfoBase database, the countries in the region with the largest gender gaps in heavy drinking are Kiribati (22 percent among males versus 1 percent among females), Samoa (22 percent versus 1 percent), Lao PDR (22 percent versus 5 percent), Japan (18 percent versus 3 percent), Mongolia (14 percent versus 0 percent), and Micronesia (13 percent versus 1 percent). The level of annual per capita consumption of pure alcohol is especially high for males

FIGURE 2.16   Men are more likely to smoke than women

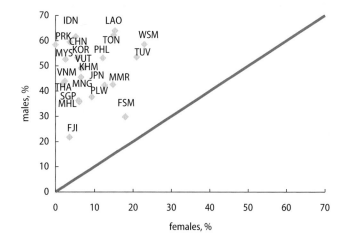

prevalence of currently smoking any tobacco product among adults (≥15 years) (%), male vs. female, 2006

Source: World Health Organization (WHO) Global InfoBase.

FIGURE 2.17   Men are more likely to drink than women

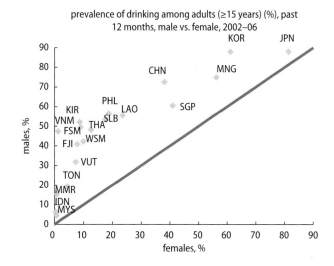

prevalence of drinking among adults (≥15 years) (%), past 12 months, male vs. female, 2002–06

Source: WHO Global InfoBase.

in Tonga (38 liters per capita consumption), Malaysia (32 liters), Thailand (29 liters), and Korea (29 liters).

These behaviors pose substantial risks to men's health and can translate into high costs for productivity and economic growth. Globally, 6 percent of all male deaths are related to alcohol, compared to 1 percent of female deaths (WHO 2011a). Half of today's smokers are likely to die from tobacco-related causes. A simple cross-country relationship (shown in figures 2.18 and 2.19) between these behaviors and morbidity and premature mortality related to smoking and alcohol use also indicates positive correlations (other

factors may also play a role, such as obesity and heart conditions, which are particularly high in the Pacific countries). In China, the rising risks of noncommunicable diseases, partly linked to men's smoking and drinking behaviors, put greater pressure on the size of the working-age population, already a concern because of its aging population (World Bank 2011a).

These behaviors are influenced by norms about masculinity, cultural beliefs about health, and the surrounding environment, and they can be slow to change. For men, smoking and drinking alcohol are commonly viewed as masculine behaviors, and studies show that men and boys feel substantial pressure to accept gender stereotypes that they should be strong and tough, and the opposite for women. A recent national survey in Vietnam found that the primary reason women did not use tobacco was the belief that women should not smoke. In a country where 50 percent of men but just over 3 percent of women smoke, 76 percent of the 2,020 young urban Vietnamese women surveyed said that this low female prevalence could be attributed to gender norms (that is, social disapproval of women who smoke). Only 20 percent said that the low prevalence was due to health concerns (WHO 2003). In addition, these patterns of smoking behaviors stayed very stable in Vietnam over the period from 1993 to 2006.

The poor tend to be slightly more likely to report "ever smoking," but the relationship is not strong among those countries where such data is available, except in Cambodia. As shown in figure 2.20, individuals ages 15 and above in the poorest quintile in Cambodia were almost twice as likely to engage in smoking cigarettes or chewing tobacco as those in the richest quintile. Among those who smoke, the intensity can vary because of the affordability of cigarettes. In Cambodia, particularly in rural areas, the richer smokers consume more cigarettes per day than the poorer smokers. However, no clear pattern between income and the intensity of smoking is observed in Mongolia and Vietnam.

FIGURE 2.18   Tobacco use is positively correlated with mortality due to lung cancer

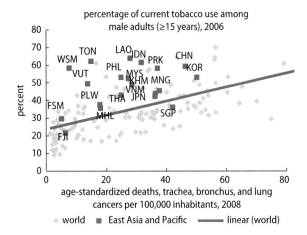

Source: WHO Global InfoBase.

FIGURE 2.19   Alcohol consumption is positively correlated with mortality due to alcohol use disorders

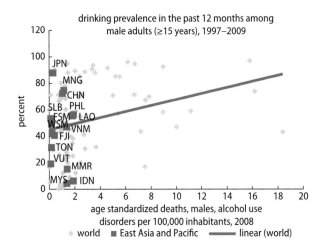

Source: WHO Global InfoBase.

Gender equality in productive assets: An unfinished agenda

Promoting gender equality in the control of productive assets (such as land, financial capital, social capital, and information and technology) is likely to enhance development, through both economic and empowerment benefits. Asset ownership can influence men's and women's income and their voice and influence within the household and within society. This effect can happen through strengthening their ability to take advantage of economic opportunities; for example, evidence shows that clear land-ownership rights have positive effects on agricultural productivity and access to credit (Deininger 2003). In the agricultural sector, evidence from Africa and Latin America suggests that ensuring equal access to productive assets and technologies raises agricultural production (Goldstein and Udry 2008; Quisumbing 1995; Udry 1996). As discussed throughout this report, income in the hands of women has been shown to positively affect children's education and health outcomes (Duflo 2003; Lundberg, Pollak, and Wales 1997). Women's assets prior to marriage have been shown to have positive effects on education expenditures on children in Indonesia and other countries (Quisumbing and Maluccio 2003). Beegle et al. (2001) showed that, in Indonesia, women with a higher share of household assets made more use of prenatal care. Asset ownership can lead to women's empowerment, such as reducing vulnerability to domestic violence in India (Panda and Agarwal 2005).

Gender equality in assets has been less responsive to growth and development than has equality in education and health. Over time, economic growth can promote access to financing that benefits women and men as well as improves economic opportunities and women's income. However, development impacts on gender equality in this domain are constrained by the complex legal, social, and economic factors that shape the control of productive assets. Gender disparities in access to and control of productive assets

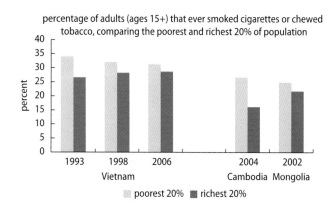

FIGURE 2.20 **The poor are slightly more likely to engage in smoking**

percentage of adults (ages 15+) that ever smoked cigarettes or chewed tobacco, comparing the poorest and richest 20% of population

Source: World Bank estimates using household income and expenditure surveys of various countries and years.

still persist around the world and in some parts of the East Asia and Pacific region. For example, household surveys in Bangladesh, Ethiopia, Indonesia, and South Africa indicate that women bring fewer assets into marriage (Quisumbing and Maluccio 2003). Evidence from the Philippines and other developing countries shows that the husband-wife asset difference at the time of marriage has not changed over time and favors the husband, even though gaps in age and education have been closing (Quisumbing and Hallman 2005).

This section focuses mainly on land and credit for the following two reasons. First, they are the major types of assets that strongly influence well-being. Land and property are usually the most valuable assets for a poor person. Aside from itself being a productive asset, land can be used as collateral to acquire credit. Access to financing is also very important because it usually presents a major barrier to realizing economic opportunities in developing countries. Second, data and rigorous quantitative evidence on gender and assets in the East Asia and Pacific region are very limited, particularly for assets other than land and credit. Individual-level data of asset ownership are sometimes available for land holdings but rarely for household durables, since many are considered jointly owned. In the face of scarce individual-level data on asset ownership, this section often resorts to comparing female-headed households and male-headed ones.[12] This section also draws on qualitative evidence and research from outside the region to help complement the limited quantitative evidence from East Asian and Pacific countries.

Persistent gender disparities in access to productive assets

Analysis of the East Asian and Pacific countries shows persistent gender disparities in productive assets that result in women having lower rates of ownership of and access to land, fewer agricultural inputs such as livestock and less access to extension services, and limited access to credit in some countries and subregions.

Gender-differentiated ownership of land
Female-headed households tend to own less land than male-headed households. Even though the probability of owning land is not substantially lower among female-headed households (figure 2.21), figure 2.22 indicates that male-headed households own much more land in terms of land size. This gap exists even among richer households. Over time, there have been small improvements in women's land ownership as well as men's, although the gender gap has not necessarily narrowed or disappeared except for the case of Vietnam in 2006. Similarly in China, most male-headed and female-headed households had access to rural land in 2008, but the amount of land per capita in female-headed households was roughly 70 percent of that in male-headed households (de Brauw et al. 2011). The pattern that female-headed households tend to own less land is similar to other developing countries (Agarwal 1994; Deere and Leon 2003; FAO 2011). Exact comparisons of the gender gaps across regions are difficult, however, since the incidence of land ownership varies drastically across countries,

even within the same region, and data are available for only a few countries in each region.

Evidence on individual land ownership suggests different ownership rates and different ownership composition by gender, depending on the context. In post-tsunami Aceh, Indonesia, women have fewer land holdings than men (Bell 2010). Vietnamese men owned more agricultural land plots than women in 2008. Figure 2.23 demonstrates the gender composition of ownership among the agricultural land plots with identified owners on long-term user right certificates. Overall, and among those plots owned by rural households, less than 17 percent of the plots are owned by a woman, but more than 65 percent are owned by a man. Plots jointly owned by a male and a female represent a nonnegligible share, in part because of Vietnam's recent land reform (discussed later in the chapter). However, women are still clearly at a disadvantage in terms of having their name on the land title. The same 2008 household survey data suggest that the gender gap is even starker among ethnic minorities. In some contexts, the allocation of land assets between men and women can depend on the type of land. For example, in matrilineal parts of West Sumatra, Indonesia, wives own more paddy land, and husbands own more forest land (Quisumbing and Maluccio 2003).

Gender disparities in agricultural inputs
Female-headed households tend to own less livestock, and female-run farms tend to have less access to extension services. In many countries, at least among rural households, livestock is one of the most valuable agricultural assets. It represents a source of income, wealth accumulation, and buffer against shocks. Yet, female-headed households tend to own less livestock across developing regions (FAO 2011). Figure 2.24 shows livestock ownership of female-headed and male-headed households in five countries in the region, for households of different wealth quintiles. The gender gap in Lao PDR appears irrespective of income,

but this observation does not apply in other contexts. In places such as Vietnam or Timor-Leste, gender gaps are more visible among poorer households. In many countries, this same pattern is observed in urban areas as well as rural areas. The data show that gaps also persist over time.

Extension service provision remains low in developing countries, and women tend to have less access to extension services than men (FAO 2011). In Cambodia, few women benefit from agricultural extension services or credit made available to rural people, despite the fact that they make up the majority of farmers and informal sector workers. Agricultural research and extension efforts usually do not consider women's activities—seed preparation and planting—or take into account the fact that men and women tend to specialize in different rural tasks. Distance to the point of service provision, lack of female agents, and insensitivity to illiterate customers (the majority of whom are often women) are other reasons for this lower access of female farmers to extension services (UNIFEM, World Bank, ADB, UNDP, and DFID/UK 2004).

Gender gaps in access to credit
Evidence on gender differentials in access to credit is mixed. Female-headed households are slightly less likely to borrow from financial institutions, as illustrated by figure 2.25. In most of the East Asian and Pacific countries for which data are available, the gap between households headed by females and males varies widely both across and within countries. The most recent data available indicate that gaps have been small except for Timor-Leste, rural Lao PDR, rural Cambodia, and urban Mongolia. Using survey data from 2000, de Brauw et al. (2011) found no difference between female-managed farms and male-managed farms in China in terms of access to credit. As discussed in chapter 3, female-run and female-owned firms in East Asia, at least those in the formal sector, do not appear to be systematically more constrained in accessing finance than male-run firms. However, a joint study by FAO/UNDP (2002) in Vietnam revealed that female-

FIGURE 2.21   The probability of owning land is not substantially lower for female-headed households than for male-headed households

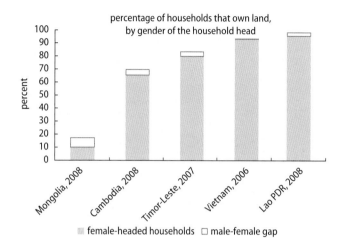

Source: World Bank estimates using household income and expenditure surveys of various countries and years.

FIGURE 2.22   Female-headed households own less land in terms of land size

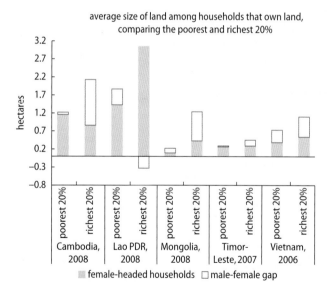

Source: World Bank estimates using household income and expenditure surveys of various countries and years.

headed households borrowed less, had less access to formal credit, and paid higher interest on loans than male-headed households.

Individual-level data on access to financing, by gender, suggest that the gender gap

FIGURE 2.23   Vietnamese men owned more agricultural land plots than did women in 2008

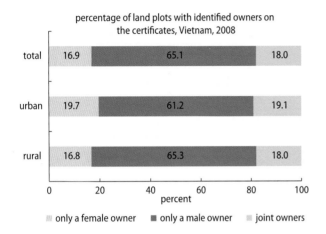

percentage of land plots with identified owners on the certificates, Vietnam, 2008

only a female owner   ▪ only a male owner   ▪ joint owners

Source: World Bank estimates using VHLSS (GSO Vietnam), 2008 data.

FIGURE 2.24   Female-headed households are less likely to own livestock

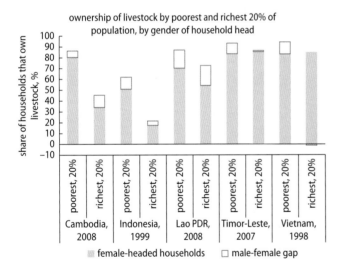

ownership of livestock by poorest and richest 20% of population, by gender of household head

▪ female-headed households   ☐ male-female gap

Source: World Bank estimates using household income and expenditure surveys of various countries and years.

in access to formal finance is likely to be small in East Asia. Evidence from Gallup surveys conducted since 2011 suggests that men and women have similar access to formal finance in East Asian countries (Demirgüç-Kunt and Klapper 2012). As figure 2.26 shows, of the nine East Asian countries surveyed, women were as likely as men to report having an account in a formal financial institution in Cambodia and Thailand; less likely to have an account in China, Indonesia, Lao PDR, Malaysia, and Vietnam (although the difference is statistically significant only in Malaysia and Vietnam); and more likely to have an account in Mongolia and the Philippines. A small-scale study of Vietnam's rural credit market in 2002 indicated that credit rationing depended on education and credit history, but found no evidence of bias against women (Barslund and Tarp 2003). The Thailand 2005 Household Socio-Economic Panel data show a similar rate between men and women holding a savings account in a financial institution (46 percent and 49 percent, respectively). Although this similarity is observed in Thailand's urban as well as rural areas, the situation may vary between rural and urban areas of other countries and might be very different in the Pacific, for which individual data are lacking.

Limited effects of economic growth on gender gaps in assets

Economic growth could, in principle, increase women's asset holdings by increasing income, but this factor alone is insufficient to close the gender gap in asset holdings. Market transactions are an important way to accumulate assets, and evidence from Latin America and the Caribbean suggests that, after inheritance, markets are the second most important channel for women in that region to acquire land. In that sense, income plays an important role. However, as shown in chapter 3, substantial gender gaps in income still persist in the East Asia and Pacific region, despite economic growth. Other complex legal and social factors also make it very challenging to close the gender gap in asset holdings with economic growth alone. Evidence from South Asia indicates that better employment opportunities and progressive legislation do not necessarily lead to gender equality in access to and control of land, because of social factors

(Agarwal 1994). Actually, wealthier house-holds with more valuable assets are not nec-essarily willing to give women more owner-ship rights. Analysis of the Vietnam 2008 household survey data by World Bank staff shows that richer households are less likely to have a female name in the title of their agricultural land plots, accounting for fac-tors such as land size and basic household characteristics.

As part of the development process, the expansion of microfinance coverage has been argued to improve women's access to credit. In terms of coverage, microfinance has greatly expanded all over the world, reach-ing many poor clients (Daley-Harris 2009). In China, microcredit has been used in vari-ous instances to support women. The Tianjin Women's Association for Business Develop-ment and Promotion, the Guangxi Provincial Women's Federation, and Liuzhou Municipal Women's Federation are examples of micro-credit schemes that target poor, laid-off, and unemployed women (ADB 2006a). Not all countries provide such targeted support. In Indonesia, although women are considered to be an important market for microfinance, the Indonesian microfinance industry has never made targeting of women a hallmark of their business. The average proportion of female clients in Indonesia served by major microfinance institutions has remained fairly constant over the past 20 years (Asia Foun-dation, ADB, CIDA, NDI, and World Bank 2006).

However, rapid expansion of microfinance does not necessarily imply a de facto control of resources. Goetz and Gupta (1996) show that women's access to microcredit has not been matched by an increase in their control of these funds. Microcredit facilities have been established for women in Lao PDR, but women's role in decision making has not improved (GRID 2005). Although women make up a high proportion of membership in credit schemes in Cambodia, they tend to be excluded from the decision-making processes and receive smaller amounts of credit than men (UNIFEM, World Bank, ADB, UNDP, and DFID/UK 2004).

FIGURE 2.25 Female-headed households are slightly less likely to borrow from a financial institution

percentage of households that borrowed from a financial institution (among those that borrowed money) by gender of the household head

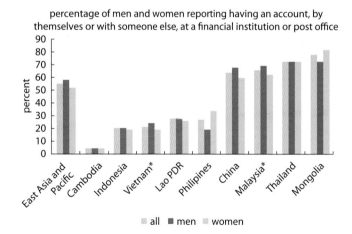

■ female-headed households □ male-female gap

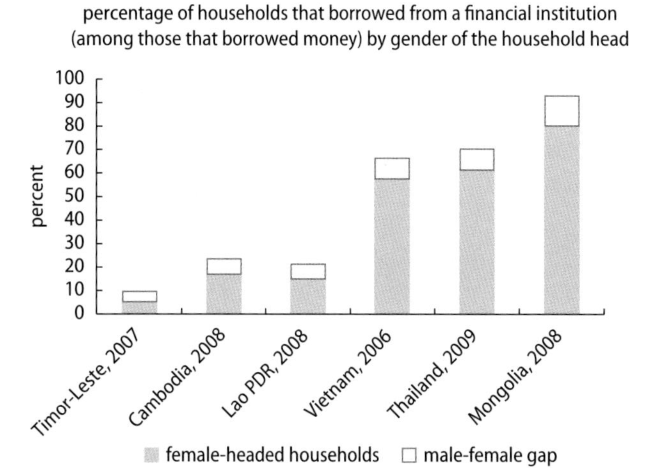

Source: World Bank estimates using household income and expenditure surveys of various countries and years.

FIGURE 2.26 Women are slightly less likely than men to report having an account at a formal financial institution

percentage of men and women reporting having an account, by themselves or with someone else, at a financial institution or post office

■ all   ■ men   ■ women

Sources: World Bank Global Financial Inclusion (Global Findex) database; Demirgüç-Kunt and Klapper 2012.
* Denotes a statistically significant difference between men and women at the 1 percent level.

Legal and social constraints to equalizing access to assets

To understand the observed gender dispari-ties in asset holdings, one must understand how assets are accumulated and the factors determining asset accumulation. Individuals can accumulate, or lose, assets in several ways. First, individuals can make market transactions—such as buying land or livestock, or acquiring a bank loan. Second, assets are acquired through inheritance or through allocation or acquisition by the state, for example, through land redistribution. Inheritance is one of the main mechanisms for asset accumulation (Deere and Doss 2006), and equality in asset endowments can be limited by differences in the right to inherit property. Third, life-cycle events such as marriage, including marriage payments, or separation also alter asset holdings.

Complex legal, social, and economic factors determine or constrain asset accumulation: (a) formal institutions—particularly the legal framework for property and inheritance rights, family laws, and law enforcement; (b) informal institutions—social norms and customary laws affecting women's preferences and ability to acquire and accumulate assets; and (c) human capital and other economic factors, such as income and the rate of returns on productive assets. The impact of economic growth on promoting more equal asset holdings is limited, as discussed earlier, because legal and social barriers often act as binding constraints. Despite positive changes in the legal framework in the East Asia and Pacific region, the interactions between formal and informal institutions still leave women at a disadvantage with respect to the control of assets.[13] The discussion below first describes the legal framework and then highlights the challenges in practice due to weak implementation and the influence of norms and customary laws.

The legal framework

The majority of countries in East Asia no longer differentiate by gender in statutory law. In that respect, the East Asia and Pacific region differs from some other regions: inheritance rights are still unequal in the Middle East and North African countries and half of the countries in South Asia (World Bank 2011c). As shown in table 2.4, most East Asian countries have legislation for property and inheritance rights with no discrimination against women. These countries—Cambodia, China, Lao PDR, Mongolia, Thailand, and Vietnam—do not have plural legal systems (for example, customary or religious laws), which means that all citizens adhere to civil law. For example, in China, many advances toward ensuring equal treatment and the protection of the rights of women under the law occurred as early as the 1950s. Women's property and inheritance rights were protected through the enactment of the 1982 constitution, which protects the right of citizens to inherit private property (Article 13),[14] and the 1985 Law of Succession of the People's Republic of China, which states that males and females are equal in their right to inheritance (Article 9).[15] The 1992 Law of the People's Republic of China on the Protection of Rights and Interests of Women further promotes gender equality; Article 28 declares that the state shall guarantee that women enjoy the equal right, with men, to property.[16]

In several of the countries examined, males and females are not treated equally, mostly under inheritance laws. Plural legal systems exist in Indonesia, Malaysia, the Philippines, and Singapore. Muslim laws govern the majority of the populations in Indonesia and Malaysia and a small minority of the population in the Philippines and Singapore. For example, according to the Islamic Law Compilation in Indonesia, when a married person dies, each son is entitled to receive a share twice as large as each daughter (Asia Foundation, ADB, CIDA, NDI, and World Bank 2006). The autonomous Muslim region of Mindanao in the Philippines can independently promulgate its own legislation following Islamic law (as allowed in the constitution). Although the Philippines is a community property regime,[17] the Muslim family code reflects the husband having the final say concerning the handling of joint property.[18] Among Pacific Island countries, Kiribati and Tuvalu have unequal statutory legislation. Equal inheritance laws exist in Fiji, Papua New Guinea, Samoa, the Solomon Islands, and Vanuatu; however, customary law in relation to land has constitutional status in these countries and may lawfully

TABLE 2.4 **Most East Asian and Pacific countries do not differentiate by gender in inheritance and property laws**

Question	Country	Answer	Law
Do sons and daughters have equal inheritance rights?	Cambodia	Yes	The Inheritance Law
	China	Yes	Law of Succession (Article 9)
	Indonesia	No	Islamic Law Compilation, Book II
	Lao PDR	Yes	The Inheritance Law
	Malaysia	No	Constitution
	Mongolia	Yes	Civil Code (Part V)
	Philippines	No[a]	Presidential Decree No. 1083
	Singapore	No[a]	Intestate Succession Act; Administration of Muslim Law Act
	Thailand	Yes	Civil and Commercial Code (Sections 1599–1710)
	Vietnam	Yes	Civil Code (Art. 635)
	Fiji	Yes[b]	Succession, Probate and Administration Act [Cap 60] 1970
	Kiribati	No	Laws of Kiribati Act 1989, Schedule 4
Do men and women have equal rights over property?	Cambodia	Yes	Constitution (Arts. 31 and 45); Law on Marriage and Family (Arts. 29, 32–37)
	China	Yes	Law on Protection of Women's Rights (Arts. 30 and 47)
	Indonesia	Yes	Marriage Law No. 1 of 1974
	Lao PDR	Yes	Law of Property (Arts. 20, 26)
	Malaysia	Yes	Constitution
	Mongolia	Yes	Civil Code (Ch. 12)
	Philippines	No[a]	Presidential Decree No. 1083
	Singapore	Yes	Women's Charter (Arts. 51, 52, 56)
	Thailand	Yes	Constitution (Section 30); Civil and Commercial Code (Book IV)
	Vietnam	Yes	Land Law; Civil Code (Sec. 8)
	Fiji	No	Constitution (Amendment) Act 1997
	Kiribati	No	Laws of Kiribati Act 1989, Schedule 4; Magistrates Court Act; Gilbert and Phoenix Islands Land Code

Sources: Women, Business and the Law database; Jivan and Forster 2007.
a. Implies that unequal legal systems apply only to a minority population.
b. Denotes that despite equal inheritance in legislation, Fijian custom in relation to land has constitutional status and may lawfully discriminate against women.

discriminate against women (Jivan and Forster 2007).

Beyond the protection of equal inheritance rights, several countries in the region have recently adopted legal changes that actively promote better gender equality in access to land. Since concerns have been raised about promoting gender equity in land titling programs, places such as Indonesia, Lao PDR, and Vietnam have recently adopted gender-sensitive reforms in land titling. Since the 2004 Land Law in Vietnam, all new land tenure certificates must include the names of both spouses, a provision intended to reduce gender inequality in access to land, protect families against unilateral decisions by one spouse, and protect women in case of divorce or disputes. Qualitative analysis of impacts in three provinces suggests that joint titling improves procedures and opportunities for

women to access loans, empowers women in case of disputes, and leads to higher mutual decision making (World Bank 2008).

Weak implementation and enforcement of the law

Although women and men may be equal under the law, these legal rights do not always translate into equal access to land in practice. Progress toward gender equality in assets is still limited because of weak implementation and enforcement of the law.

Implementation of the laws may be imperfect and ineffective as a result of challenges in incorporating existing cultural norms and practices. Even where the legal system supports equal access to land, traditional values and norms create difficulty with respect to enforcing the legislation. After the 2004 Land Law requiring joint titling was introduced in

Vietnam, the results have been varied across ethnic majority and minority groups. Analysis of the Vietnam Household Living Standards Survey 2008 shows that ethnic minority households are much less likely to have a female name in the title of their agricultural land plots, even when accounting for factors such as land size and basic household characteristics. Studies of existing kinship systems in China and Vietnam show resistance to endowing daughters with land, as land is considered lost when daughters get married (Bélanger and Li 2009). In Northern Liaoning, China, women are more likely to become landless at marriage because the population control and land tenure policies reinforce traditional forms of gender bias: a daughter is expected to marry out, whereas a son is expected to reside with or near his parents (Chen and Summerfield 2007). Therefore, during the redistribution of farmland to households based on household size, township and village officials allocated a larger share of land for each son in the household than for each daughter.

The Lao PDR land titling program 1997–2010 offered useful implementation lessons to account for context-specific social norms. An early assessment during the first phase of the program noted that traditional family roles dedicated the man to handling taxes and, thus, to having only his name on land-related tax documents. Mostly men interacted with government officials and participated in information meetings and titling activities. As a response to this assessment, Lao PDR introduced a stronger gender inclusion program, engaging the Lao Women's Union, to raise community awareness of land titling, to include special training for women on their rights, and to ensure their participation during titling. Following this active engagement, a higher number of titles went to women than to men (Lao Land Titling Project II ICR 2010).

Qualitative research, more broadly, has stressed that a lack of information about legal entitlement is a key barrier to enforcing women's land rights throughout the region. Women are generally less aware of the laws governing land ownership. They are also less likely to know about land registration requirements (including whether the land owned by their households is registered and whether their names are included in land titling documents). Dissemination activities on rights and entitlements can exclude women, especially in rural areas.[19] Even if they have information about their legal land rights, women tend not to pursue formal complaints in cases of land grabbing or disputes regarding inheritance or division of property. A 2008 AusAID report on land tenure in the Pacific region stresses the following factors constraining women's access to the formal legal system to resolve disputes regarding land ownership and use: (a) the system was culturally unfamiliar and based on "adversarial" methods rather than on the consensus building usually preferred by communities; (b) women also faced more "practical" obstacles to accessing courts (limited access to transport, lack of time, income); (c) the "technical nature" of the procedures and inadequate support from court staff were also noted as important barriers (AusAID 2008).

Social norms and practices

In some contexts, customary practices, rather than statutory laws, directly govern land ownership and land use. Most land in the Pacific region is under customary authority (approximately 80 percent of total land area) (AusAID 2008). Although the rules vary widely within the Pacific region, usually in customary systems, (a) land can only be transferred within networks of social/political relationships, (b) land use is governed by reciprocal relations within the kinship or customary land groups, and (c) social hierarchies and status are important factors determining one's rights to land. In these customary systems, women have access to land primarily through their kinship relations with men. Women also have less voice in public decisions about the use of land (AusAID 2008). For example, Fijian women in most parts of the Fiji Islands are excluded from inheritance rights in customary land and have no rights to land other than those permitted by their fathers or husbands.

Nor do they normally receive land rents. Most Indo-Fijians with land also practice father-to-son inheritance (ADB 2006b).

Practices within the region and within countries vary considerably. Inheritance practices based on norms can be patrilineal, matrilineal, or bilateral and, as such, are not always tilted against women. In the matrilineal society of Sumatra, Indonesia, together with the shift from communal to individual tenure, the inheritance system became more egalitarian in that sons and daughters inherit the type of land that is most intensive in their own work effort, and gender bias in land inheritance is either nonexistent or small (Quisumbing and Otsuka 2001). In the Philippines, sons are preferred in land inheritance but daughters are favored in education investments (Estudillo, Quisumbing and Otsuka 2001). In Lao PDR, land inheritance and ownership are important elements of women's autonomy in lowland areas, with daughters customarily inheriting land. However, women in the midlands and highlands, such as those in the Khmu and Hmong ethnic groups, face important barriers to controlling land (Ireson-Doolittle 1999). Understanding the specifics of a particular context is thus important for designing appropriate policies and interventions.

In summary, given the limited evidence in the East Asia and Pacific region, understanding gender inequality in assets is an important research agenda. Gender disparities in access to productive assets can hinder women's ability to participate and benefit from economic opportunities as well as constrain women's voice and representation in the society. Further research and better data are called for to disentangle the complex mechanisms influencing men's and women's control of assets and to shed light on policies.

Policies to promote gender equality in endowments

The analysis thus far has identified the factors influencing or constraining gender equality in endowments. This analysis also sheds light on where policies may be used and can

have an impact. What follows is an initial discussion about policy priorities for promoting gender equality in human capital: (a) closing persistent gender gaps in human development, (b) reducing gender streaming in education, (c) promoting balanced sex ratios at birth, and (d) addressing male-specific gender issues. Policy approaches to close gender gaps in assets are important for gender equality in endowments, and they also enable more equal economic opportunity and agency. These policies are discussed within the general framework of policies to promote gender equality in economic opportunity, which is discussed in chapter 3 and chapter 6. A more detailed discussion will follow in chapter 6 on policies to promote overall gender equality and more effective development.

Closing persistent gender gaps in human development

For countries with overall low and unequal gender outcomes in education and health, the priority remains to improve these outcomes. Actions to strengthen the education and health systems are called for to improve overall outcomes, in addition to any focus on gender. Interventions may be needed at the national level and may yield high economic and social returns. For countries with localized gender disparities among certain ethnic groups or low-income regions, interventions may be targeted to these groups. Though the exact constraints vary by country context, the analysis in this chapter has shown that both demand-side and supply-side factors are responsible for these poor human capital outcomes. Policies can have an impact through improving service delivery (infrastructure, staffing, incentives, use of information and communication technology) and demand-side interventions (conditional cash transfers, information campaigns, accountability). For education, policies to improve education outcomes in general are expected to also improve gender equality. For health, the slow improvements in health outcomes in several East Asian and Pacific countries underscore the importance of improving service delivery in general, including efforts to account for gender norms that affect service utilization and effectiveness.

Reducing gender streaming in education

For many countries in the East Asia and Pacific region, addressing education quality—specifically, gender streaming in education—requires policy attention. Many aspects of gender issues regarding basic access in education and health have abated with growth and development. However, concerted efforts in education and labor market policies are needed to break the traditional patterns of females going into certain fields of study and, consequently, jobs in lower-paying occupations and sectors. Within the education system, possible approaches in this agenda include both curriculum reforms to reduce gender stereotyping and active interventions—financial and nonfinancial incentives as well as information campaigns—to promote entrance into nontraditional fields.

Promoting balanced sex ratios at birth

In the few countries with "missing girls" at birth, rooted in the prevalence of son preference, continuing efforts are needed. A promising strategy is to adopt policy approaches that aim to enhance the relative value of daughters as perceived by families. General policies to promote economic development may play a role, but Korea's recent experience suggests that interventions to influence social norms and facilitate the spread of new values may be very important, rather than simply relying on raising female education and labor force participation. Information campaigns, financial incentives, and improved social security for the elderly are worthwhile efforts. China has been adopting many of these programs, and they can be expected to reduce the imbalance in the sex ratio at birth.

Addressing male-specific gender issues

Attention to male gender issues is crucial in many country contexts since they may also hamper growth and development. First, the initial signs of the reversed gender gap in education need to be monitored closely where applicable. Second, the excessive tobacco and alcohol consumption among males in many parts of the East Asia and Pacific region deserves policy attention because the social costs, passed on as externalities to other members of the society, are usually higher than private costs. Possible measures to tackle this challenge include providing information about the health risks of excessive tobacco and alcohol consumption, taxation, regulatory measures on advertisement, and restrictions on smoking in public sites.

Notes

1. See Malhotra, Pande, and Grown (2003) on impacts of investments in female education on gender equality.
2. See *Engendering Development* (World Bank 2001) and *World Development Report 2012* (World Bank 2011c) for similar literature. However, few studies rigorously identify the causal effects as opposed to simple correlation. And female education or income might not always have dominant, widespread benefits over male education as commonly perceived. For example, controlling for household average education, Breierova and Duflo (2004) found no impact of female education on infant mortality in Indonesia. Edmonds (2006) found that in South Africa, pension money going to grandmothers improved children's health while that going to grandfathers improved children's schooling.
3. Enrollment in different types of education also shows gender differences that vary across countries in the region. Females' completion rate in vocational training has been increasing in Thailand and Vietnam. In recent years, this rate among females is still lower than that among males in Indonesia, Thailand, Vietnam, and Mongolia, even though the reverse tends to be true for completion of general secondary education. Cambodia experiences the opposite patterns, that is, women have lower general secondary completion rates but higher vocational completion rates than men (Sakellariou 2011).

4. Conflict and postconflict areas also suffer poor education outcomes for both boys and girls, for different reasons, such as the risk for boys of being taken out of school to join the military and the risk of safety for girls at the school.

5. Girls do not always lag boys in poor rural areas. Evidence from China's Gansu province shows no significant gender disadvantage (Hannum and Adams 2002).

6. The returns to education for women have not increased uniformly: the returns for women increased relative to those of men in Vietnam but decreased in Indonesia and Thailand since the late 1990s (Sakellariou 2011).

7. See Kobia (2009) and Lloyd, Mensch, and Clark (1998), for example, for analysis of similar gender stereotyping in Kenya.

8. Prior empirical evidence on the extent to which income causally affects health status has been controversial, partly owing to methodological challenges such as possible omitted variables and the reversed feedback from health to income (Deaton 2006; Filmer and Pritchett 1999; Pritchett and Summers 1996).

9. Das Gupta, Chung, and Shuzhuo (2009) argue that the recent provincial sex ratios in China suggest an incipient turnaround of the "missing girls" phenomenon in East Asia. However, the concern is far from over because child sex ratios in China are still high.

10. See World Bank 2011c, *World Development Report 2012*, chapter 3 technical annex for detailed methodology.

11. Prevalence estimates of current smoking of any tobacco product result from the latest adult tobacco use surveys, which have been adjusted according to the WHO regression method for standardizing described in the Method of Estimation. "Tobacco smoking" includes cigarettes, cigars, pipes, or any other smoked tobacco products. "Current smoking" includes both daily and nondaily or occasional smoking.

12. Households headed by women and those headed by men may be very different as a result of unobserved factors. The observation that female-headed households have fewer assets does not necessarily mean that women have less access to assets in general (many are widows or single-parent families that face economic difficulties anyway).

Looking at gender of the household head as an approximate measure presumes that females in female-headed households own and control most of the assets while females in male-headed households control and own relatively few assets. This measure could be misleading in cases where male heads of households are temporarily absent.

13. Recent analysis of women and land in Association of Southeast Asian Nations (ASEAN) countries identifies the following supply constraints to women's equal access to land in the region: unfavorable legal framework, pro-male customs, limited opportunities, and lack of data (ASEAN 2008).

14. http://english.people.com.cn/constitution/constitution.html.

15. http://www.chinaembassycanada.org/eng/lsfw/Relevant%20Chinese%20Laws%20and%20Regulations/t37737.htm.

16. http://www.womenofchina.cn/html/report/515-1.htm.

17. In a community property jurisdiction, most property acquired during the marriage (except for gifts or inheritances) is owned jointly by both spouses and is divided upon divorce, annulment, or death.

18. http://www.law.emory.edu/ifl/legal/philippines.htm.

19. In China, for example, it is a common practice for new policies, regulations, and programs to be discussed in village meetings in which heads of households (primarily men) participate. Information on changes to land laws, for example, might reach women later (or partially, as they may rely on other household members to convey the information) (Liaw 2008). In Indonesia, land acquired during marriage tends to be registered primarily under the name of the male head of household because most landowners are not aware of the possibility of registering land in more than one person's name (Brown 2003).

References

ADB (Asian Development Bank). 2005. *Country Gender Assessment Timor-Leste*. Pacific Regional Department and Regional and Sustainable Development Department, Asian Development Bank East and South East Asia Regional Office United Nations Development Fund for Women. Manila, Philippines: ADB.

———. 2006a. *ADB Country Gender Assessment: People's Republic of China.* Manila, Philippines: ADB.

———. 2006b. *Republic of the Fiji Islands: Country Gender Assessment.* Manila, Philippines: ADB.

Agarwal, Bina. 1994. *A Field of One's Own: Gender and Land Rights in South Asia.* Cambridge, U.K.: Cambridge University Press.

Albanesi, Stephania, and Claudia Olivetti. 2009. "Production, Market Production and the Gender Wage Gap: Incentives and Expectations." *Review of Economic Dynamics* 12 (1): 80–107.

ASEAN (Association of Southeast Asian Nations) 2008. "ASEAN Continues to Empower Women." http://www.aseansec.org/Bulletin-Feb-08.htm#Article-2.

Asia Foundation, ADB (Asian Development Bank), CIDA (Canadian International Development Agency), NDI (National Democratic Institute), and World Bank. 2006. *Indonesia: Country Gender Assessment.* Manila, Philippines: Asia Foundation, ADB, CIDA, NDI, and World Bank.

Atkin, David. 2010. "Endogenous Skill Acquisition and Manufacturing in Mexico." Yale University, Department of Economics, New Haven, CT. http://www.econ.yale.edu/~da334/Endogenous_Skill_Acquisition_Mexico.pdf.

AusAID. 2008. *Making Land Work: Vol. 1. Reconciling Customary Land and Development in the Pacific.* Canberra: Australian Agency for International Development.

Bailey, B., and M. Bernard. 2003. "Establishing a Database of Gender Differentials in Enrolment and Performance at the Secondary and Tertiary Levels of the Caribbean Education Systems." Canada-Caribbean Gender Equality Fund Programme II (CARICOM), Georgetown, Guyana.

Baird, Sarah, Jed Friedman, and Norbert Schady. 2007. "Aggregate Income Shocks and Infant Mortality in the Developing World." Policy Research Working Paper 4346, World Bank, Washington, DC.

Barslund, Mikkel, and Finn Tarp. 2003. "Rural Credit in Vietnam." Discussion Paper, University of Copenhagen, Denmark.

Beegle, Kathleen, Elizabeth Frankenberg, and Duncan Thomas. 2001. "Bargaining Power within Couples and Use of Prenatal and Delivery Care in Indonesia." *Studies in Family Planning* 32 (2): 130–46.

Behrman, Jere Richard, and James C. Knowles. 1999. "Household Income and Child Schooling in Vietnam." *World Bank Economic Review* 13: 211–56.

Bélanger, Danièle, and T. H. Oanh Khuat. 2009. Second-Trimester Abortions and Sex-Selection of Children in Hanoi, Vietnam. *Population Studies* 63 (2): 163–71.

Bélanger, Danièle, and Xu Li. 2009. "Agricultural Land, Gender and Kinship in Rural China and Vietnam: A Comparison of Two Villages." *Journal of Agrarian Change* 9 (2): 204–30.

Bélanger, Danièle, and Giang Linh Tran. 2011. "The Impact of Transnational Migration of Gender and Marriage in Sendign Communities of Vietnam." *Current Sociology* 59 (1): 59–77.

Bélanger, Danièle, Thi Hai Oanh Khuat, Jianye Liu, Thanh Thuy Le, and Viet Thanh Pham. 2003. "Are Sex Ratios at Birth Increasing in Vietnam?" *Population (English Edition)* 58 (2): 231–50.

Bell, Keith Clifford. 2010. "Study on Gender Impacts of Land Titling in Post-Tsunami Aceh, Indonesia." Photocopy, World Bank, Washington, DC.

Bharadwaj, Prashant, and Leah K. Nelson. 2010. "Discrimination Begins in the Womb: Evidence of Sex-Selective Prenatal Investments." Photocopy, University of California, San Diego.

Blumberg, Rae Lesser. 2007. "Gender Bias in Textbooks: A Hidden Obstacle on the Road to Gender Equality in Education." Paper commissioned for the *Education for All Global Monitoring Report 2008, Education for All by 2015: Will We Make It?* http://unesdoc.unesco.org/images/0015/001555/155509e.pdf.

BPS (Badan Pusat Statistik). Indonesia National Socioeconomic Survey (SUSENAS). Jakarta, Indonesia.

Breierova, Lucia, and Esther Duflo. 2004. "The Impact of Education on Fertility and Child Mortality: Do Fathers Really Matter Less Than Mothers?" NBER Working Paper 10513, National Bureau of Economic Research, Cambridge, MA.

Brown, Jennifer. 2003. "Rural Women's Rights in Java, Indonesia: Strengthened by Family Law but Weakened by Land Registration." *Pacific Rim Law and Policy Journal* 12 (631): 51.

Cambodia Development Resource Institute. 2001. "Social Assessment of Land in Cambodia—A Field Study." Cambodia Development Resource Institute, Phnom Penh, Cambodia.

Cameron, L., and C. Worswick. 2001. "Education Expenditure Responses to Crop Loss in Indonesia: A Gender Bias." *Economic Development and Cultural Change* 49 (2): 351–63.

Chattier, Priya. 2011. "Gender and Economic Choice." Background paper for the *World Development Report 2012*. University of the South Pacific Laucala Campus, Suva, Fiji.

Chen, Junjie, and Gale Summerfield. 2007. "Gender and Rural Reforms in China: A Case Study of Population Control and Land Rights Policies in Northern Liaoning." *Feminist Economics* 13: 63–92.

Chung, W., and Monica Das Gupta. 2007. "The Decline of Son Preference in South Korea: The Roles of Development and Public Policy." *Population and Development Review* 33 (4): 757–83.

Cutler, D., and G. Miller. 2005. "The Role of Public Health Improvements in Health Advances: The Twentieth Century United States." *Demography* 42 (1): 1–2.

Daley-Harris, Sam. 2009. *State of the Microcredit Summit Campaign Report 2009*. Washington, DC: Microcredit Summit Campaign.

Das Gupta, Monica. 2005. Explaining Asia's "Missing Women": A New Look at the Data. *Population and Development Review* 31 (3): 529–35.

———. 2009. "Family Systems, Political Systems, and Asia's 'Missing Girls.'" Policy Research Working Paper 5148, World Bank, Washington, DC.

Das Gupta, Monica, W. Chung, and L. Shuzhuo. 2009. "Is There an Incipient Turnaround in Asia's 'Missing Girls' Phenomenon?" Policy Research Working Paper 4846, World Bank, Washington, DC.

Davis, James. 2002. "Boys to Men: Masculine Diversity and Schooling." Paper presented at the School Leadership Centre of Trinidad and Tobago, Port of Spain, Trinidad, August 6–8.

Deaton, Angus. 2006. "Global Patterns of Income and Health: Facts, Interpretations, and Policies." WIDER Annual Lecture, Helsinki, September 29.

de Brauw, Alan, Jikun Huang, Linxiu Zhang, and Scott Rozelle. 2011. "The Feminization of Agriculture with Chinese Characteristics." Background paper for the *World Development Report 2012*, World Bank, Washington, DC.

Deere, Carmen Diana, and Cheryl Doss. 2006. "The Gender Asset Gap: What Do We Know and Why Does It Matter?" *Feminist Economics* 12 (1–2): 1–50.

Deere, Carmen Diana, and Magdalena Leon. 2003. "The Gender Asset Gap: Land in Latin America." *World Development* 31 (6): 925–47.

Deininger, Klaus. 2003. *Land Policies for Growth and Poverty Reduction*. Washington, DC: World Bank.

Demirgüç-Kunt, Asli, and Leora Klapper. 2012. "Measuring Financial Inclusion: The Global Findex Database." Policy Research Working Paper 6025, World Bank, Washington, DC.

Duflo, Esther. 2000. "Child Health and Household Resources in South Africa: Evidence from the Old Age Pension Program." *American Economic Review* 90 (2): 393–98.

———. 2003. "Grandmothers and Granddaughters: Old-Age Pensions and Intrahousehold Allocation in South Africa." *World Bank Economic Review* 17 (1): 1–25.

Dunne, M., and F. Leach. 2005. "Gendered School Experiences: The Impact on Retention and Achievement in Botswana and Ghana." Researching the Issues 56, Department for International Development, London.

Ebenstein, A. 2010. "The Missing Girls of China and the Unintended Consequences of the One Child Policy." *Journal of Human Resources* 45 (1): 87–115.

Eccles, J. S., and P. Blumenfeld. 1985. *Classroom Experiences and Student Gender: Are There Differences and Do They Matter?* In *Gender Influences in Classroom Interaction*, edited by L. C. Wilkinson and C.B. Marrett. Orlando, FL: Academic Press.

Edlund, L., and C. Lee. 2009. "Son Preference, Sex Selection and Economic Development: Theory and Evidence from South Korea." Discussion Paper 0910-04, Columbia University, Department of Economics, Chicago, IL.

Edmonds, E. 2006. "Child Labor and Schooling Responses to Anticipated Income in South Africa." *Journal of Development Economics* 81 (2): 386–414.

Eskes, T., and C. Haanen. 2006. "Why Do Women Live Longer Than Men?" *European Journal of Obstetrics Gynecology and Reproductive Biology* 133 (2): 126–33.

Estudillo, J., Agnes Quisumbing, and Keijiro Otsuka. 2001. "Gender Differences in Land Inheritance, Schooling and Lifetime Income: Evidence from the Rural Philippines." *Journal of Development Studies* 37 (4): 23–48.

FAO (Food and Agriculture Organization)/UNDP (United Nations Development Programme). 2002. "Gender Differences in the Transitional Economy of Viet Nam." Hanoi. http://www .fao.org/docrep/005/ac685e/ac685e00.htm.

FAO. 2011. *The State of Food and Agriculture 2010–11: Women in Agriculture: Closing the Gender Gap for Development*. Food and Agriculture Organization, Rome.

Figueroa, M. 2000. "Making Sense of the Male Experience: The Case of Academic Underachievement in the English-Speaking Caribbean." *IDS Bulletin* 31 (2).

Filmer, Deon, Jed Friedman, and Norbert Schady. 2008. "Development, Modernization, and Son Preference in Fertility Decisions." Policy Research Working Paper 4716, World Bank, Washington, DC.

Filmer, Deon, and Lant Pritchett. 1999. "The Impact of Public Spending on Health: Does Money Matter?" *Social Science and Medicine* 49 (10): 1309–23.

Filmer, Deon, and Norbert Schady. 2008. "Getting Girls into School: Evidence from a Scholarship Program in Cambodia." *Economic Development and Cultural Change* 56 (3): 581–617.

———. 2009. "School Enrollment, Selection and Test Scores." Policy Research Working Paper 4998, World Bank, Washington, DC.

Fiszbein, Ariel, and Norbert Schady. 2009. "Conditional Cash Transfers: Reducing Present and Future Poverty." Policy Research Series, World Bank, Washington, DC.

Flabbi, Luca. 2011. "Gender Differences In Education, Career Choices and Labor Market Outcomes on a Sample of OECD Countries." OECD Background Paper for *World Development Report 2012*, World Bank, Washington, DC.

Frankenberg, Elizabeth, A. Buttenheim, B. Sikoki, and W. Suriastini. 2009. "Do Women Increase Their Use of Reproductive Health Care When It Becomes More Available?" *Evidence from Indonesia, Studies in Family Planning* 40 (1): 27–38.

Frankenberg, Elizabeth, and Duncan Thomas. 2001. "Women's Health and Pregnancy Outcomes: Do Services Make a Difference?" *Demography* 38 (2): 253–65.

Global Findex (Global Financial Inclusion Database). World Bank, Washington, DC. http://data.worldbank.org/data-catalog/ financial_inclusion.

Goetz, A. M., and R. Gupta. 1996. "Who Takes the Credit? Gender, Power, and Control over Loan Use in Rural Credit Programs in Bangladesh. *World Development* 24 (1): 45–63.

Goldstein, Markus, and Christopher Udry. 2008. "The Profits of Power: Land Rights and Agricultural Investment in Ghana." *Journal of Political Economy* 116 (6): 981–1022.

GRID (Gender Resource Information and Development Center). 2005. "Lao PDR Gender Profile." GRID, with support of the World Bank, Washington, DC. http://siteresources .worldbank.org/INTLAOPRD/Resources/ Lao-Gender-Report-2005.pdf.

GSO (General Statistics Office) Vietnam. Household Living Standards Surveys, http:// www.gso.gov.vn.

Guo, L., and Z. Zhao. 2002. "Children, Gender, and Language Teaching Materials." *Chinese Education and Society* 35 (5): 34–52.

Hanna, Rema, and Leigh Linden. 2009. "Measuring Discrimination in Education." NBER Working Paper 15057, National Bureau of Economic Research, Cambridge, MA.

Hannum, E., and J. Adams. 2002. "Girls in Gansu, China: Expectations and Aspirations for Secondary Schooling." In *Exclusion, Gender and Education*, edited by M. Lewis and M. Lockheed, 71–96. Washington, DC: Center for Global Development.

HNPStats (Health Nutrition Population Statistics) database. World DataBank, World Bank, Washington, DC. http://data.worldbank.org/ data-catalog/health-nutrition-population- statistics.

IDN CGA. (2006). "Indonesia: Country Gender Assessment." Asian Development Bank, Canadian International Development Agency, National Democratic Institute, and the World Bank.

Ireson-Doolittle, C. 1999. "Gender and Changing Property Rights in Laos." In *Women's Rights to House and Land: China, Laos, Vietnam*, edited by I. Tinker and G. Summerfield, 145– 52. Boulder, Colorado: Lynne Rienner.

Jayachandran, S., and A. Lleras-Muney. 2009. "Life Expectancy and Human Capital Investments: Evidence from Maternal Mortality Declines." *Quarterly Journal of Economics* 124 (1): 349–97.

Jensen, R. 2010. "Economic Opportunities and Gender Differences in Human Capital: Experimental Evidence for India." NBER

Working Paper 16021, National Bureau of Economic Research, Cambridge, MA.

Jha, Jyotsna, and Fatimah Kelleher. 2006. *Boys' Underachievement in Education: An Exploration in Selected Commonwealth Countries*. London, U.K.: Commonwealth Secretariat, Gender Section, Social Transformation Programmes Division, and the Commonwealth of Learning.

Jivan, Vedna, and Christine Forster. 2007. *Translating CEDAW into Law: CEDAW Legislative Compliance in Nine Pacific Countries*. Suva, Fiji: UNDP Pacific Centre and UNIFEM Pacific Regional Office.

Johnson, Kiersten, Sovanratnak Sao, and Darith Hor. 2000. *Cambodia 2000 Demographic and Health Survey: Key Findings*. Calverton, MD: ORC Macro.

Jung, Kyung, and Haesook Chung. 2005. *Gender Equality in Classroom Instruction: Introducing Gender Training for Teachers in the Republic of Korea*. Bangkok, Thailand: UNESCO.

Kevane, M., and D. Levine. 2003. "Changing Status of Daughters in Indonesia." Working Paper C03-126, University of California, Berkeley, Center for International Development, Berkeley, CA.

King, E., and D. van der Walle. 2007. *Girls in Lao PDR: Ethnic Affiliation, Poverty and Location*. In *Exclusion, Gender and Education: Case Studies from the Developing World*, edited by M. Lewis and M. Lockheed, 31–70. Washington, DC: Center for Global Development.

Klasen, Stephan, and Francesca Lamanna. 2009. "The Impact of Gender Inequality in Education and Employment on Economic Growth: New Evidence for a Panel of Countries." *Feminist Economics* 15 (3): 91–132.

Knowles, Stephen, A. K. Lorgelly, and Dorian Owen. 2002. "Are Educational Gender Gaps a Brake on Economic Development? Some Cross-Country Empirical Evidence." *Oxford Economic Papers* 54 (1): 118–49.

Kobia, J. 2009. "Femininity and Masculinity in English Language Textbooks in Kenya." *The International Journal of Language Society and Culture*, 28: 57–71.

Korea National Statistical Office. Daejeon, Republic of Korea. http://kostat.go.kr.

Lao Land Titling Project II ICR. 2010. "Implementation Completion and Results Report (Ida-38010) on a Credit in the Amount of Sdr 10.8 Million (US$14.82 Million Equivalent) to the Lao People's Democratic Republic for the Second Land Titling Project." World Bank, Washington, DC.

Lavy, Victor. 2008. "Do Gender Stereotypes Reduce Girls' or Boys' Human Capital Outcomes? Evidence from a Natural Experiment," *Journal of Public Economics*. 92 (10–11): 2083–2105.

Li, H., and H. Zheng. 2009. "Ultrasonography and Sex Ratios in China." *Asian Economic Policy Review* 4: 121–37.

Liaw, H. Ray. 2008. "Women's Land Rights in Rural China: Transforming Existing Laws into a Source of Property Rights." *Pacific Rim Law and Policy Journal* 17 (1): 237–64.

Lloyd, C., B. Mensch, and W. Clark. 1998. "The Effects of Primary School Quality on the Educational Participation and Attainment of Kenyan Girls and Boys." Population Council Paper 116. http://www.popcouncil.org/pdfs/wp/16.pdf.

LSB (Lao Statistics Bureau). Expenditure and Consumption Survey, LAO People's Democratic Republic. http://www.nsc.gov.la/.

Lundberg, S. J., R. Pollak, and T. J. Wales. 1997. "Do Husbands and Wives Pool Their Resources? Evidence from the United Kingdom Child Benefit." *Journal of Human Resources* 32 (3): 463–80.

Malhotra, A., R. Pande, and C. Grown. 2003. "Impact of Investments in Female Education on Gender Equality." Commissioned by the World Bank Gender and Development Group and International Center for Research on Women, Washington, DC.

Marks, J. 2001. "Girls Know Better: Educational Attainment of Boys and Girls." CIVITAS— The Institute for the Study of Civil Society, London, U.K.

Martino, W., and D. Berrill. 2003. "Boys, Schooling and Masculinities: Interrogating the 'Right' Way to Educate Boys." *Educational Review* 55 (2): 99–117.

Naudeau, Sophie, Naoko Kataoka, Alexandria Valerio, Michelle Neuman, and Leslie Elder. 2011. *Investing in Young Children: An Early Childhood Development Guide for Policy Dialogue and Project Preparation*. Washington, DC: World Bank.

Niehof, A. 2003. "Women and the Social Context of Fertility Under the New Order." In *Two is Enough: Family Planning in Indonesia*

Under the New Order 1968–1998, edited by A. Niehof and F. Lubis, 163–183. Leiden: KITLV Press.

Nielsen, Harriet Bjerrum, and Bronwyn Davies. 2010. "Discourse and the Construction of Gendered Identities in Education." In *Encyclopedia of Language and Education: Volume 3, Discourse and Education*, edited by Marilyn Martin-Jones, Anne-Marie de Mejía, and Nancy H. Hornberger. Berlin: Springer.

NIS (National Institute of Statistics) Cambodia. Cambodia Socio-Economic Survey (CSES). National Institute of Statistics, Phnom Penh, Cambodia. http://www.nis.gov.kh/index.php/social-statistics/cses.

NSD (National Statistics Directorate) Timor-Leste. Timor-Leste Survey of Living Standards. Ministry of Planning and Finance, National Statistics Directorate, Dili, Timor-Leste. http://dne.mof.gov.tl/TLSLS/AboutTLSLS/index.htm.

NSD (National Statistics Directorate) Timor-Leste, Ministry of Finance, and ICF Macro. 2010. *Timor-Leste Demographic and Health Survey 2009–10*. Dili, Timor-Leste: NSD and ICF Macro.

NSO Mongolia (National Statistical Office of Mongolia). Living Standards Measurement Survey (LSMS). Ulaanbaatar, Mongolia: National Statistical Office. http://www.nso.mn.

NSO Philippines (National Statistics Office, Republic of the Philippines). 2006. *Family Income and Expenditure Survey 2006*. Manila, the Philippines: National Statistics Office.

NSO Thailand (National Statistical Office Thailand). Thailand Socio-Economic Survey (SES). Bangkok, Thailand: National Statistical Office. http://web.nso.go.th/tnso.htm.

Oster, Emily, and Bryce Millet. 2010. "Do Call Centers Promote School Enrollment? Evidence from India." Draft, University of Chicago Booth School of Business.

Panda, Pradeep, and Bina Agarwal. 2005. "Marital Violence, Human Development and Women's Property Status in India." *World Development* 33 (5): 823–50.

Papua New Guinea Department of Education. 2009. "Achieving Universal Education for a Better Future: Universal Basic Education Plan 2010–2019." National Executive Council, Port Moresby, Papua New Guinea.

PISA (Programme for International Student Assessment). Organisation for Economic Co-operation and Development, Paris, France. www.pisa.oecd.org/.

Pritchett, Lant, and Lawrence H. Summers. 1996. "Wealthier Is Healthier." *Journal of Human Resources* 31 (4): 841–68.

Qian, Nancy. 2008. "Missing Women and the Price of Tea in China: The Effect of Sex-Specific Earnings on Sex Imbalance." *Quarterly Journal of Economics* 123 (3): 1251–85.

Quisumbing, Agnes. 1995. "Gender Differences in Agricultural Productivity: A Survey of Empirical Evidence." FCND Discussion Paper 5, International Food Policy Research Institute, Washington, DC.

Quisumbing, Agnes, and Kelly Hallman. 2005. "Marriage in Transition: Evidence on Age, Education, and Assets from Six Developing Countries." In *The Transition to Adulthood in Developing Countries: Selected Studies*, edited by Jere Behrman, Barney Cohen, Cynthia B. Lloyd, Nelly Stromquist. Washington, DC: National Academies Press.

Quisumbing, Agnes R., and John Maluccio. 2003. "Resources at Marriage and Intra-household Allocation: Evidence from Bangladesh, Ethiopia, Indonesia, and South Africa." *Oxford Bulletin of Economics and Statistics* 65: 283–327.

Quisumbing, Agnes R., and Keijiro Otsuka. 2001. "Land Inheritance and Schooling in Matrilineal Societies: Evidence from Sumatra." *World Development* 29 (12): 2093–2110.

Ross, Heidi, and Jinghuan Shi, eds. 2003. "Entering the Gendered World of Teaching Materials, Part II." *Chinese Education and Society* (May–June) 36: 3.

Sakellariou, Chris. 2011. "Determinants of the Gender Wage Gap and Female Labor Force Participation in EAP." Paper commissioned for *Toward Gender Equality in East Asia and the Pacific: A Companion to the* World Development Report. Washington, DC: World Bank.

Schleicher, A. 2008. "Student Learning Outcomes in Mathematics from a Gender Perspective: What Does the International PISA Assessment Tell Us?" In *Girls' Education in the 21st Century*, edited by M. Tembon and L. Fort. Washington, DC: World Bank.

Schultz, T. P. 1993. "Returns to Women's Schooling." In *Women's Education in Developing Countries: Barriers, Benefits and Policy.* Baltimore, MD: Johns Hopkins University Press.

———. 1997. "The Demand for Children in Low-Income Countries." In *Handbook of Population and Family Economics*: Vol. 1A, edited by M. R. Rosenzweig and O. Stark. Amsterdam: North-Holland.

———. 2006. "Does the Liberalization of Trade Advance Gender Equality in Schooling and Health?" IZA Discussion Papers 2140, Institute for the Study of Labor, Bonn.

Schultz, T. P., and Z. Yi. 1997. "The Impact of Institutional Reform from 1979 to 1987 on Fertility in Rural China." Photocopy. August 29.

Sen, Amartya. 1992. "Missing Women." *British Medical Journal* 304: 587–88.

Shel, T. A. 2007. "Gender and Inequity in Education Literature Review." Background paper prepared for the *Education for All Global Monitoring Report 2008 Education for All by 2015: Will We Make It?* http://unesdoc .unsco.org/imagers/0015/001555/155580e. pdf.

Shi, Jinghuan, and Heidi Ross, eds. 2002. "Entering the Gendered World of Teaching Materials, Part I." *Chinese Education and Society* 35 (5): 3–13.

Tararia, A. 2011. "WDR Rapid Qualitative Assessment on Gender and Economic Choice." Country Synthesis Report, Papua New Guinea. World Bank, Washington, DC.

Thomas, Duncan. 1995. "Like Father, Like Son, Like Mother, Like Daughter, Parental Resources and Child Height." Papers 95-01, RAND Reprint Series, *RAND* corporation, Santa Marica, CA.

Thomas, Duncan, and John Strauss. 1992. "Prices, Infrastructure, Household Characteristics and Child Height." *Journal of Development Economics* 39 (2): 301–33.

TIMSS (Trends in International Mathematics and Science Study). International Association for the Evaluation of Educational Achievement, Amsterdam, the Netherlands. http://rms.iea-dpc.org/.

Udry, Christopher. 1996. "Gender, Agricultural Production, and the Theory of the Household." *Journal of Political Economy* 104 (5): 1010–14.

UNESCO (United Nations Educational, Scientific and Cultural Organization) and the Vietnam Ministry of Education and Training. 2011. "National Textbook Review and Analysis from a Gender Perspective: Report of Findings." UNESCO, Hanoi, Vietnam.

UNFPA (United Nations Population Fund). 2009. "Recent Change in the Sex Ratio at Birth in Viet Nam." UNFPA Viet Nam, Hanoi.

UNICEF (United Nations Children's Fund). 2004. "What About the Boys?" In *The State of the World's Children 2004*, ed. UNICEF, 58–69. New York: UNICEF.

———. 2009. "Education for All Mid-Decade Assessment, East Asia and Pacific." UNICEF East Asia and Pacific Regional Office, Bangkok, Thailand.

UNIFEM, World Bank, ADB, UNDP, and DFID/ UK. 2004. *A Fair Share for Women: Cambodia Gender Assessment.* UNIFEM, WB, ADB, UNDP, DFID/UK. Phnom Penh.

UN DESA (United Nations Department of Economic and Social Affairs), Population Division. 2009. *World Urbanization Prospects: The 2009 Revision.* New York: United Nations. http://esa.un.org/unpd/wup/ index.html.

Velasco, E. 2001. "Why Are Girls Not in School: Perceptions, Realities and Contradictions in Changing Cambodia." UNICEF/Cambodia, Phnom Penh.

Waldron, Ingrid. 1998. "Factors Determining the Sex Ratio at Birth." In *Too Young to Die: Genes or Gender?*, 53–63. New York: United Nations.

WHO (World Health Organization) Global Infobase. World Health Organization, Geneva, Switzerland. https://apps.who.int/infobase/ report.aspx.

———. 2003, November. "Gender, Health and Tobacco." *Gender and Health.* http:// www.who.int/gender/documents/Gender_ Tobacco_2.pdf.

———. 2010. *World Health Statistics.* Geneva: WHO.

———. 2011a. "Action Needed to Reduce Health Impact of Harmful Alcohol Use." News Release, February 11. http://www .who.int/mediacentre/news/releases/2011/ alcohol_20110211/en/.

———. 2011b. *Global Status Report on Alcohol and Health.* Geneva: WHO. http://www .who.int/substance_abuse/publications/ global_alcohol_report/en/.

Williamson, N. 1976. *Sons or Daughters: A Cross-Cultural Survey of Parental Preferences.* Beverley Hills, CA: Sage.

Women, Business and the Law (database). World Bank, Washington, DC. http://wbl.worldbank .org/data.

World Bank. 2001. *Engendering Development: through Gender Equality in Rights, Resources, and Voice.* Washington, DC: World Bank.

———. 2008. "Analysis of the Impact of Land Tenure Certificates with Both the Names of Wife and Husband in Vietnam." World Bank, Washington, DC. https://openknowledge .worldbank.org/handle/10986/7810.

———. 2010. *"... and Then She Died": Indonesia Maternal Health Assessment.* Washington, DC: World Bank. https://openknowledge .worldbank.org/handle/10986/2837.

———. 2011a. "Toward a Healthy and Harmonious Life in China: Stemming the Rising Tide of Non-Communicable Diseases. Human Development, East Asia and the Pacific." World Bank, Washington, DC.

———. 2011b. *Vietnam Country Gender Assessment.* Washington, DC: World Bank. http://www-wds.worldbank.org/external/ default/WDSContentServer/WDSP/IB/2011 /11/14/000333038_20111114003420/ Rendered/PDF/655010WP0P12270sessment .0Eng.0Final.pdf.

———. 2011c. *World Development Report 2012: Gender Equality and Development.* Washington, DC: World Bank.

World Development Indicators (WDI) database. World DataBank, World Bank, Washington, DC. http://data.worldbank.org/data-catalog/ world-development-indicators.

Zeng, Yi, Tu Ping, Gu Baochang, Xu Yi, Li Bohua, and Li Yongqing. 1993. "Causes and Implications of the Recent Increase in the Reported Sex Ratio at Birth in China." *Population and Development Review* 19 (2): 283–302.

Gender and Economic Opportunity | 3

In the dynamic East Asia and Pacific region, many countries have undergone structural transformations that have shifted the balance of economic opportunities away from rural areas and toward urban areas. These growth processes have opened up nonfarm economic opportunities for men and women in the region, particularly among younger cohorts. The educational attainment and health outcomes of women, particularly younger women, have been catching up to those of men during this period of growth, as discussed in chapter 2. Along some dimensions of gender equality in economic opportunities, there has been substantial progress over the last two decades. For example, the evidence suggests that the labor market participation decisions of younger women resemble those of their male counterparts in the region. However, many other indicators demonstrate the substantial challenges to be overcome to close the gender gap in access to economic opportunities. The type of work women do remains very different from that of men, and their remuneration for these tasks is lower. Women of all ages are more likely than men to be in poorly remunerated occupations and sectors, women are paid less than men for similar work, and gender norms in the

division of labor within the household imply that women work longer hours than men, although fewer of those hours are devoted to remunerated activities.

Reducing gender inequalities in economic opportunities can improve economic outcomes in multiple ways. First, reducing employment segregation (the unequal distribution of male and female workers across occupations and sectors) will reduce efficiency losses associated with the misallocation of talent (Anker 1998; Morrison, Raju, and Sinha 2007). Men and women often choose occupations on the basis of norms, gender stereotypes, and sometimes prejudice, rather than on the basis of earnings or job match (Klasen and Lamanna 2009). Encouraging workers and employers to make labor choices on the basis of their skills, competencies, and inherent ability is likely to raise productivity and may have a positive impact on economic growth by increasing the size of the labor force as well as by expanding the pool of managerial and innovative talent in the economy (Esteve-Volart 2004). As noted in chapter 1, estimates for East Asian and Pacific countries suggest that output per worker could be 7 to 18 percent higher if female entrepreneurs and workers were to work in the same sectors, types of jobs, and activities as men (Cuberes and Teignier-Baqué 2011). Second, empirical evidence from other developing regions suggests that reducing gender inequalities in access to productive inputs can increase overall production by increasing the productivity of female-run farms and enterprises (Goldstein and Udry 2008; Quisumbing 1995; Udry 1996).

This chapter examines differences in productive opportunities for men and women within and across countries in the East Asia and Pacific region and explores the economic and institutional factors that have determined how those opportunities have evolved over time. Because education enrollment and attainment are growing and labor markets are changing, this chapter differentiates between the experiences of older and younger generations. It also uses examples from both high-income and low-income countries across the globe to understand these trends.

Gender differences in the economic sphere manifest themselves in several indicators, including differences in labor force participation, in the time spent on productive and reproductive activities, in the sectors in which men and women work, in the tasks and occupations that they do within those sectors, and in the types of firms that employ them. Differences in these indicators contribute to and are themselves determined by gaps in the earnings of men and women. Women are often paid less than men for the same work, and female-run enterprises and farms typically produce less than those of men.

Gender inequalities in economic opportunities are driven by multiple interacting factors. In this chapter we examine how gender differences in access to human and physical capital, technologies, and government services; gender stereotypes; and gender roles explain gender inequality in economic opportunities. These factors are, in themselves, determined by the household, market, and institutional environment in which preferences and gender roles are learned and in which education and time allocation choices are determined.

Three main messages for the East Asia and Pacific region emerge in this chapter:

- In some dimensions, such as labor force participation, gender inequalities in the economic sphere in the region have improved in recent decades and are narrower than those in other regions; however, multiple dimensions of inequalities remain that will require concerted effort to change, such as persistent gender wage and productivity gaps. In several areas, policies and public investment can help to ease the constraints on women and support them in their multiple roles as entrepreneurs, farmers, wage and salaried workers, mothers, and caregivers.
- The constraints faced by women vary across sectors of the economy and also by country according to the institutional environment. Some common themes emerge, however. In the agricultural sector, female farmers' access to productive capital, technologies, and governmental services is lower than that of male farmers; improving access to these inputs is tantamount to increasing productivity. In the nonfarm sector, female-run enterprises are smaller and in different sectors than male-run enterprises. The constraints and productivity differences of female enterprises are predominantly attributable to their small size and to the sectors in which they are found. Constraints in all countries are likely to be greatest at start-up, when access to finance and entrepreneurial skills are likely to be important determinants of sector and initial scale of enterprise. In the labor market, gender-based employment segmentation—or sorting across types of firms, industries and sectors—affects both the wages women earn as well as productivity in the economy, particularly when men and women sort into occupations on the basis of gender rather than skill. Policies that encourage both men and women to think outside of gender-based occupational norms will be productivity enhancing and likely have positive repercussions for female empowerment.

- As in other parts of the world, women in the region have multiple roles and will require more support to manage competing demands for their time from productive, reproductive, and community management activities as development proceeds and greater nonfarm sector opportunities emerge. Nearly all countries in the region see declines in the female labor force participation of young mothers. Furthermore, to manage their dual roles, women are often obliged to enter into different occupations and work fewer hours than men, both of which are found to have negative implications on their wages and earnings. Policies that support women in juggling the competing demands of home and market work will be required, particularly as women start moving into "male" occupations that have not traditionally allowed them the flexibility to lead their dual lives.

The chapter is structured as follows. The first section explores whether growth is sufficient to reduce gender inequalities in productive activities, drawing upon evidence from high-income countries within and outside the region. The second section describes the current situation with regard to gender inequalities in economic opportunities in East Asian and Pacific countries. The third section examines the determinants of the most persistent gender inequalities, and the fourth section concludes by briefly examining policy directions, a discussion that is taken up again in chapter 6.

Limited effects of growth on gender gaps in economic opportunity

The empirical literature suggests that economic development alone is not sufficient to narrow gender differences in earnings (Blau and Kahn 2003; Hertz et al. 2009; Tzannatos 1999). Evidence from East Asian and Pacific countries suggests instead that social, political, and cultural factors are as important as economic development in

determining gender wage gaps and, indeed, that these factors interact with the development process to determine the degree to which growth narrows gender inequalities (Meng 1996).[1]

In high-income countries, the age profile of women's economic participation and the sensitivity of their participation to life-cycle factors have changed with development. Between 1950 and 2010, female labor force participation in Hong Kong SAR, China; Japan; the Republic of Korea; and Singapore increased substantially across all age cohorts. For example, in Korea, women's labor force participation has increased monotonically over time (figure 3.1). Similar changes were seen in the United States, particularly among married women (Juhn and Potter 2006). Furthermore, the decline in labor force participation by women in their early 30s has become less severe in these high income countries. Rising female labor force participation with development has been attributed to a number of demographic factors, including later marriage and childbirth and lower fertility rates.

However, income growth in these countries has not been enough to eliminate gender

FIGURE 3.1 Female labor force participation in the Republic of Korea rose for women of all ages between 1960 and 2005

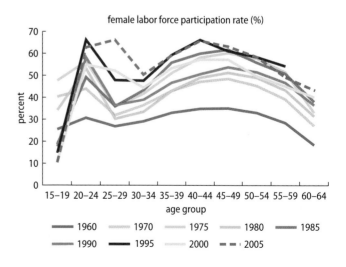

Source: ILO Key Indicators of the Labour Market (KILM) database: Korea Labor Force Survey and Population Census.

inequalities in all dimensions of economic opportunities. Despite a substantial increase in young women's labor force participation, sharp gender differences in employment status, occupational status, and wage rates continue to exist. In Japan, women are more likely than men to be in nonregular employment, including temporary work, contract work, and part-time work (Hill 1996; Yu 2002). Women continue to leave the workforce in substantial numbers during their childbearing years, even if only temporarily. The greatest drop in female labor force participation rates comes for married women in their mid-20s to early 30s when they start having children and assuming greater family responsibilities (Hill 1996; Lee, Jang, and Sarkar 2008; Miller 1998, 2003; Sasaki 2002).

Welfare, child care, and tax reforms; legislative changes; and women's movements have helped to narrow gender differences in economic opportunities in high-income countries in the East Asia and Pacific region and across the world. The United States provides a good example of how the combination of economic forces, social changes, and legal reforms has transformed women's labor force participation. Explanations of changes in female labor force participation include demand-side factors that shifted women's market wage as well as supply-side factors that reduced women's opportunity costs of working (Fang and Keane 2004; Galor and Weil 1996; Greenwood, Seshadri, and Yorukoglu 2005; Weinberg 2000). Since the 1980s, Korean labor and family law has sought to improve women's status within marriage and the family and to eliminate gender bias in other areas, including in labor law (Kim 2005). In addition, the Korean government has undertaken several measures to support married women with their child-care responsibilities, such as reforming maternity and paternity leave and expanding expenditures on child-care facilities. Women's and labor organizations have played an important role in advocating for legislation that reduces gender discriminatory practices (Kim 2005). Legislative

reforms to level the playing field for women in the labor market have been difficult to enforce in Japan, however (Lam 1992; Liu and Boyle 2001; Miller 2003).

Gender differences in economic activity

Gender differences in the economic sphere manifest themselves in a number of indicators. This section focuses on examining gender differences in labor force participation, in earnings, and in the labor market segregation of men and women.

Labor force participation

The East Asia and Pacific region is characterized by high female labor force participation on average, but also by substantial variation across and within countries. Labor force participation is defined as all productive work, whether as wage workers or as unpaid family workers. Female labor force participation in the East Asia and Pacific region is the highest in the developing world—70 percent of females were participating in labor market activities in 2008 (see figure 1.14 in chapter 1). The participation gap (the difference between the participation of men and women in paid or unpaid market-oriented work) was approximately 15 percentage points in the region. In comparison, the female labor force participation rate in the Latin America and the Caribbean region and Europe and Central Asia region was 55 and 58 percent in 2008, respectively, and their gender gaps in labor force participation were 27 and 16 percentage points, respectively. Box 3.1 discusses the ways in which high female labor force participation may, or sometimes may not, be an indicator of enhanced welfare for women.

The average rates of female labor force participation vary substantially within the region (see figure 1.15 in chapter 1). Participation and gender gaps in some parts of the region are among the highest in the world, whereas in others they are among the lowest. For example, in Fiji, Samoa,

BOX 3.1   Is higher female labor force participation always a good thing?

Studies from across the world indicate that increased female labor force participation and control by women over household resources are associated with their enhanced well-being and status. For example, a study on India shows that female mortality is lower where female labor force participation rates and earnings are higher (Murthi, Guio, and Drèze 1995). At the same time, although higher household income is associated with higher levels of welfare for household members, the marginal impacts are considerably greater when income is in the hands of the mother. As women gain more control over expenditure decisions, a larger share of household resources tends to be devoted to children's education, health, and nutrition (Thomas 1990; Thomas and Strauss 1997). Women's borrowing from microfinance programs also appears to improve child welfare more substantially than does borrowing by men, including significant improvements in children's nutritional levels and increases in the school enrollment levels of both boys and girls. However, labor force participation is not always empowering and may be a consequence of poverty, which pushes women into low-paying jobs with poor working conditions and job insecurity. Jobs may also be exploitative and may reinforce traditional gender roles (Elson and Pearson 1981).

and Malaysia, participation rates in 2008 were an order of magnitude lower than the average for the region, with approximately 55 percent of working-age females in these countries participating in the labor force. Female labor force participation is highest in countries where the state has put a priority on gender equality, for example, in socialist or formerly socialist countries, as well as in agrarian and rural economies where high rates of participation may be a consequence of poverty.

Substantial gender differentials in participation have also been noted within countries in the region. National averages of labor force participation fail to capture important differences in participation across rural and urban areas, as well as regional variation within countries. In all countries within the region, rural labor force participation is higher than urban labor force participation, for both males and females (figure 3.2). Participation gaps in all countries, apart from Mongolia and the Philippines, are substantially greater in urban areas than in rural areas. In Vietnam, higher levels of female participation and lower gender differentials are found in poorer and often low-productivity areas, where female participation in employment to supplement household income is necessary (Pierre 2011).

Even in countries with high overall participation rates, female labor force participation rates decline during childbearing years and old age. Average female labor force participation rates hide important variation over women's life cycles due to factors such as marriage and childbearing, whereas male participation rates remain fairly constant throughout their life cycles. The gaps between men and women in labor force participation, earnings, and job composition increase after marriage and childbearing in many East Asian and Pacific countries, as they do globally. In the United States, for example, young single men and women have more similar labor participation, earnings, and career profiles than do married men and women. In many East Asian and Pacific countries, life-cycle patterns of labor force participation by birth cohorts reveal specific trends that are not apparent in the average rates of participation, especially during periods of rapid growth and structural transformation. Rapid changes in education levels across cohorts, the availability of new economic opportunities in growing sectors, and urbanization mean that young women are more likely than older women to participate in the labor force, and

FIGURE 3.2   Labor force participation is greater in rural areas than in urban areas for both males and females

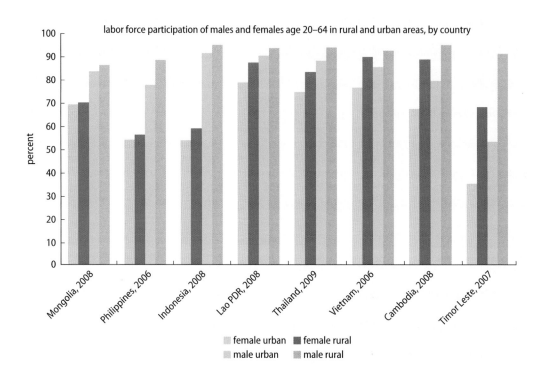

Source: Data generated using country level household data, with the exception of Indonesia which uses labor force data.
Note: The countries are sorted by the gap between females in rural and urban areas.

to participate in different sectors (Mammen and Paxson 2000).

Female labor force participation is more sensitive to life-cycle factors, such as marriage and childbearing, in some countries than others. The birth cohort life-cycle participation patterns of women vary substantially within the East Asia and Pacific region and have also exhibited considerable change over time. Three patterns can be distinguished in the region (Horton 1996).[2] The "plateau" pattern exhibits relatively flat female labor force participation until women reach their early 50s, and then declines into old age. The "double-peaked" or "M" pattern is generally observed in more industrialized countries: it is characterized by high participation in the labor market prior to marriage and childbearing, with a subsequent return to the labor force once children

are older. A "single-peaked" pattern exhibits higher rates of participation at younger ages, which then fall after the peak. This is a more extreme version of the double-peaked pattern and characterizes a labor market in which few women return to work after marriage and childbearing. In the region, this pattern can be seen only in Malaysia, where it continues despite increases in female participation between 1980 and 2010 (figure 3.3). The plateau pattern, in contrast, describes a situation in which female labor force participation is less sensitive to life-cycle effects than in the single- and double-peaked patterns. The plateau pattern can be seen in countries with political regimes that encouraged women's economic independence, such as China, Mongolia, and Vietnam.

Labor force participation decisions of rural and urban women in many countries

FIGURE 3.3   Female labor force participation rates in Malaysia have risen over time among 20- to 55-year-olds, but continue to decline during child-bearing years

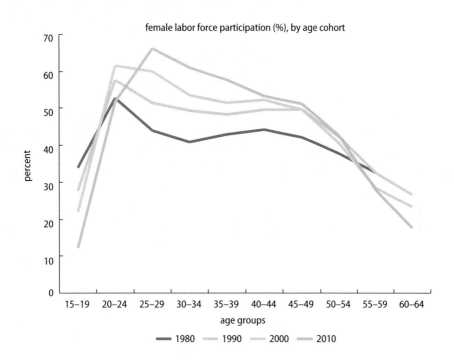

Source: LABORSTA Internet (1980, 1990, and 2000 data), CEIC Data (2010 data).

display different responses to life-cycle changes, and their sensitivity to life-cycle changes may evolve differently over time. In Indonesia, for example, urban women are more likely than rural women to leave the labor force when they have children, and they also leave the labor force for longer. Patterns of participation have changed substantially since 1990, however, and have evolved differently in rural and urban areas. Although urban women with young children in 2008 are still less likely to work than their rural counterparts, urban women display less sensitivity to childbearing decisions than they did in 1990. In contrast, rural women in 2008 are *more* likely to reduce their labor supply when having children and to leave the labor force for longer than rural women did in 1990.

East Asian and Pacific countries with higher rates of female labor force participation overall have smaller declines in participation during childbearing years, although predominantly in urban settings. In all countries in the region, women with young children are substantially less likely to participate in the labor market than men with young children and than women without young children. In some countries, such as Mongolia and Vietnam, a reduction in labor market participation is seen in either urban or rural areas alone, whereas in others, such as Indonesia and the Philippines, it is seen in both rural and urban areas. Data from the 2009 Vietnam Labor Force Survey (GSO [Vietnam]) show that rural Vietnamese women between ages 25 and 35 with a child under the age of 2 are a third less likely to participate in the labor market than men of the same age with a child. In comparison, the participation of women between ages 25 and 35 without a child is fairly similar to that of men in this age group (figure 3.4). In Vietnam, the effect of children on labor force participation is smaller and of shorter duration than in Mongolia. In Vietnam, the participation gap disappears by the time the child reaches school age, whereas in Mongolia the participation gap continues to be substantial even for older children.

In countries where female labor force participation is low on average, it also tends to be highly sensitive to marriage and childbearing years. Indonesia, Japan, Korea, and Malaysia have historically displayed a single- or double-peaked pattern; that is, women participate in the labor market prior to marriage and childbearing and, in the case of a double-peaked pattern, eventually return to the labor force once the children are older. These countries also have lower average female labor force participation than countries in the region that exhibit a plateau pattern. In Indonesia and

the Philippines, in marked contrast to the examples of Mongolia and Vietnam above, female-male participation gaps are substantial even among women who do not have children, and the gaps widen with age.

Earnings

Gender gaps in wages and earnings are found in all countries in the region and in all sectors, with rare exceptions. Male and female wages differ in multiple sectors and settings across the region, with female agricultural wage workers earning less than male agricultural wage workers, and female urban wage workers earning less than their male counterparts.[3] Female entrepreneurs and farmers often display lower revenues and profits than their male counterparts.

A substantial body of evidence has accumulated over the past three decades to suggest that women are paid less for similar work across the world as well as in the region. A cross-country comparison of wages indicates that women earn between 70 and 80 percent of the wages men receive for similar work (Hausmann, Tyson, and Zahidi 2010).[4] Evidence from across the world suggests that economic development is not sufficient to reduce the gender earnings gaps (Blau and Khan 2003). Even in high-income countries in the East Asia and Pacific region—notably Japan and Korea—the average woman earns less than half the wage of the average man. Lower-income countries, including Mongolia, Lao People's Democratic Republic, Papua New Guinea, and Vietnam, have lower gender wage gaps, on average, than many richer countries.

Within the agricultural, manufacturing, service, and government sectors, women earn less than men on average (figure 3.5), although the ratio of male-to-female earnings varies substantially across sectors as well as across countries. Female-to-male wage ratios tend to be lower in the agricultural and manufacturing sectors and higher in the service sector and in government, with some exceptions.

FIGURE 3.4   Women in rural Vietnam with children under age 2 are substantially less likely to participate in the labor market than those without young children

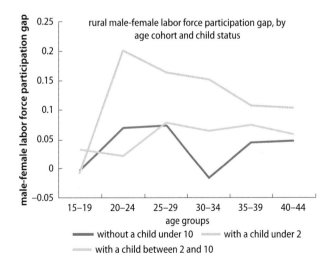

Source: World Bank staff estimates using Vietnam Labor Force Survey (GSO Vietnam), 2009 data.

Gender wage gaps in the East Asia and Pacific region are greatest between men and women with the lowest educational endowments and in the lowest paying occupations. Several studies have indicated that the gap is wider at the bottom end of the wage distribution than at the top (Chi and Li 2007; Li and Song 2011; Sakellariou 2011). In contrast, in Organisation for Economic Co-operation and Development (OECD) countries, gender wage gaps tend to be wider at the top than at the bottom of the wage distribution (Albrecht, Björklund, and Vroman 2003; Arulampalam, Booth, and Bryan 2007; de la Rica, Dolado, and Llorens 2005).

In the nonfarm sector, the performance of male and female firms around the world varies substantially, as measured by total factor productivity, labor productivity, profitability, and capital intensity (Sabarwal, Terrell, and Bardasi 2011). In the East Asia and Pacific region, formal sector firms with at least one female owner do not have significantly lower sales than those with no female owners, with the exception of firms in the Philippines (figure 3.6). However, in the informal sector, output per worker in female-owned enterprises is substantially lower than that in male-owned enterprises. Data from Indonesia and Vietnam suggest that in the informal sector, gender earnings gaps are more pronounced than in the formal sector: female-owned enterprises with fewer than five employees generate only approximately 60 to 70 percent of the output per worker generated by male-owned enterprises (figure 3.7).

Labor market segregation

Multiple studies find that men and women work in different sectors, industries, occupations, and types of firms across the world (Anker 1998; Boserup 1970). As can be seen in figures 3.8 and 3.9, across the region women are overrepresented in unpaid family work, particularly in rural areas, and are slightly more likely to be employed in the informal sector.[5,6] Within the informal sector, women are more likely to be

FIGURE 3.5   Women earn less than men in the majority of East Asian and Pacific countries and in all sectors of the economy

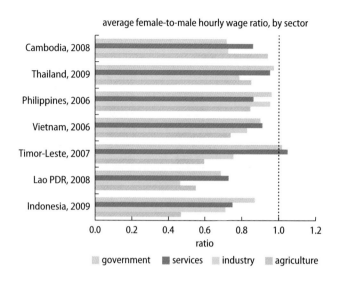

Source: World Bank estimates using household income and expenditure surveys. Cambodia Socio-Economic Survey (CSES) (NIS Cambodia), 2008 data; Indonesia National Socioeconomic Survey (SUSENAS) (BPS Indonesia), 2009 data; Lao Expenditure and Consumption Survey (LSB [Lao PDR]), 2008 data; Philippines Family Income and Expenditures Survey (NSCB 2006); Thailand Socio-Economic Survey (SES) (NSO Thailand), 2009 data; Timor-Leste Survey of Living Standards (NSD Timor-Leste), 2007 data; Vietnam Household Living Standards Surveys (VHLSS) (GSO Vietnam), 2006 data.

FIGURE 3.6   Male- and female-owned firms in the formal sector do not display substantial differences in productivity

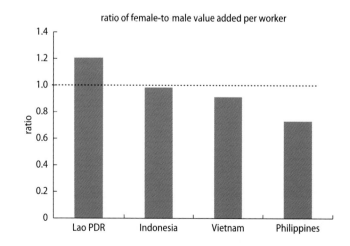

Source: World Bank estimates using Enterprise Surveys, 2006–11 data.
Note: Data are for small (19 employees or less), medium (20–99 employees), and large (100 and more employees) firms in the formal sector. Female-owned firms are defined as those with at least one female owner. Value added is measured as revenues minus material expenses.

FIGURE 3.7   In the informal sector, gender-based differences in productivity are more pronounced than in the formal sector

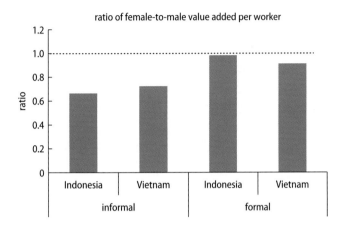

Source: Data for informal firms are based on Indonesia Family Life Survey 2007/2008; VHLSS (GSO Vietnam), 2008 data. Data on formal firms are from Enterprise Surveys.
Note: Based on a sample of firms with fewer than five employees.

FIGURE 3.8   Women are more likely than men to work as unpaid family workers

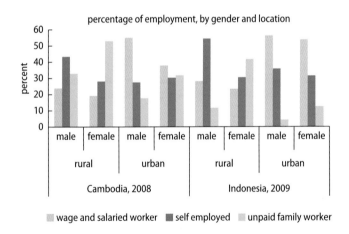

Source: World Bank estimates using socioeconomic surveys: CSES (NIS Cambodia), 2008 data; Indonesia SUSENAS (BPS Indonesia), 2009 data.

own-account workers (self-employed workers working by themselves) and subcontracted workers, whereas men are more likely to be employers or paid employees of informal enterprises (World Bank 2011g). Globally, women are more likely than men to work part-time (ILO 2010).

Across the world, women in the nonagricultural workforce are more likely to work in social and communal services (such as education and health) and in commerce and restaurants, whereas men are more likely to work in transport, construction, public administration, and manufacturing (ILO 2010). Similar trends are found in East Asian and Pacific countries. For example, men in Cambodia and Indonesia are disproportionately found in the manufacturing, transport, construction, and public administration sectors, and women are found in manufacturing, education, health and social services, and commerce (figure 3.10). Estimates from household survey data suggest that, in Fiji and Vietnam, men are more likely to work as professionals and managers and as plant and machine workers, whereas women are more likely to work as technicians, sales workers, and clerks, and in elementary occupations.[7] Women also make up a smaller fraction of the public sector workforce than men in Cambodia, Thailand, Timor-Leste, and Vietnam. The Philippines, however, displays much higher levels of occupational segregation and also has a higher fraction of women in the public sector than men, a trend that is likely to be related to the relatively high investment in education by Filipino women.

Within the manufacturing sector, women are more likely to be found in industries such as textiles and food processing, and are also found in large and export-oriented firms. Data from country enterprise surveys (2002–05) indicate that, in all developing regions, the fraction of full-time female workers is greater in export-oriented firms. The East Asia and Pacific region has the second-largest fraction of full-time female workers, after Europe and Central Asia. However, within export-oriented firms, women are also more likely than men to be temporary workers in Cambodia, China, and Indonesia, although in Thailand men are more likely to be temporary workers in all firms (see figure 1.18 in chapter 1).

Women are underrepresented in managerial positions and positions of power in all sectors, from government to manufacturing (Anker 1998). The East Asia and Pacific region performs relatively well compared to other regions in terms of the fraction of firms with a top female manager (see figure 1.16 in chapter 1). However, women remain less likely than men to serve as managers and directors. The share of female directors ranges from 10 percent in the Philippines and 7 percent in China and Thailand to 5 percent in Indonesia and under 2 percent in Korea and Japan (CWDI 2010). Globally, only one country in the world has succeeded in having more than 30 percent female representation on corporate boards (namely, Norway), and one-third of countries have female board representation over 10 percent (CWDI 2010).

Segregation goes beyond the wage labor market, with smaller female-owned and female-managed firms located in less-capital-intensive sectors than male-owned and male-managed firms. Female-owned and female-run enterprises are, on average, smaller than male-run enterprises in terms of the number of employees hired, sales, and profits (Aterido and Hallward-Driemeier 2009; Badiani and Posadas 2011; Costa and Rijkers 2011; Sabarwal, Terrell, and Bardasi 2011). Among formal sector firms in the East Asia and Pacific region, enterprise survey data suggest that small firms are more likely than medium and large firms to have a top female manager (see figure 1.19 in chapter 1). Evidence by ownership displays a more mixed picture. In Indonesia, Lao PDR, Mongolia, Timor-Leste, Tonga, and Vanuatu, this pattern still holds, but evidence from the Philippines, Samoa, and Vietnam suggests that female owners are not disproportionately represented among small firms (figure 3.11).[8] This pattern may be the result of more open cultural norms regarding women's role in business. However, the lower levels of female management relative to ownership suggest that women may still have less control or representation within firms.

Female-run enterprises are also more likely to be found in labor-intensive sectors, such as

FIGURE 3.9 **Women are slightly more likely to be employed in the informal sector than men. Rural-urban differences in informality are greater than gender differences within rural or urban areas**

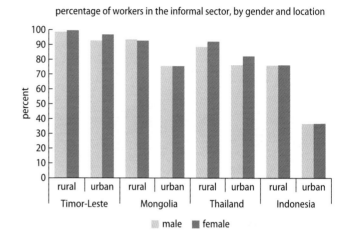

Source: World Bank estimates using Indonesia SUSENAS (BPS Indonesia), 2009 data; Mongolia Living Standards Measurement Survey (LSMS) (NSO Mongolia), 2007–08 data; Thailand SES (NSO Thailand), 2009 data; Timor-Leste Survey of Living Standards (NSD Timor-Leste), 2007 data.
Note: The informal sector is defined using information on an individual's occupation and sector of employment.

services and trade, than in capital-intensive sectors (Klapper and Parker 2010). For example, a study commissioned by this report finds that women are more likely to be found in manufacturing, food sales, and food preparation in Indonesia and less likely to be found in transport, construction, and other services (Badiani and Posadas 2011). Similar patterns have been observed in other countries in the region, including Lao PDR (Davies and Record 2010), Mongolia (World Bank 2011c), and Vietnam (Bjerge and Rand 2011).

The enterprises that women work in are also less productive and capital intensive. In Indonesia, the sectors that women are more likely to be employed in—food and garments production—are among the least capital intensive and productive sectors (figure 3.12). By contrast, the transportation and other service sectors—where male entrepreneurs are most likely to be found—has higher productivity and capital intensity.

Having a female presence in management may have positive implications for workers, however, even if productivity per worker is lower. Female-run firms have been found to

FIGURE 3.10   Men and women work in different sectors throughout the East Asia and Pacific region

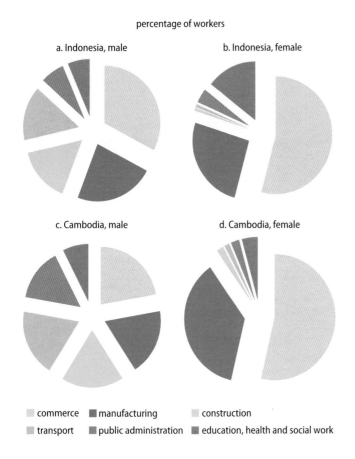

percentage of workers

a. Indonesia, male

b. Indonesia, female

c. Cambodia, male

d. Cambodia, female

- commerce
- manufacturing
- construction
- transport
- public administration
- education, health and social work

Source: CSES (NIS Cambodia), 2008 data; World Bank estimates using Indonesia SUSENAS (BPS Indonesia), 2009 data.

influence factors beyond productivity—from the provision of benefits to gender-sensitive policies.[9] In Vietnam, workers in female-run small and medium enterprises are more likely to receive fringe benefits in addition to wage compensation (Rand and Tarp 2011). Female owners are more likely to provide health and social insurance and to allow sick, vacation, and maternity leave with pay.

Labor market sorting has been found to contribute to the gender wage and earnings gaps.[10] Differences in occupational and industrial sorting explain a greater fraction of gender wage gaps in the East Asia and Pacific region and across the world than differences in human capital (Ñopo, Daza,

and Ramos 2011; Sakellariou 2011). Among entrepreneurs, differences in industry and size of enterprise have been found to explain a substantial fraction of the raw performance gaps in profits and revenues among male and female firms.[11]

What determines gender inequality in economic opportunities?

Men and women differ in their economic activities in a number of respects. These differences in activities may represent differences in choices, but they may also represent differences in the constraints that men and women face in their working lives. This section explores the economic, demographic, and cultural factors that determine current gender inequalities in access to economic opportunities in the East Asia and Pacific region.

Determinants of labor market participation

Female labor force participation is affected by growth processes and development. As development occurs, changes in household income, education, wages, marital, and fertility choices affect participation. The evolution of market opportunities alters the types of jobs present in the economy, as well as the relative demand for skills. The institutional framework of society affects and is affected by the economic participation of women during the process of development; that is, gender norms, expectations, and perceptions within the household and society affect female labor force participation and are also likely to be affected by its evolution.

Female labor force participation and its determinants vary over the life cycle. Early in their careers, women are similar to men in terms of their decisions to join the labor market, but their participation begins to differ as their domestic responsibilities increase. This change is partly due to the increased time that women devote to household activities, such as housework and child rearing, as their marital and familial status changes, but it is also a reflection of differences in the roles of and expectations of married and single women. Younger women in the region are investing more in education and hence have delayed their entry into the labor market, in both rural and urban areas. Married women appear to take substantial time out of the workforce for raising children, but the same is not true for men. Furthermore, large labor force participation gaps open up toward the end of women's careers, in part as a result of labor market regulations such as gender-differentiated retirement policies.

The stage of development of countries and their institutions

Evidence from across the world suggests that, as countries develop, female labor force participation displays a U-shaped trajectory.[12] Female labor force participation usually declines as incomes rise and opportunities in the labor market become less attractive to female workers; it then increases again when more attractive employment opportunities emerge (Bloom et al. 2009; Chaudhuri 2009; Goldin 1995; Sinha 1967, cited in Mammen and Paxson 2000; Tam 2011). In poor, agricultural economies, female participation tends to be high because agricultural work and family responsibilities can easily be combined. However, in middle-income countries dominated by the manufacturing and service sectors, female participation declines in part because most new jobs are difficult to combine with family responsibilities. Female participation rates are higher in high-income countries that have large service sectors and a highly educated workforce. This finding holds both across and within countries over time (Fatima and Sultana 2009; Fuwa 2004; Juhn and Ureta 2003; Tansel 2001).[13]

The stylized U-shaped relationship between female labor force participation and economic development holds for countries in the East Asia and Pacific region, as well as globally. Figure 3.13 depicts the relationship between economic development (as captured by income per capita) and female labor force participation across the globe between 1980

FIGURE 3.11   The pattern of female ownership by firm size varies across countries

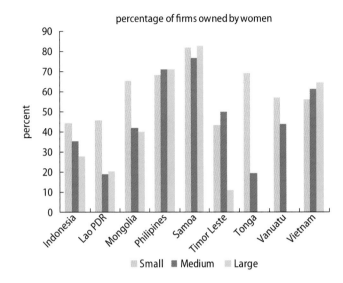

Source: World Bank estimates using Enterprise Surveys (database), 2006–11 data.
Note: Survey data are for small (less than 20 employees), medium (20–99 employees), and large (100 and more employees) firms in the formal sector. Female-owned firms are defined as those with females among the owners.

FIGURE 3.12   In Indonesia, female-led enterprises are clustered in lower-productivity and capital-intensive industries

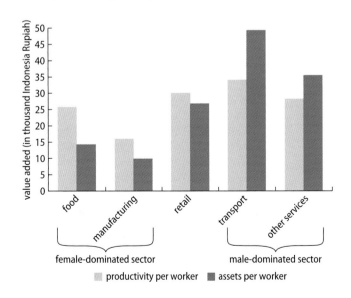

Source: World Bank staff estimates using Indonesia Family Life Survey 2007/2008.
Note: The graph shows productivity and assets per worker in five industries for firms with fewer than five workers. Productivity is measured by value added.

FIGURE 3.13   **Female labor force participation is high by global standards but also varies substantially across the region**

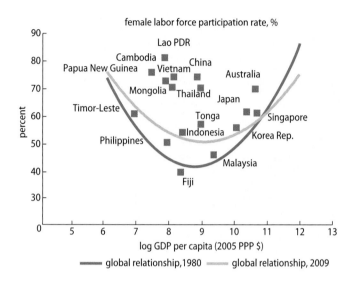

Source: World Bank estimates using labor force data from the ILO KILM database and purchasing power parity (PPP)-adjusted GDP per capita from the Penn World Table.
Note: GDP = gross domestic product, PPP = purchasing power parity. The data shown for each country reflect the 2009 data, and the estimated U-shaped relationships use data from across the world.

and 2009. The position of the East Asian and Pacific countries relative to the rest of the world is depicted in red. Two key features emerge. First, the U-shape pattern holds across countries for both periods of time. Second, the participation rate associated with each level of development has increased between 1980 and 2009, as can be seen by the upward shift in the U-shaped pattern over time. In the East Asia and Pacific region, Vietnam and China have substantially higher levels of female labor force participation than the world average relative to their income levels, whereas participation is near the world average in countries such as Indonesia and the Philippines. The pattern of female labor force participation seen in parts of the Pacific partly reflects economic structure, whereby labor force participation is higher in agriculture-based economies. For example, although Fiji has lower rates of participation relative to its income level, the rates of participation in Papua New Guinea and Vanuatu are substantially higher.

Gender norms and expectations
An important question is why labor force participation varies so much between countries with similar per capita income levels, as can be seen in figure 3.13. Gender norms strongly influence labor force participation rates and help to explain the variation in participation across countries that is not explained by the level of economic development. Societal perceptions of women in the workplace and gender norms strongly influence labor force participation decisions (Antecol 2000; Fernández 2010; Fernández and Fogli 2005; Fernández, Fogli, and Olivetti 2004). Countries in which strong socioreligious views exist about women's role in the public sphere, including the workplace, have been found to have lower female labor force participation rates (Psacharopoulos and Tzannatos 1989).

Whether a woman is entitled to make her own choices in the economic sphere, such as entering the workforce or starting a business, varies substantially from country to country. For example, in Vietnam, young single women are increasingly choosing to migrate for employment and to commute long distances (World Bank 2011a). However, in rural Morobe Province in Papua New Guinea, young women are not allowed by their families to migrate out of the village because of the fear of early marriages, which are regarded as taboo (World Bank 2011a).

In countries with large male-female participation gaps throughout the life cycle—such as Indonesia, Malaysia, and the Philippines—female labor force participation is considered socially and culturally acceptable as long as it does not interfere with women's primary role as wives and mothers. For example, public gender discourse in Indonesia and Malaysia places particular emphasis on motherhood and child care as a "woman's true vocation" (Blackburn 2001, 2004; Stivens 2006; White 2006). Political and institutional evolution also affects the scope for women's participation in education and in the workplace. Women's labor force participation is substantially higher in socialist and ex-socialist countries, such as China, Mongolia, and Vietnam, than in others.

Legislation can codify social norms into discriminatory labor practices. In Korea, marriage bars to private and public employment were common until the 1980s (Hill 1996). In Mongolia, women retire approximately 10 years earlier than men; this practice is, in part, attributable to a lower retirement age for women (World Bank 2011c). In Mongolia, these differences in retirement ages have contributed to the female-male participation gap rising by approximately 20 percentage points between the ages of 50 and 60.

However, gender roles and relations within households do change over time, particularly in evolving environments. In China, the economic reforms of recent decades have increased the range of opportunities available to both men and women in paid employment, as the structure of the economy has moved away from predominantly agrarian with a capital-intensive heavy industry sector toward labor-intensive light industry and services (Hughes, Maurer-Fazio, and Zhang 2007). However, this transition has been argued to have created new obstacles for women: the state has retreated from its former commitments to gender equality and to strong enforcement of workplace protections for women, thus allowing the reemergence of traditional patriarchal values (Entwisle and Henderson 2000).

Individual and household-level factors: Income, education, marriage and children
Changes in the characteristics of women and households—notably, changes in educational attainment, changes in the demographic profile of the population, and growth in household incomes—do not, for the most part, explain changes in the male-female gap in labor force participation in many countries in the East Asia and Pacific region during the past decade (Sakellariou 2011).

Rising education levels among girls in countries across the region have led to a substantial decline in girls' workforce participation rates. For example, women's participation in rural

Vietnam has declined from almost 89 percent in 1998 (similar to men) to about 81 percent in 2008 (versus 84 percent for men). Most of the decline occurs in rural areas and is accounted for by rising participation in education. Given the corresponding though smaller decline in male participation, the overall male-female participation gap increased from approximately 1.5 percentage points in 1998 to 4.5 percentage points in 2008 (Sakellariou 2011). Similarly, in China, the population census indicates that the labor force participation of 15- to 22-year-old urban males and females dropped from 70.6 and 72.7 percent, respectively, in 1982 to 43.8 and 46.4 percent, respectively, in 2000, largely because of an increase in educational enrollment (Hughes, Maurer-Fazio, and Zhang 2005). In rural areas, participation of the female population ages 15 to 22 declined from 84.3 percent to 67.3 percent between 1982 and 2000 for the same reason. Where education acquisition varies by gender, a gender gap in participation can be observed.

Although changes in education have not explained a large fraction of changes in participation, variation in education contributes to explaining differences in female labor force participation within a country. In Indonesia, women with higher levels of education are more likely to enter the labor market, particularly in urban areas, which may reflect their higher wage premiums and higher opportunity cost of being inactive (Ogawa and Akter 2007; World Bank 2010a). Literacy in Indonesia is also strongly associated with both participation in the labor market and occupational segregation among women; thus, being illiterate poses a double barrier to labor force outcomes (Gallaway and Bernasek 2004). In Vietnam, women with no educational qualifications are more likely than their male counterparts to be inactive, those with primary or secondary education are slightly less likely than their male counterparts to be working, and those with higher levels of education are more likely than their male counterparts to be working (Pierre 2011).

Declines in fertility have been found to exert a large positive effect on the labor force participation rate across the world. The effect is strongest for women ages 20 to 39, with an additional child being associated with a reduction of approximately four years of paid work over a woman's lifetime (Bloom et al. 2009). In Korea, the reduction in the total fertility rate, from 5.6 children per woman in 1962 to 1.2 in 2002, has been suggested to have increased per capita income by approximately 36 percent over the period, because of an increase in the size of the workforce and higher female labor force participation as well as a longer-term increase in the capital-to-labor ratio (Bloom et al. 2009).

> What will stop us from getting a job is having plenty of children and having nobody to mind them.
>
> *Young woman, Indonesia*
> (World Bank 2011a)

Child-care costs have a negative impact on female labor force participation, particularly in families without the support of familial networks.[14] Reductions in child-care provisions in China and Mongolia have had a negative impact on labor force participation. In China, participation of women of childbearing age has declined in urban areas, a trend that is partly attributable to higher child-care costs now that child care is no longer subsidized by the state (Chi and Li 2007; Du and Dong 2010; Li and Song 2011; Maurer-Fazio et al. 2011). Similar patterns have been observed in Mongolia, where state-funded early child care and education were rolled back in the 1990s (World Bank 2011c; World Bank and ADB 2005).

The labor force participation of older women is also affected by the presence of children, since they are often responsible for caring for younger household members. In Mongolia and rural Vietnam, the evidence suggests that a "grandmother effect" is present. Older women in households with children under the age of 10 have a 15 percentage point larger participation gap than women in households with no children under the age of 10, a difference that may be attributable to older women staying at home to look after their grandchildren.

Retirement policies

In nearly all countries in the region, the male-female participation gap rises after the age of 50, indicating that women retire earlier than men. These differences can be partly attributed to gender-differentiated retirement policies. In 4 of 12 countries studied in the region (see table 3.1), the statutory age of retirement for women in the private sector is five years earlier than that of men, although only in China is it mandatory to retire at the statutory age. In China, gender differences in mandatory retirement ages influence the prospective value of hiring older men and women, implying that the expected return of hiring an older man is greater than that of hiring an otherwise identical older woman (Giles, Wang, and Cai 2011). Differences in retirement prospects across rural and urban areas also contribute to the explanation of differences in participation across rural and urban areas among older workers. In urban areas, where most long-term residents have had formal wage employment, residents retire at a relatively young age and receive substantial pension support. In contrast, rural residents often lack pension support and hence make their labor supply decisions in the absence of pension availability and the constraint of mandatory retirement (Giles, Wang, and Cai 2011). Since women over 60 in rural areas are less likely to be vested in formal labor markets, they are less likely than other elderly to be covered by the pension system—only 1 percent of rural women over 60 report that pension income is their most significant source of financial support, compared with 8 percent of rural men over 60 (World Bank 2012).

Differences in retirement ages are likely to affect outcomes beyond labor force participation, including educational investment, the number of women in positions of power, and the risk of poverty for elderly women. Because women anticipate having a shorter working life than men, differences in retirement ages affect education and occupation choices. Gender differences in retirement ages also imply that women are less likely to rise to the top of occupational ladders, because they have less experience than men toward the end of their careers. Fewer women in Mongolia reach higher-level managerial positions, despite women having one of the highest levels of education on average in the region. Furthermore, early retirement reduces pension payouts, which depend on the length of tenure. This disadvantage in pension receipt can increase the risk of poverty for low-income households headed by females—women accounted for approximately 70 percent of elderly single–headed households in Mongolia in 2010 (World Bank 2011c).

Determinants of gender gaps in earnings

Earnings gaps between male and female workers

The gender earnings and wage gap may reflect differences between men and women in education and other characteristics, as well as differences in the returns to these characteristics by gender. The literature separates gender gaps into parts that are "explained" by measured factors, such as education, age, experience, and marital status, and those that are "unexplained," often regarded as a measure of discrimination against female workers. Care must be taken in interpreting the remaining component of wages as discrimination, however, since this interpretation assumes that all relevant factors have been considered.[15]

The level and changes in the gender wage gap over time in the East Asia and Pacific region vary by country and across the income distribution, suggesting that the forces that lead to narrowing wage gaps over time are idiosyncratic. A study commissioned for this report found that the evolution of gender wage gaps over time has followed different paths in Cambodia, Indonesia, Mongolia, the Philippines, Thailand, and Vietnam (Sakellariou 2011). In Thailand, gender wage gaps throughout the wage distribution decreased substantially between 1996 and 2006. Changes in experience, education, and return to experience were found to have contributed significantly to this reduction. In

TABLE 3.1 Legal retirement ages in the East Asia and Pacific region

Retirement age	Men	Women
China	60	55
Hong Kong SAR, China	65	65
Indonesia	55	55
Lao PDR	60	60
Malaysia	55	50
Mongolia	60	55
Papua New Guinea	55	55
Philippines	65	65
Singapore	67	67
Taiwan, China	65	65
Thailand	55	55
Vietnam	60	55

Source: World Bank 2011f.

contrast, in the Philippines, the wage gap widened between 2000 and 2009, a change that is partly attributable to growing differences between men and women in terms of their returns to education and other characteristics. Indonesia has the widest earnings gap among the six countries examined, and the study found substantial differences between rural and urban areas, in the size of the wage gap and the factors contributing to it.

Average differences between the characteristics of men and women, such as education, experience, and the sector and occupation of employment, explain up to one-third of the male-female wage differentials across the region and across the world.[16] Occupational and industrial sorting have been found to contribute more to explaining the gender wage gaps than do differences in human capital across the world, and the same holds true for the East Asia and Pacific region (Ñopo, Daza, and Ramos 2011; Sakellariou 2011).

The narrowing of gender wage gaps over time can be partially attributed to the converging of educational attainment of men and women. Between 1985 and 2005, the average characteristics of the female wage and salaried workforce in Thailand improved over time relative to those of males. Nakavachara (2010) finds that the substantial increase in the education of females between 1985 and 2005 was the major source of the narrowing of the gender earnings gap in Thailand during this period. Dhanani and Islam (2004) find that, although females earned on average about 30 percent less than men between 1976 and 2000 in Indonesia, overall wage inequality varies with industrial affiliation as well as education and age. The wage gap in Indonesia increases substantially with age, reflecting in part growing differences in education and experience between older men and women.

Since women have higher levels of education on average than men in some countries in the region, estimates suggest that women would have earned more than men had they faced the same returns to their education and other characteristics as men. In the Philippines and Mongolia, women's levels of human capital are higher than those of men on average. In Mongolia, taking into account the different characteristics of men and women, women should have earned 22 percent more than men in 2006 (Pastore 2009). Similarly, in the Philippines, education and other characteristics of women would suggest that, on average, the wages of women should be higher than those of men (Sakellariou 2011).

The bulk of the gender wage gap within the region is due to differences in the labor market value of male and female characteristics.[17] In the East Asia and Pacific region, the fraction of the gender wage gap explained by differences in characteristics (such as education and experience)—the explained component—is smaller than the fraction of the wage gap attributable to differences in returns—the unexplained component (Sakellariou 2011). Figure 3.14 shows the percentage difference between male and female wages at deciles of the wage distribution in Indonesia in 2009, and the difference in male and female wages that is attributable to differences in characteristics and returns on those characteristics.[18] Two noteworthy points emerge from figure 3.14. First, the gender wage gap is wider at the bottom than at the top of the wage distribution, pointing toward the phenomenon of "sticky floors." Second, differences in returns to characteristics between men and women are able to account for a greater share of the gender wage gap at all points in the wage distribution—the

gender wage gap attributable to returns to characteristics is greater than the gender wage gap attributable to characteristics across the distribution. On average, differences in characteristics explain just over 35 percent of the gross gap in 2009. Differences in labor market experience and returns to labor market experience constituted the major contributor to the characteristics component in both years.

There is substantial variation in the share of gender wage gaps explained by differences in characteristics between men and women. In Vietnam, the fraction of the gap explained by differences in characteristics is only 11 percent in urban areas, whereas in rural areas the characteristics of females would suggest that, on average, the wages of women should be higher than those of men (Sakellariou 2011). In China, the mean gender pay gap increased substantially between 1987 and 2007, from 18 percent in 1995 (Chi and Li 2007) to nearly 30 percent in 2007 (Li and Song 2011). Over this period, the majority of the increase was not attributable to differences in characteristics but rather was due to unexplained differences in the returns to male and female characteristics.

Marriage and childbearing have a larger negative effect on the wages of females than on the wages of males. The trade-off for women between career, earnings growth, and family does not appear to exist for men. For some women, this trade-off is associated with increasingly stark choices. Anecdotal evidence suggests that women in richer parts of East Asia are increasingly less likely to marry and marry later when they do, in part because of the perceived incompatibility of marriage and career (*Economist* 2011). The negative effect of childbearing on earnings and employment has been found across the world as well as within countries in the East Asia and Pacific region.[19] In Cebu, in the Philippines, children have a strong negative effect on a woman's likelihood of participating in the labor force and, once in the labor force, on her earnings over time. The negative effect of children on women's earnings represents both a reduction in the number of hours worked and a shift to lower-paying and often less secure jobs that are more compatible with child-bearing responsibilities (Adair et al. 2002). In China, married women have substantially larger gender wage gaps than their unmarried counterparts, and the wage gap between married men and women grew between 1995 and 2007 while it closed for single men and women (Hughes and Maurer-Fazio 2002; Li and Song 2011). Furthermore, the proportion of the gender wage gap unexplained by differences in the productive characteristics of men and women is greater for married than for single women, although the gap is narrower in the case of more educated married women. These results may be attributable to the greater responsibilities borne by Chinese women for household chores and child care.

Gender wage paths over careers may reflect differences in male and female workplace behavior, as well as differences in the types of characteristics that men and women value in a job. Evidence from the organizational and human resources literature shows females as being less competitive, more risk averse (Croson and Gneezy 2009), less likely to push for pay rises (Babcock and Laschever 2003), and more likely to be content with lower starting salaries. This tendency may be due in part to women having different demands for job characteristics and expectations of labor market outcomes than men. For example, Bender, Donohue, and Heywood (2005) find that flexibility is an important job characteristic that appeals to women and may play a contributing role in gender-based labor market segregation.

In sum, men earn more than women across countries in the East Asia and Pacific region, although the reasons for this vary from country to country. Education differences between men and women are not able to explain the bulk of gender gaps in labor market wages. In some countries, such as Mongolia and the Philippines, education differences in fact imply that women should earn more than men. The most important differences between men and women are those related to occupational choice and family life. Marriage and childbearing are likely to affect men's and women's earnings in different ways, partly

FIGURE 3.14 **The fraction of the gender wage gap explained by differences in characteristics is smaller than the fraction of the wage gap attributable to differences in returns in Indonesia, 2009**

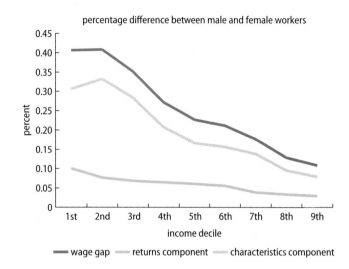

Source: Sakellariou 2011.
Note: The graph uses the approach of Firpo, Fortin, and Lemieux (2009), based on Indonesia National Labor Force Survey (BPS Sakernas), 1997 and 2009 data.

reflecting their respective responses to these life-cycle transitions.

Earnings gaps between male and female entrepreneurs

Female-run firms have fewer employees, lower sales, and lower capital stock than those run by men, as evidence from Europe and Central Asia, Africa, and Latin America and the Caribbean has shown (Amin 2010; Bardasi, Blackden, and Guzman 2007; Bruhn 2009; Costa and Rijkers 2011; Sabarwal and Terell 2008). Various conditions explain why female-run firms produce less per worker than male-run firms and, in particular, why women tend to head smaller firms in less capital-intensive sectors than men. The three predominant explanations examined here are sectoral segregation, skills, and constraints to business development.

Sectoral segregation. Firm size and industrial and occupational sorting along gender lines explain a large part of the differences in returns to capital between firms run by men and those run by women. In a number of studies comparing the productivity of male- and female-run enterprises around the world, productivity gaps diminish substantially once the industry or sector of employment is taken into account (Aterido and Hallward-Driemeier 2009; Bruhn 2009; Costa and Rijkers 2011; de Mel, McKenzie, and Woodruff 2009a, 2009b). This finding also holds for the East Asia and Pacific region. In Mongolia, substantial differences in capital and sales exist between male- and female-owned firms. These are largely explained by the sector of employment and firm size (World Bank 2011c). In Lao PDR, differences in the productivity between male- and female-owned firms in the formal sector disappear once differences in sector, capital, size, and other factors that affect productivity are considered (Davies and Record 2010). In Vietnam, a study commissioned for this report found that there were no gender differences in short-term revenue growth or survival probability among in firms in the formal sector between 1997 and 2009 (Bjerge and Rand 2011).[20] In the informal sector in Indonesia, the sector of operation and the firm's size accounts for almost 90 percent of the observed gender gap in productivity in rural areas (Costa and Rijkers 2011), although they account for substantially less of the gap in urban Indonesia (Badiani and Posadas 2011).

Informality is likely to be a greater determinant of an enterprise's success than the gender of the manager or owner. Evidence from Indonesia and Vietnam suggests that differences in profits between female-headed firms and male-headed firms are substantially smaller than the differences between female-headed firms in the formal sector and those in the informal sector (figure 3.15). This mirrors evidence from Africa (Aterido and Hallward-Driemeier 2009).

Evidence on gender constraints in informal enterprises within the East Asia and Pacific region is more limited than evidence from the formal sector, however. More data are urgently needed on the informal sector, as microenterprises constitute a substantial fraction of enterprises and employment in the region. In Indonesia, Vietnam, and the Philippines,

microenterprises make up 98.9, 98.2, and 92.3 percent of all micro, small, and medium enterprises (MSMEs), respectively, and they make up 79.3 percent of all enterprises in Malaysia (Kushnir, Mirmulstein, and Ramalho 2010). The East Asia and Pacific region has the highest ratio of MSME employment to total employment of all regions, mainly driven by China, where MSMEs account for 80 percent of total employment.

Skills. Evidence on whether female entrepreneurs have a lower level of managerial skills than male entrepreneurs is limited in the region. However, gender differences in both education and soft skills are likely to affect men's and women's respective decisions to participate in self-employment and in their sector of choice (Brush 1992). Furthermore, differences in entrepreneurial and managerial skills (such as being able to identify market niches and do bookkeeping) between men and women may explain differences in the size of the firms that they establish and earnings gaps (Bruhn, Karlan, and Schoar 2010).

Although evidence from studies of entrepreneurial skills in the region is limited, the evidence from outside the region is mixed with regard to whether a lack of skills is an important constraint for entrepreneurs. A study from Peru suggests that giving business training to female clients of a microcredit program did not lead to higher profits or revenues on average, although the clients did adopt some of the activities taught in the program, including thinking proactively about new markets and profit-making opportunities (Karlan and Valdivia 2011). In Pakistan, business training was found to increase the survival and profitability of male-run firms but had no effect on female-run firms (Giné and Mansuri 2011). However, in India, a two-day training program for female clients of an Indian microfinance institution was found to increase both the amount that they borrowed and the likelihood of the clients' receiving labor income (Field, Jayachandran, and Pande 2010).

Within the region, female entrepreneurs themselves consider that their skill limitations are a barrier to their success. In Vietnam, a survey of 500 female owners of enterprises revealed that female entrepreneurs felt the need to improve their skills, particularly in business management and leadership, through training and education (IFC and MPDF 2006). These findings are supported by another survey that indicated that females have lower general training levels than men (VCCI 2006).

Constraints to business development. In the formal sector, the constraints faced by female-run firms do not appear to be uniformly greater than those faced by male-run firms (Davies and Record 2010; IFC 2011; IFC/NORC Indonesia 2010; World Bank 2011b).[21] In the Philippines, qualitative and quantitative research has found little evidence of gender differentiation in lending or borrowing in small and medium enterprises (IFC 2011). In Indonesia, insufficient finance and financial management were found to be less of a concern for women (IFC/NORC Indonesia 2010).

Data from enterprise surveys carried out in five Pacific and four East and Southeast Asian countries in 2009 show that entrepreneurs, regardless of gender, named competition, finance, and electricity as their top three constraints in five of the nine countries. Male-led and female-led firms reported the same constraints as being the most important in all countries, except Tonga, Vanuatu, and Vietnam. The difference between self-reported access to credit constraints for male- and female-run formal sector firms is small in all countries, with the exception of Timor-Leste and Tonga (figure 3.16).

However, credit constraints are greater in the smallest firms and in the informal sector where female-run firms are concentrated. Constraints to female entrepreneurs may arise in both formal and informal institutional structures, for example, because of difficulties interacting with male officials who adhere to cultural norms of female propriety. In Indonesia, access to capital is the most important constraint reported in both male- and female-run informal firms (figure 3.17), and female-run informal firms appear to have substantially less capital than male-

FIGURE 3.15 **Differences in productivity across informal and formal firms in Vietnam are larger than differences across male- and female-led enterprises**

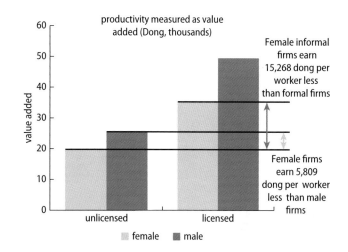

Source: World Bank estimates based on Vietnam Household Living Standards Surveys (VHLSS) (GSO Vietnam), 2009 data.
Note: unlicensed = informal.

FIGURE 3.16 **Self-reporting shows little difference in access to finance between male- and female-run firms, with the exception of Timor-Leste and Tonga**

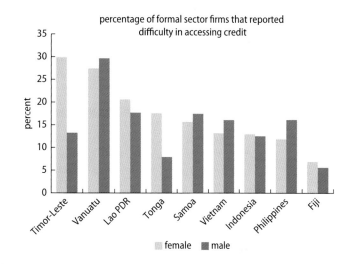

Source: World Bank estimates using Enterprise Surveys (database), 2006–11 data.
Note: Finance is considered to be an obstacle if it is reported as a major or severe constraint.

run informal firms. Female-led microenterprises also have lower levels of start-up capital than male-run microenterprises.[22]

Access to capital may be more limited—and more expensive—for female entrepreneurs

FIGURE 3.17   Male- and female-led informal firms report similar constraints in Indonesia

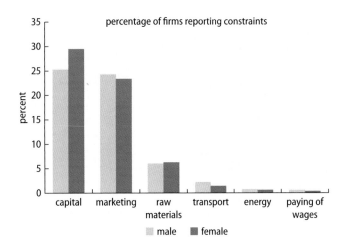

Source: World Bank estimates using SUSI Integrated (Survey of Cottage and Small-Scale Firms) 2002.

because they lack collateral (World Bank 2011g). According to calculations from the World Bank Enterprise Surveys database of 2006–11, in all countries in the region for which data were available, land and buildings were the predominant forms of collateral used to acquire a loan for production purposes. This practice puts female entrepreneurs at a disadvantage since they have fewer of these assets as well as less secure access to land or other immobile assets. However, the most important determinant of access to credit is the size of the firm (Beck et al. 2006). Credit institutions tend to regard small firms as a bigger risk than large firms, and, since women tend to manage smaller firms, this is likely to lead to female-run businesses being more constrained in the credit market than businesses that are run by men (Simavi, Manuel, and Blackden 2010).

Earnings gaps between male and female farmers

The agricultural sector continues to be the major sector of employment in many countries in the region. In Vanuatu, 80 percent of the population rely upon small-scale agriculture (IFC 2010), and in Cambodia and

Lao PDR over 70 percent of the workforce is employed in agriculture.

As the countries across the world have developed, women have been playing a greater role in agricultural activities as men move to the cities in search of non-farm work (Deere and Leon 2003; Ganguly 2003; Mu and van de Walle 2009). This "feminization" of agriculture can be seen in multiple countries in the region. In China, the number of households in which women participated in agricultural activities and performed all the farm work rose from 14 percent in 1991 to just under 30 percent in 2009, whereas the fraction of households in which men did all the work remained fairly stable (de Brauw et al. 2011; Rawski and Mead 1998). In Cambodia, women make up the majority of farmers—1.4 million female farmers compared to 1.2 million male farmers (UNIFEM, World Bank, ADB, UNDP, and DFID/UK 2004). In Vietnam, agriculture accounted for 64 percent of working women in rural areas in 2008 compared to 53 percent of working men (World Bank 2011e).

Despite the important role played in agriculture by women in the region, little evidence is available regarding gender differences in productivity, access to inputs, or agricultural services. Remedying this information gap is likely to become increasingly important if the feminization of agriculture continues within the region.

The evidence from other regions suggests that female farmers earn less than male farmers.[23] For example, evidence from Africa suggests that, within a household, yields on male-run plots are higher than those on female-run plots, predominantly because of increased male access to fertilizer and labor inputs (Goldstein and Udry 2008; Udry 1996). In comparison, the available evidence from China suggests that female farmers are just as productive and efficient as male farmers, despite differences in mechanization and fertilizer use (de Brauw et al. 2011). Although female farmers are able to produce similar amounts per hectare to men, they do not produce as much overall owing to differences in land and nonland inputs between male and female farmers.

Agricultural production varies in part between male and female farmers because of differences in access to land. Female farmers across the world own and operate less land, and often have lower-quality land than male farmers (Deere and Doss 2006; Quisumbing 1998; Quisumbing, Estudillo, and Otsuka 2004).[24] Evidence from the East Asia and Pacific region suggests that there are substantial differences in access to land between male and female farmers, as discussed in chapter 2. In China, female-headed households own 30 percent less land per capita than male-headed households (de Brauw et al. 2011). In Lao PDR, male-headed agricultural households own approximately 16 percent more land than female-headed households on average, although in the north of the country men have access to 28 percent more land (FAO/Sida Partnership Cooperation 2010). A World Bank study of women's landholdings in post-tsunami Aceh found that women's landholdings were considerably less than men's (World Bank 2010b).

Female farmers have also been found to have less access to technological inputs such as fertilizers and high quality seeds (Peterman, Behrman, and Quisumbing 2010). An important theme in the literature on the use of productivity-enhancing technology is that *access* to inputs, not the propensity to use inputs or productivity once the inputs have been used, is the key factor for many female farmers. In a review of 24 studies of technological input use, access, and adoption of fertilizer, seed varieties, tools, and pesticides in Africa and South Asia, Peterman, Behrman, and Quisumbing (2010) found that men have greater access to technological resources in the majority of cases. To know whether the same constraints are pertinent in the region requires further evidence on this topic.

Gender differences in access to extension services are also likely to contribute to differences in the adoption of new technologies and farming practices and to perpetuate existing gender inequalities in access to inputs. In Cambodia, access to agricultural extension services was found to be substantially lower for female farmers than for male farmers in the early 2000s, despite the fact that there were more female farmers than male farmers (UNIFEM, World Bank, ADB, UNDP, and DFID/UK 2004).[25] Reasons for the lower access to extension services among female farmers include the extension services' focus on activities normally conducted by men, the focus on literate farmers, and the need to travel several kilometers to access services (UNIFEM, World Bank, ADB, UNDP, and DFID/UK 2004). Efforts to reduce these constraints in the Cambodian context are discussed in chapter 6.

Finally, gender-based differences in social and political capital have been found in a number of rural areas, reducing women's access to information on farming techniques and their ability to protect and regulate local resources and their marketing channels. In the Philippines, men and women do not differ in their level of participation in local groups, but they do differ in the types of groups that they join. Men are more likely than women to be members of production-based groups, whereas women are more likely to be members of civic groups, which include women's groups, village youth associations, school committees, and village officials (Godquin and Quisumbing 2008). In Indonesia, men have been found to participate in civil society organizations related to community-level governance, physical infrastructure, environmental improvements, and neighborhood security, whereas women participate in organizations that focus on family welfare, economics, and health (Beard 2005).

Determinants of labor market segregation

The determinants of persistent gender employment segregation across countries and over time lie at all levels of the economy and society, from gender norms within households and communities to economic signals in the marketplace and ideological predilections at a societal level. Factors that influence labor market segregation along gender lines include (a) differences between men and women in the level and composition of their education and experience, (b) differences in their preferences for types of work and job characteristics, (c) their comparative advantages due to their differences in physical endowments, (d) employers' perceptions about male and female employees (sometimes to the extent of discrimination), and (e) labor market institutions that limit or restrict occupational choices. These factors are, in themselves, determined by the economic signals, institutions, and norms facing individuals, households, and communities.

This section examines the effects of three key factors: (a) gender differences in education levels and labor market skills; (b) gender differences in time-use patterns and gender roles within households, both of which affect the characteristics and types of work women are able to do; and (c) labor market and institutional barriers to women's choices.

Education and skills

Education levels and the type of education acquired (for example, vocational versus general education) affect the skills that individuals bring into the labor market and hence are likely to contribute to their sectoral and occupational choices.[26] As discussed in chapter 2, male and female education has been converging over the past two decades in the East Asia and Pacific region, and in some countries, such as Mongolia and the Philippines, education acquisition by women has overtaken that of men. At the tertiary level of education (universities, technical training schools, and so forth), a student's field of specialization was likely to be closely related to his or her subsequent sectoral choices. Any gender differences in education choices are therefore likely to be mirrored in sectoral dissimilarities among men and women. If women tend to choose different fields than men, sectoral and occupational segregation is likely to increase as the number of men and women continuing to higher and tertiary education rises.

In the lower- and middle-income countries in the East Asia and Pacific region,

only a small fraction of men and women continue to higher education. Therefore, the nexus of education choice and occupational segregation is not currently a primary driving force for labor market segregation in the region. However, it is likely to play a larger role among younger cohorts and future generations as education continues to expand. For example, in Taiwan, China, gender wage gaps declined substantially between 1979 and 1995, a period during which there was a rapid increase in average educational attainment and a shift from academic general-curriculum education toward vocational training. Among university graduates, differences in degree type between men and women are able to explain between 20 percent and 50 percent of the gender wage gap, although the link between occupation and degree type is lower among women than men (Baraka 1999).

Time use and gender roles within households

Differences in the ways men and women spend their time are informative in illuminating why they may invest differently in human capital, how gender differences in labor market participation may arise, and what their occupational and industry choices will be. Time is a valuable resource that is endowed equally across men and women—everyone has just 24 hours in a day. Time can be devoted to a number of uses: labor market work; unpaid work within the household, such as conducting domestic chores and caring for children and the elderly; and personal activities, including sleeping, eating, and leisure (Becker 1965, 1981).

Women across the world work more than men. This stylized fact holds true in multiple low- and middle-income countries across the world (Berniell and Sanchez-Páramo 2011; World Bank 2011g), as well as in several OECD countries (Burda, Hamermesh, and Weil 2007; Slootmaekers-Miranda 2011). Gender differences in time-use patterns exists at all ages. Whereas men are able to focus predominantly on their single productive role and conduct their other roles sequentially, women are more likely to play these roles simultaneously and have to balance competing uses of their time (Blackden and Wodon 2006).

However, time differences in hours worked by men and women diminish when growth in gross domestic product (GDP) is combined with gender-neutral social norms. A study of 25 countries across the world finds that men and women did the same amount of total work in rich northern countries (Burda, Hamermesh, and Weil 2007). The gender difference in total working time is close to zero for countries with relatively high female employment and with more gender-neutral social norms.

In the East Asia and Pacific region, women work more hours and devote more time to caregiving and housework than men, and men specialize in market-oriented activities. Since women's share of unpaid work is higher than that of men, this translates into shorter time in paid work.

> Women indeed work harder. We go to the field and return home at the same time as our husband. Afterwards, we still have to cook, do laundry and do other household chores. The men after they return home from the field do not want to work anymore.
>
> *Adult woman in rural Batu Palano, Indonesia* (World Bank 2011a)

A number of stylized facts emerge from time-use patterns by gender in the region. First, women work more hours than men; this holds true at all ages. In Lao PDR, women work on average 2 hours extra per day than men, and in Cambodia, they work on average 1.2 hours more per day.

Second, gender differences in time-use patterns are starker during the childbearing years. In Cambodia, men and women spend similar amounts of time in market work until they are 20 years old, but after that point, women devote less time to market work than men and more time to domestic activities (CSES [NIS Cambodia], 2004 data). The greatest difference between male and female hours worked is during childbearing years. In households with young children, women work on average 2 hours extra per day in

FIGURE 3.18 **In Cambodia, women—particularly those with young children—balance household work commitment with market work**

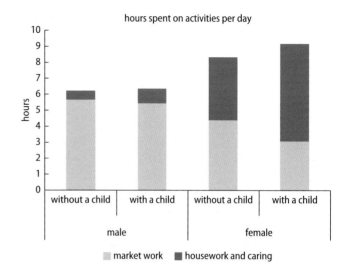

Source: World Bank estimates using CSES (NIS Cambodia), 2004 data.

Cambodia and 1 hour extra per day in Lao PDR (figure 1.20 and 3.18).

Third, gender differences in the time devoted to household activities are smaller in high-income households, although this represents a reduction in female working time rather than an increase in male working time. In Timor-Leste, rural women in the highest income quintile spend more time on domestic activities than men (figure 3.19). The difference between men and women narrows with rising income. The narrowing of time-use differences, however, arises from a reduction in the time devoted by women to these activities, thus reflecting increased access to technologies and household help rather than an increase in male participation.

Finally, gender differences in the time devoted to household activities start at an early age. In the Philippines, Pörtner (2009) studied time use of children between the ages of 7 and 16 and found that, although participation rates and time spent in school by boys and girls were similar, time spent on work and household chores is closely related to the gender of the child. Boys participate in market-related activities more than twice

FIGURE 3.19   Women in Timor-Leste spend more time on domestic activities than men, and these differences are found among richer as well as poorer households

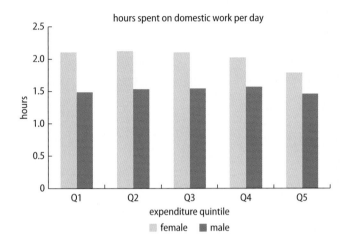

Source: World Bank estimates using Timor-Leste Survey of Living Standards (NSD [Timor-Leste] 2007 data).

as often as girls, whereas from the age of 14, girls spend on average twice as much time on household chores than boys.

Spending greater time on household activities has implications for women's labor market outcomes. Because women do a greater proportion of household work and child care than men, they have less time for productive activities. This pattern makes them less likely to enter the labor market and implies that, when they do, they are more likely to choose occupations that give them the flexibility to conduct their domestic responsibilities (Anker 1998; Becker 1965).

Women may regard the informal sector as more compatible with childbearing and household responsibilities than the formal sector, since it offers greater flexibility in hours, although flexibility comes at a price—less security, lower status, and a lack of nonwage benefits such as social security, health insurance, and paid sick leave. Female entrepreneurs may similarly prefer to keep their businesses small because doing so allows them to combine household responsibilities with work (Sabarwal, Terrell, and Bardasi 2011). In Thailand, single women without children are more likely

to be found in the formal sector, whereas married women with children are found in the informal sector, particularly in self-employment. The evidence suggests that the movement of married women with children into self-employment is a choice rather than an effect of discrimination (Bosch and Maloney 2011).

Differences in the time-use patterns of men and women also reflect society's norms regarding gender roles and "appropriate work." These social norms and customs affect the ability of both men and women to participate in the economic, social, or political domain. In China, despite strong governmental support for gender equality on multiple levels in the workplace, cultural norms have still emphasized the role of women as *xianqi liangmu* (a good wife and a good mother) and as the bearer of responsibility for household work (Chen 2005; Honig 2000). These norms persist even as development progresses and women's share of income in household economies rises. In Beijing, even when the wife's income represents a larger share of the couple's combined income, women still do the majority of the household work (Zuo and Bian 2001). Both men *and* women justified this in terms of their gender roles in the household.[27] Similar patterns are found in Nanjing (Kim et al. 2010).

Social norms may be reinforced at the level of the nation-state, if the role of women as homemakers and mothers and of men as breadwinners is underlined in political discourse. In Malaysia, the Nation of Character project focused on 25 key values important for the development of good character in children. This project very clearly put forward that women's most important tasks were related to the home and to strengthening of the family. The division of labor between men and women in the public and private sphere is further emphasized from an early age in parts of the region (White 2006). In Cambodia, a book of moral codes—the *Chba'p srey*—emphasizes the "proper" behavior and conduct of women, and limits their opportunities outside the household (Dasgupta and Williams 2010).

Labor regulations and informal institutions

Labor laws and regulations can directly affect the demand for female workers and constrain these workers in their choice of occupation. In many countries, restrictions on women's working hours or industries were introduced as measures to protect the health of women (particularly those who are pregnant or lactating) or women in potentially hazardous jobs. However, with improvements in labor market conditions in dangerous industries and with the passing of employment legislation designed to protect the health and safety of workers, many of these restrictions may no longer be relevant and could be changed. Measures that limit women to work only daytime hours or restrict their work to a subset of industries limit their employment options and also drive employers to hire only men for jobs that women might otherwise do.

A study on labor laws commissioned for this report found that protective legislation prohibits women from working in industries and occupations in 9 of the 12 low- and middle-income countries examined in the region (China, Fiji, Kiribati, Korea, Malaysia, Papua New Guinea, the Philippines, Thailand, and Vietnam) (World Bank 2010f). In Thailand, women are prohibited from working in certain occupations, including mining; working on a scaffold more than 10 meters high; and producing or transporting explosives or flammable materials. In Vietnam, the Labor Code prohibits assigning female workers to "heavy or dangerous work, or work requiring contact with toxic substances, which has adverse effects on her ability to bear and raise a child." The code also provides that women, regardless of age, cannot be employed in mines or in deep water. No similar provisions are applicable to men. In Mongolia, until 2008, extensive labor market regulations limited female participation in multiple sectors of the economy, including mining, transport, and construction (World Bank 2011c). Pregnant women are even more limited in their choices: in 6 out of 10 countries in the region, they have

more restrictions on industry choices than other men and women (World Bank 2010f).

These protective restrictions may increase the cost of employing women. For example, in Kiribati, Papua New Guinea, and the Philippines, women are not allowed to work the same night hours as men. In the Philippines, the law states that women are unable to work between 10 at night and 6 in the morning of the following day. The restrictions on women's work have been criticized as particularly restrictive by the call center industry, which employs a large proportion of women who are required to work at night (Keitel 2009). Paid maternity leave also increases the cost to employers of hiring female workers rather than male workers, particularly in countries where paternity leave policies are not in place. In contexts where employers bear the burden of this cost, this is particularly likely to raise the cost of hiring female workers.

Discriminatory laws in the area of family and marriage also affect women's economic opportunities. Laws relating to family, marriage, and inheritance play a key role in influencing women's economic rights, including access to land, housing, and other forms of property (Ellis, Kirkwood, and Malhotra 2010). The law of succession in Tonga's constitution, for example, allows only males to inherit. In Indonesia, the civil code prevents women from entering into contracts on their own behalf, whether to sell or buy property, which enables men to influence women's access to collateral. Access to land is dependent on a woman's married status, and her control and ownership can be lost upon divorce, widowhood, migration, or desertion by her husband (IFC 2010).

Women also have less access to information on opportunities and job networks. Personal connections are recognized as facilitating job search, but women's domestic responsibilities make it difficult for them to forge useful social connections (Timberlake 2005). In Nanjing city, China, a case study suggests that women secured less-attractive jobs through their own networks than through those of their husbands. Women who have fewer social contacts also were more vulnerable to redundancy and experienced more hardship in finding a new job after having been displaced from their old job (Hiroko, Liu, and Tamashita 2011).

Finally, occupational segregation may be related to persistent stereotypes about what are appropriate occupations for men and women. Stereotypes of women as homemakers and men as breadwinners can translate into perceptions about their skills in the labor market and, therefore, the type of market work they do (Anker 1998). For example, positive stereotypes of women— having a caring nature, manual dexterity, skill at conducting household work, and greater honesty—could lead to the hypothesis that women would be better qualified to work as nurses, teachers, clerks, and sales assistants, among other occupations (Anker 1998). In qualitative work undertaken in Fiji, Indonesia, Papua New Guinea, and Vietnam, gendered beliefs about appropriate employment for men and women were remarkably consistent across these countries

and among genders (World Bank 2011a). See box 3.2 for more detail.

Policies to promote gender equality in economic opportunity

The analysis in this chapter has identified several indicators of gender inequalities in the productive sphere where challenges remain and where policy may be able to reduce persistent gender-based differences in economic opportunities and outcomes. A detailed discussion of policy examples and their impacts will follow in chapter 6.

Mitigating trade-offs between women's household and market roles

Promoting gender equality in economic opportunity requires policies and investments that address the trade-offs between women's household and market roles. Women often face stark time trade-offs between household chores and market work, particularly in

---

BOX 3.2   Gender-related beliefs on appropriate employment in Fiji, Papua New Guinea, Indonesia, and Vietnam

Gender-related beliefs about appropriate employment for men and women were remarkably consistent across East Asian and Pacific countries where qualitative fieldwork was undertaken (Fiji, Papua New Guinea, and Vietnam), and among both men and women, although some change in gender norms was observed among younger cohorts. Comments largely revolved around the concept of "heavy" (physically demanding) versus "light" work, with the former ascribed to men and the latter to women. In Vietnam, an adult man from Hanoi said that "In general, men are better with heavy jobs like taxi driver. Women are better with lighter jobs like sewing, and selling things. . . . Men are better with big business because it is hard work—men are more decisive than women. Job[s] like killing pigs, in principle women can do, but it would be strange to see

a woman with a knife in her hand." Differing skill levels could be conflated with conceptions of heavy and light work, for example, when one young male respondent said that men do heavy work such as that of an auto mechanic, while women do light work, "such as accounting." A focus group discussion with adult women in the urban National Capital District of Papua New Guinea revealed further gender stereotypes and fears held by these women about what they considered to be gender-specific employment. They said that secretaries should be women (so that they are not "aggressive to their bosses"), and trading stores should be run by men as they have the "ideas, mindset and business knowledge" and would not be "targeted by criminals."

Source: World Bank 2011a.

rural areas. In such contexts, programs and policies targeted at reducing women's time on chores—for example, through investment in infrastructure—are likely to increase their ability to engage in market-based, income-earning opportunities. Policies that support women in balancing their caregiving and market roles are also critical to strengthening their access to economic opportunity. Access to affordable and accessible child care can be critical in this regard. Community child-care centers, particularly those targeted at low-income neighborhoods, have been found to increase maternal employment in a number of Latin American countries.

Parental and paternity leave can promote greater parity between the sexes by facilitating a more equitable division of child-rearing responsibilities, thus allowing women to have the same opportunities as men for advancing their careers in the formal sector. Within the region, only Cambodia, Indonesia, and the Philippines currently have provisions for paternity leave. Evidence from the OECD on the take-up of paternity leave is mixed, however, suggesting that providing paternity leave alone is not sufficient to change the current gender division of child-rearing responsibilities within households. Rather, it needs to be combined with approaches to breaking down gender norms regarding household caregiving roles.

Breaking down gender silos in the labor market

A critical element of breaking down gender silos in the labor market involves investing in skills on the basis of productivity rather than on gender norms and perceptions of "appropriate" occupations. Beyond efforts to reduce gender streaming in education, discussed above, programs that encourage both women and men to think outside of gender stereotypes in the job market will likely improve the allocation of talent toward jobs in a way that enhances both equality of economic opportunity and productivity.

Breaking down social norms and perceptions is an area where the public sector can also lead by example. Even if women enter more "male" occupations, stereotypes are likely to persist with respect to women as leaders and managers. The public sector is in a unique position to establish good practice in this regard by encouraging women's professional advancement, either through direct measures such as targets or quotas, or through specialized training programs. In this context, the government of Malaysia put in place a system of quotas for female managers in the public sector; the approach has recently been extended to private sector firms to encourage women to assume leadership roles. In Mexico, the government initiated a system of grants to firms to improve gender-based employment issues in their workplace and also to improve the gender distribution in management.

Eliminating resource constraints on female-led farms and enterprises

Despite progress, the existing evidence suggests that women continue to have less access to a range of productive resources than do men, as a function of their gender as opposed to their innate productive capabilities. Public policies may thus have an important role to play in promoting gender equality in the control of productive inputs—whether land, agricultural extension, technology, or financial capital. Policies aimed at promoting equal access to assets, particularly land, requires careful thought, since complex legal, social, and economic factors are at play and the evidence base is thin. Levelling the legal playing field is usually a good start, but it is even more important to work with the informal institutions and take into account norms and customary practices in order to remove barriers in that domain.

Improving women's access to productive assets can play an important role in raising enterprise productivity, whether in the farm or nonfarm sector, as in the following examples:

- Several countries in the region have made headway in recent years in increasing ownership and control of land. In response to concerns about persistent gender inequalities in land, several countries—including Indonesia, Lao PDR, and Vietnam—recently adopted gender-sensitive reforms in land titling. Because the reasons for women's lower access to land differ across the region—from unfavorable legal frameworks to cultural norms and practices that deem land a male asset—effective policies to increase female landholdings need to account for context-specific constraints in developing context-specific solutions.

- Increasing women's access to information and training, extension services, and other productive inputs can also play a key role in enhancing the productivity of female-led enterprises, both within and outside agriculture.
- Although evidence on access to finance in East Asian and Pacific countries is mixed, women do face particular challenges in accessing credit, given their weaker access to land, which is an important source of collateral.

Where evidence is thin, greater resources should be invested in uncovering the greatest constraints. For example, investments could target the collection of information, by gender, on access to inputs in the agricultural sector in the region, as well as to inputs and productivity in the informal sector.

As in the case of education and health, broader systemic weaknesses—whether in the form of cumbersome registration procedures, weak systems of financial intermediation, or lack of electricity—affect both female- and male-led enterprises. Evidence suggests that such constraints may be more onerous among small and informal firms than among larger firms and, as such, may constrain female-led enterprises disproportionately. Thus, interventions that focus on improving the overall investment climate and, in particular, promoting small business development will be a critical part of a strategy to promote gender equality in access to economic opportunity. In sum, strategies to promote gender equality in economic opportunity should address systemic as well as gender-specific constraints.

Creating an enabling environment for gender equality in employment

Creating an enabling environment is a key component of efforts to promote gender equality in economic opportunity in the long term in East Asia and the Pacific. An important starting point for promoting equal opportunity in employment is to ensure that women and men face a level legal playing field with respect to jobs and sectors. Labor regulations that result in asymmetries in the costs of hiring male and female workers can be found in countries across the region. Ostensibly protective legislation—in the form of restrictions on women working at night, working overtime, and working in so-called dangerous sectors—in practice inhibits women's economic participation. Priority should be given to reducing labor market restrictions that limit women's employment options. Where the original concerns motivating these policies continue to be valid—for example, health and safety concerns—measures should be taken to ensure that these concerns are addressed more directly and for both men and women, whether through workplace safety codes or through the provision of safe and reliable transport infrastructure.

Although formal sector employment is still small as a share of total employment in most East Asian and Pacific countries, an important role for public policy is to strengthen the enabling environment for gender equality in economic opportunity. Active labor market policies are one means of overcoming gender differences in access for formal employment. For example, wage subsidies may allow individuals, albeit temporarily, to signal their abilities to future employers and make it cheaper for employers to hire female workers whom they may not otherwise have considered. This approach provides the opportunity to reduce stereotypes through directly observing their skills, and also gives women valuable labor market experience. Skills training programs may enable women and men to move into professions outside of gender silos, particularly when paired with apprenticeship opportunities. Although evidence on the impact of active labor market policies in East Asia and the Pacific is limited, studies from Latin America and the Middle East suggest that well-designed active labor market policies can help to improve women's employment outcomes.

Affirmative action policies have also been used to overcome gender-specific barriers to employment, whether those barriers are due to implicit or overt discrimination in hiring and promotion. Although the literature continues to debate the benefits and costs of affirmative action, the evidence (largely from high-income countries) suggests that carefully designed policies can help break down barriers to female employment with few or no adverse effects on firm productivity (World Bank 2011g). Affirmative action hiring and promotion in the public sector can also have important demonstration effects. For example, in 2004, the government of Malaysia introduced a public sector gender quota of 30 percent female representation across all decision-making levels, including positions from department head to secretary general (ASEAN 2008).

Notes

1. Using cross-country data for selected Asian economies, Meng (1996) found no significant relationship between economic development and the relative earnings of men and women. In fact, gender inequality in earnings within the East Asia and Pacific region was worse in high-income countries such as Japan and the Republic of Korea than in the low- and middle-income countries.
2. Unfortunately, data constraints prevent looking at birth age-cohort patterns within the majority of countries in the region.
3. Gender wage gaps do not capture earnings differences among all men and women of working age. First, they miss a large fraction of the workforce, notably those in unpaid work or self-employed workers. Second, since there may be differential selection between males and females into the labor force—and into wage employment rather than entrepreneurship and agriculture—gender wage gaps are also likely to reflect these selection decisions.
4. Aggregate relative wage data should, however, be treated with caution since it confounds

differences in human capital and experience, occupational and sectoral selection, underlying ability, selection into the labor market, and discrimination.
5. Household surveys in Thailand and Vietnam corroborate this. In urban areas, women are approximately 10 percentage points less likely to receive benefits than men in Vietnam, whereas they are 3 percentage points less likely to do so in Thailand. In the postreform period in China, a growing number of women and urban workers have been pushed into temporary, part-time, insecure, or low-paying work in the informal sector (Yuan and Cook 2010).
6. Notably, an individual is defined as working in the informal sector if he or she works in agriculture or is self-employed, is working in the household enterprise, or is working as an unpaid family worker. However, an individual is classified as working in the formal sector if he or she works as a legislator or manager, professional, technician or associate professional, or plant machine operator or assembler, or is in the armed forces.
7. Elementary occupations consist of simple and routine tasks that mainly require the use of hand-held tools and often some physical effort. For a more detailed explanation, please refer to the International Standard Classification of Occupations (ISCO) by the International Labour Organization.
8. Care should be taken when interpreting the data from Samoa, Timor-Leste, Tonga, and Vanuatu because of very small sample sizes.
9. Evidence from across the world suggests that firms with greater female representation in management display lower levels of gender inequalities, including wage gaps and inequalities within firms (Cohen and Huffman 2007; Graves and Powell 1995; Huffman, Cohen, and Pearlman 2010).
10. Occupational and industrial segregation by gender is detrimental for labor market efficiency and welfare for four principal reasons (Anker 1998). First, men and women are not working in occupations or industrial sectors to which they are best suited and most productive but are rather choosing their work based on other factors. This trend reduces overall incomes and aggregate productivity. Second, gender-based segregation increases labor market rigidity and reduces the ability of labor markets to respond to economic reforms and labor market shocks, such as those

related to globalization. Third, segregation reinforces and perpetuates negative gender stereotypes, consequently reducing women's status, income, education, and skills. Finally, segregation of the current generation has negative effects on future generations.

11. In Indonesia, controlling for sector of operation and firm size accounts for 17 percent of differences in profits in rural areas, and 50 percent in urban areas (Badiani and Posades 2011), and industry accounts for 9 to 14 percent of earnings among self-employed individuals in the United States (Hundley 2001).

12. Several hypotheses suggest why female labor force participation first falls before rising with economic development. Boserup (1970) suggests that men's greater access to education and technologies implies that they displace women from the labor force during the early stages of a country's development. As development continues and women gain more access to education and technologies, female labor force participation increases. Another well-established hypothesis for this phenomenon focuses on income and substitution effects (Goldin 1995; Mammen and Paxson 2000). As development occurs, households' unearned incomes rise, reducing the incentive of women to work outside the home. The negative impact of rising incomes on women's labor force participation is termed the "income effect," since greater household income implies that households are able to afford more female leisure time. The substitution effect works in the opposite direction—as female wages rise, more women are enticed to enter the workforce (Goldin 1995; Mammen and Paxson 2000).

13. In the East Asia and Pacific region, evidence of declining female labor force participation as incomes rise has been found in Thailand (Mammen and Paxson 2000).

14. In the OECD, a number of studies find that childcare costs negatively impact rates of female labor force participation and that the provision of subsidized child care raises participation (Anderson and Levine 1999; Blau and Currie 2006; Chevalier and Viitanen 2002; Del Boca 2002; Gelbach 2002; Gustafsson and Stafford 1992).

15. For example, even if differences in average human capital have been considered, the residual is likely to still contain differences between the composition and quality of education of males and females that may help

to explain gender earnings gaps, as well as other unobserved characteristics that may vary between males and females, such as the intensity of work conducted, workplace characteristics, and unobserved measures of ability. Furthermore, if the occupation choice or tenure trajectories within a firm reflect the impact of discrimination, then measuring discrimination as the unexplained component of wage gaps will underestimate its true extent.

16. In the OECD, Blau and Kahn (2003) find that 38 percent of the total gender wage gap is unexplained by differences in occupational and sectoral choice, education, and experience. In addition, women have less labor market experience—in OECD countries as well as developing countries (Goldin, Katz, and Kuziemko 2006)—in part because they are more likely both to take time out from the work force and to work part-time.

17. The evidence suggests a similar pattern in the United States. Bertrand, Goldin, and Katz (2010) find that female master of business administration (MBA) graduates earn less, even if they choose similar professional paths as men. Bayard et al. (2003) find that a large portion of the gender wage gap in the United States can be explained by pay differences between males and females within narrowly defined occupations and within establishments.

18. The gender wage gap is decomposed at different points in the earnings distribution using the decomposition method based on unconditional quantile regression as outlined in Firpo, Fortin, and Lemieux (2009). In this method, the estimates from the unconditional quantile regression constitute average partial effects of a small location shift of an independent variable on the unconditional quantile of the dependent variable.

19. Evidence from across the world suggests that marriage and childbearing have a large impact on the gender wage gap. In the United States, marital status and young children account for approximately half the gender wage gap faced by young women. A large component of the gender gap in earnings is attributable to women having more career interruptions and shorter work hours, including more work in part-time positions and self-employment (Becker 1981; Bertrand, Goldin, and Katz 2010; Korenman and Neumark 1992; Mincer and Polachek 1974; Sasser 2005; Wood, Corcoran, and Courant 1993). These estimates may, however, underestimate the effect of parenting and

gender divisions of household labor on wage gaps, since the demands placed on women at home can influence education, career, and work choices (O'Neill 2003).

20. Indeed, household survey data in Vietnam, which also cover microenterprises, suggest that female-run household enterprises were more likely to survive compared to male-run enterprises between 1993 and 1998 (Vijverberg and Haughton 2004). The higher rates of female-run firms' survival are linked to women's predominance in certain sectors.

21. Evidence from the enterprise surveys conducted in 2009 in Lao PDR suggests that female-run firms consistently report business environment constraints to be less severe than male-run firms (Davies and Record 2010). Similarly, in rural Indonesia, there are no differences in the severity of reported business constraints between female- and male-run household enterprises (Costa and Rijkers 2011).

22. In Indonesia, start-up capital is 3.4 times smaller in female-managed nano- and micro-enterprises than in male-managed enterprises (Badiani and Posades 2011).

23. Data on female-managed or female-owned plots and on female farmers are extremely limited in the world and, in particular, in the East Asia and Pacific region. Therefore, wherever necessary, this report examines differences between farms run by female-headed households and those run by male-headed households. Because female-headed households are different from male-headed households in a number of respects, this comparison is likely to exaggerate gender differences in agricultural productivity and access to inputs because it will confound differences in socioeconomic characteristics with differences in productivity.

24. In Latin America, women constitute between 13 and 27 percent of landowners in Mexico, Nicaragua, Paraguay, and Peru and own smaller plots of land than men in all of these countries (Deere and Doss 2006).

25. It should be noted that the fraction of farmers overall receiving agricultural extension services is very low.

26. Gender differences in "soft skills," which are acquired within and outside of school, also contribute to gender differences in occupations. Personality traits have wage returns that are both occupation and gender specific (Cobb-Clark and Tan 2011). Women have been found to be employed in safer or lower-risk jobs, which is consistent with evidence that women are more risk averse than men (Eckel and Grossman 2008).

27. Women tended to consider men's "over engagement" in household chores as "unmanly" and "non ambitious." Women often explained the fact that they carried out most domestic work (while having a full-time job) as fulfilling their obligation to care for their families. In addition, the view was expressed that women with extreme career ambitions—who did not assume much household responsibilities—would be criticized as "selfish" or "non feminine." Interestingly, wives with higher income and "occupational prestige" than their husbands often retained the primary responsibility for taking care of domestic work. This was seen as a "counter balance" to their violation of cultural values about who should be the primary breadwinner.

References

Adair, Linda, David Guilkey, Eilene Bisgrove, and Socorro Gultiano. 2002. "Effect of Childbearing on Filipino Women's Work Hours and Earnings." *Journal of Population Economics* 15 (4): 625–45.

Albrecht, James, Anders Björklund, and Susan Vroman. 2003. "Is There a Glass Ceiling in Sweden?" *Journal of Labor Economics* 21 (1): 145–77.

Amin, Mohammad. 2010. "Gender and Firm-Size: Evidence from Africa." *Economics Bulletin* 30 (1): 663–68.

Anderson, Patricia M., and Phillip B. Levine. 1999. "Child Care and Mothers' Employment Decisions." NBER Working Paper 7058, National Bureau of Economic Research, Cambridge, MA.

Anker, Richard. 1998. *Gender and Jobs: Sex Segregation of Occupations in the World.* Geneva: International Labour Organization.

Antecol, Heather. 2000. "An Examination of Cross-Country Differences in the Gender Gap in Labor Force Participation Rates." *Labour Economics* 7: 409–26.

Arulampalam, Wiji, Alison L. Booth, and Mark L. Bryan. 2007. "Is There a Glass Ceiling over Europe? Exploring the Gender Pay Gap across the Wage Distribution." *Industrial and Labor Relations Review* 60 (2): 163–86.

ASEAN (Association of Southeast Asian Nations). 2008. "ASEAN Continues to Empower Women." *ASEAN Bulletin.* http://www.aseansec.org/Bulletin-Feb-08.htm#Article-2.

Aterido, Reyes, and Mary Hallward-Driemeier. 2009. "Whose Business Is It Anyway?" Policy Research Working Paper 5571, World Bank, Washington, DC.

Babcock, Linda, and Sara Laschever. 2003. *Women Don't Ask: Negotiation and the Gender Divide*. Princeton, NJ: Princeton University Press.

Badiani, Reena, and Josefina Posadas. 2011. "Gender and Entrepreneurship in Household Firms in Indonesia: Characteristics and Performance." Paper commissioned for *Toward Gender Equality in East Asia and the Pacific: A Companion to the* World Development Report. Washington, DC: World Bank.

Baraka, Jessica. 1999. "Does Type of Degree Explain Taiwan's Gender Gap?" Research Program in Development Studies, Princeton University, Princeton, NJ.

Bardasi Elena, Mark Blackden, and Juan Carlos Guzman. 2007. "Gender, Entrepreneurship, and Competitiveness in Africa." *Africa Competitiveness Report 2007*. Geneva, Switzerland: World Economic Forum.

Bayard, Kimberly, Judith Hellerstein, David Neumark, and Kenneth Troske. 2003. "New Evidence on Sex Segregation and Sex Differences in Wages from Matched Employee-Employer Data." *Journal of Labor Economics*. 21 (4): 887–922.

Beard, Victoria. A. 2005. "Individual Determinants of Participation in Community Development in Indonesia." *Environment and Planning C: Government and Policy* 23 (1): 21–39.

Beck, Thorsten., Aslı Demirgüç-Kunt, Luc Laeven, and Vojislav Maksimovic. 2006. "The Determinants of Financing Obstacles." *Journal of International Money and Finance* 25 (6): 932–52.

Becker, Gary S. 1965. "A Theory of the Allocation of Time." *Economic Journal* 75 (299): 493–517.

———. 1981. *A Treatise on the Family*. Cambridge, MA: Harvard University Press.

Bender, Keith, Susan Donohue, and John Heywood. 2005. "Job Satisfaction and Gender Segregation." *Oxford Economic Papers* 57 (3): 479–96.

Berniell, M. I., and Carolina Sanchez-Páramo. 2011. "Closing the Access Gap: Recent Advances in Female Labor Force Participation." Background paper for the *World Development Report 2012*, World Bank, Washington, DC.

Bertrand, Marianne, Claudia Goldin, and Lawrence F. Katz. 2010. "Dynamics of the Gender Gap for Young Professionals in the Financial and Corporate Sectors." *American Economic Journal: Applied Economics* 2 (3): 228–55.

Bjerge, Benedikte, and J. Rand. 2011. "Gender Differences in the Vietnamese Business Sector: Evidence from the SME Survey." Paper commissioned for *Toward Gender Equality in East Asia and the Pacific: A Companion to the World Development Report*. Washington, DC: World Bank.

Blackburn, Susan. 2001. "Gender Relations in Indonesia: What Women Want." In *Indonesia Today, Challenges of History*, edited by Grayson Lloyd and Shannon Smith, 270–82. Pasir Panjang, Singapore: Institute of Southeast Asian Studies.

———. 2004. *Women and the State in Modern Indonesia*. Cambridge, U.K.: Cambridge University Press, Cambridge.

Blackden, C. Mark, and Quentin Wodon, eds. 2006. *Gender, Time Use, and Poverty in Sub-Saharan Africa*. Washington, DC: World Bank.

Blau, David, and Janet Currie. 2006. "Preschool, Day Care, and Afterschool Care: Who's Minding the Kids?" In *Handbook of the Economics of Education, Volume 2*, edited by Eric A. Hanushek and Finis Welch. Amsterdam: North-Holland.

Blau, Francine, and Lawrence M. Kahn. 2003. "Understanding International Differences in the Gender Pay Gap." *Journal of Labor Economics* 21 (1): 106–44.

Bloom, David, David Canning, Günther Fink, and Jocelyn Finley. 2009. "Fertility, Female Labor Force Participation and the Demographic Dividend." *Journal of Economic Growth* 14: 79–101.

Bosch, Mariano, and William Maloney. 2011. "Sectoral Choice and Family Formation: Evidence from Labor Market Transitions in Ghana, Mexico, Serbia, and Thailand." Background paper for the *World Development Report 2012* and commissioned for *Toward Gender Equality in East Asia and the Pacific: A Companion to the* World Development Report. Washington, DC: World Bank.

Boserup, Ester. 1970. *Woman's Role in Economic Development*. London: Earthscan.

BPS (Badan Pusat Statistik). Indonesia National Labor Force Survey (SAKERNAS). Jakarta, Indonesia.

———.BPS (Badan Pusat Statistik). Indonesia National Socioeconomic Survey (SUSENAS). Jakarta, Indonesia. http://dds.bps.go.id/.

Bruhn, Miriam 2009. "Female-Owned Firms in Latin America. Characteristics, Performance, and Obstacles to Growth." Policy Research Working Paper 5122, World Bank, Washington, DC.

Bruhn, Miriam, Dean Karlan, and Antoinette Schoar. 2010. "What Capital Is Missing in Developing Countries?" *American Economic Review: Papers & Proceedings* 100 (2): 629–33.

Brush, Candida G. 1992. "Research on Women: Past Trends, a New Perspective and Future Directions." *Entrepreneurship Theory and Practice* 16 (4): 5–30.

Burda, Michael, Daniel S. Hamermesh, and Philippe Weil. 2007. "Total Work, Gender and Social Norms." NBER Discussion Paper 2705, National Bureau of Economic Research, Cambridge, MA.

CEIC Data. New York. http://www.ceicdata.com/.

Chaudhuri, Sanjukta. 2009. "Economic Development and Women's Empowerment." Working paper, University of Wisconsin–Eau Claire.

Chen, Feinian. 2005. "Employment Transitions and the Household Division of Labor in China." *Social Forces* 84 (2): 831–51.

Chevalier, Arnaud, and Tarja Viitanen. 2002. "The Causality between Female Labour Force Participation and the Supply of Childcare." *Applied Economics Letters* 9: 915–18.

Chi, Wei, and Bo Li. 2007. "Glass Ceiling or Sticky Floor? Examining the Gender Pay Gap across the Wage Distribution in Urban China, 1987–2004." MPRA Paper 3544, University Library of Munich, Germany.

Cobb-Clark, Deborah A., and Michelle Tan. 2011. "Noncognitive Skills, Occupational Attainment, and Relative Wages." *Labour Economics* 18 (1): 1–13.

Cohen, Philip, and Matt Huffman. 2007. "Working for the Woman? Female Managers and the Gender Wage Gap." *American Sociological Review* 72 (5): 681–704.

Costa, Rita, and Bob Rijkers. 2011. "Gender and Rural Non-Farm Entrepreneurship." Background paper for the *World Development Report 2012*, World Bank, Washington, DC.

Croson, Rachel, and Uri Gneezy. 2009. "Gender Differences in Preferences." *Journal of Economic Literature* 47(2): 448–74.

Cuberes, D., and Marc Teignier-Baqué. 2011. "Does Gender Inequality Hinder Growth? The Evidence and Its Policy Implications." Background paper for the *World Development Report 2012*, World Bank, Washington, DC.

CWDI (Corporate Women Directors International). 2010. "CWDI/IFC 2010 Report: Accelerating Board Diversity." CWDI and International Finance Corporation, Washington, DC.

Dasgupta, Sukti, and David Williams. 2010. "Women Facing the Economic Crisis—The Garment Sector in Cambodia." In *Poverty and Sustainable Development in Asia: Impact and Responses to the Global Economic Crisis*, edited by A. Bauer and M. Thant. Manila: Asian Development Bank.

Davies, Simon, and Richard Record. 2010. "Background Paper for the Second Lao PDR Investment Climate Assessment: Gender and Entrepreneurship in the Lao PDR. Evidence from the 2009 Enterprise Survey." World Bank., Washington, DC.

de Brauw, Alan, Jikun Huang, Linxiu Zhang, and Scott Rozelle. 2011. "The Feminization of Agriculture with Chinese Characteristics." Background paper for the *World Development Report 2012*, World Bank, Washington, DC.

Deere, Carmen Diana, and Cheryl Doss. 2006. "The Gender Asset Gap: What Do We Know and Why Does It Matter?" *Feminist Economics* 12 (1–2): 1–50.

Deere, Carmen Diana, and Magdalena Leon. 2003. "The Gender Asset Gap: Land in Latin America." *World Development* 31 (6): 925–47.

de la Rica, Sara, Juan J. Dolado, and Vanesa Llorens. 2005. "Ceiling and Floors: Gender Wage Gaps by Education in Spain." IZA Discussion Paper 1483, Institute for the Study of Labor (IZA), Bonn, Germany.

Del Boca, Daniela. 2002. "The Effect of Child Care and Part Time Opportunities on Participation and Fertility Decisions in Italy." *Journal of Population Economics* 15 (3): 549–73.

de Mel, Suresh, David McKenzie, and Chris Woodruff. 2009a. "Are Women More Credit Constrained? Experimental Evidence on Gender and Microenterprise Returns." *American Economic Journal: Applied Economics* 1 (3): 1–32.

———. 2009b. "Measuring Microenterprise Profits: Must We Ask How the Sausage Is Made?" *Journal of Development Economics* 88 (1): 19–31.

Dhanani, Shafiq, and Iyanatul Islam. 2004. *Indonesian Wage Structure and Trends, 1976–2000*. Geneva: International Labour Organization.

Du, Fenglian, and Xiao-Yuan Dong. 2010. "Women's Labor Force Participation and Childcare Choices in Urban China During the Economic Transition." World Paper 2010-04, University of Winnipeg, Department of Economics, Manitoba.

Eckel, Catherine C., and Philip J. Grossman. 2008. "Men, Women and Risk Aversion: Experimental Evidence." In *Handbook of Experimental Economic Results*, edited by C. Plott and V. Smith. Amsterdam: Elsevier.

Economist. 2011. "Asian Demography: The Flight from Marriage." August 20.

Ellis, Amanda, Daniel Kirkwood, and Dhruv Malhotra. 2010. *Economic Opportunities for Women in the East Asia and Pacific Region*. Directions in Development Series. Washington, DC: World Bank.

Elson, Diane, and Ruth Pearson. 1981. "The Subordination of Women and the Internationalisation of Factory Production." In *Of Marriage and the Market: Women's Subordination in International Perspective*, edited by Kate Young, Carol Wolkowitz, and Roslyn McCullagh. London: CSE Books.

Enterprise Surveys (database). World Bank, Washington, DC. http://data.worldbank.org/data-catalog/enterprise-surveys

Entwisle, Barbara, and Gail E. Henderson, eds. 2000. *Re-Drawing Boundaries: Work, Household, and Gender in China*. Berkeley: University of California Press.

Esteve-Volart, Berta. 2004. "Gender Discrimination and Growth: Theory and Evidence from India." Development Economics Discussion Paper Series 42, Suntory and Toyota International Centres for Economics and Related Disciplines, London School of Economics and Political Science, London.

Fang, Hanming, and Michael P. Keane. 2004. "Assessing the Impact of Welfare Reform on Single Mothers." *Brookings Papers on Economic Activity* 35 (1): 1–116.

FAO (Food and Agricultural Organization)/Sida Partnership Coooperation. 2010. *National Gender Profile of Agricultural Households, 2010: Report based on the Lao Expenditure and Consumption Surveys, National Agricultural Census and the National Population Census*. Vientiane, Lao PDR: FAO.

Fatima, Ambreen, and Humera Sultana. 2009. "Tracing Out the U-Shape Relationship between Female Labor Force Participation Rate and Economic Development for Pakistan." *International Journal of Social Economics* 36 (1/2): 182–98.

Fernández, Raquel. 2010. "Does Culture Matter?" NBER Working Paper 16277, National Bureau of Economic Research, Cambridge, MA.

Fernández, Raquel, and Alessandra Fogli. 2005. *Fertility: The Role of Culture and Family Experience*. NBER Working Paper 11569, National Bureau of Economic Research, Cambridge, MA.

Fernández, Raquel, Alessandra Fogli, and Claudia Olivetti. 2004. "Mothers and Sons: Preference Formation and Female Labor Force Dynamics." *Quarterly Journal of Economics* 119 (4): 1249–99.

Field, Erica, Seema Jayachandran, and Rohini Pande. 2010. "Do Traditional Institutions Constrain Female Entrepreneurship? A Field Experiment on Business Training in India." *American Economic Review Papers and Proceedings* 100 (2): 125–29.

Firpo, S., N. M. Fortin, and T. Lemieux. 2009. "Unconditional Quantile Regressions." *Econometrica* 77 (3): 953–73.

Fuwa, Makiko. 2004. "Macro Level Gender Inequality and the Division of Household Labor in 22 Countries." *American Sociological Review* 69 (6): 751–67.

Gallaway, Julie H., and Alexandra Bernasek. 2004. "Literacy and Women's Empowerment in Indonesia: Implications for Policy." *Journal of Economic Issues* 38 (2): 519–25.

Galor, Oded, and David N. Weil. 1996. "The Gender Gap, Fertility, and Growth." *American Economic Review* 86: 374–87.

Ganguly, Anuradha, and Talwar Swapan. 2003. "Feminization of India's Agricultural Workforce." *Labour Education* (2–3): 29–33.

Gelbach, Jonah B. 2002. "Public Schooling for Young Children and Maternal Labor Supply." *American Economic Review* 92 (1): 307–22.

Gender Statistics (database). World DataBank, World Bank, Washington, DC. http://data.worldbank.org/data-catalog/gender-statistics.

Giles, John, and Firman Witoelar Kartaadipoetra. 2011. "Education, Segregation and Labor Markets in Indonesia." Background paper for the *World Development Report 2012*, World Bank, Washington, DC.

Giles, John, Dewen Wang, and Wei Cai. 2011. "The Labor Supply and Retirement Behavior of China's Older Workers and Elderly in Comparative Perspective." Policy Research Working Paper 5853, World Bank, Washington, DC.

Giné, Xavier, and Ghazala Mansuri. 2011. "Money or Ideas? A Field Experiment on Constraints to Entrepreneurship in Rural Pakistan." Photocopy, Development Economics Research Group, World Bank, Washington, DC.

Godquin, Marie, and Agnes Quisumbing. 2008. "Separate but Equal? The Gendered Nature of Social Capital in Rural Philippine Communities." *Journal of International Development* 20 (1): 13–33.

Goldin, Claudia. 1995. "The U-Shaped Female Labor Force Function in Economic Development and Economic History." In *Investment in Women's Human Capital,* edited by T. Paul Schultz. Chicago: University of Chicago Press.

Goldin, Claudia, Lawrence F. Katz, and Ilyana Kuziemko. 2006. "The Homecoming of American College Women: The Reversal of the Gender Gap in College." *Journal of Economic Perspectives* 20 (Fall): 133–56.

Goldstein, Markus, and Christopher Udry. 2008. "The Profits of Power: Land Rights and Agricultural Investment in Ghana." *Journal of Political Economy* 116 (6): 981–1022.

Graves, Laura, and Gary Powell. 1995. "The Effect of Sex Similarity on Recruiters' Evaluations of Actual Applicants: A Test of the Similarity-Attraction Paradigm." *Personnel Psychology* 48 (1): 85–98.

Greenwood, Jeremy, Ananth Seshadri, and Mehmet Yorukoglu. 2005. "Engines of Liberation." *Review of Economic Studies* 72 (1): 109–33.

GSO (General Statistics Office) [Vietnam]. Vietnam Household Living Standards Surveys. Hanoi, Vietnam. http://www.gso.gov.vn.

———. Vietnam Labor Force Survey. Hanoi, Vietnam. http://www.gso.gov.vn.

Gustafsson, Siv, and Frank Stafford. 1992. "Child Care Subsidies and Labor Supply in Sweden." *Journal of Human Resources* 27 (1): 204–30.

Hallward-Driemeier, Mary. 2006. "Improving the Legal Investment Climate for Women in Sub- Saharan Africa." Paper prepared for the NEPAD–OECD Conference, Brazzaville, December 12, 2006.

Hausmann, Ricardo, Laura D. Tyson, and Saadia Zahidi, eds. 2010. *The Global Gender Gap Report 2010.* Geneva: World Economic Forum.

Hertz, Thomas, Anna P. de la O Campos, Esteban J. Quiñones, Benjamin Davis, Carlo Azzari, and Alberto Zezza. 2009. "Wage Inequality in International Perspective: Effects of Location, Sector, and Gender." Paper presented at the FAO-IFAD-ILO Workshop on "Gaps, Trends and Current Research in Gender Dimensions of Agricultural and Rural Employment: Differentiated Pathways out of Poverty," Rome, March 31–April 2, 2009.

Hill, Anne M. 1996. "Women in the Japanese Economy." In *Women and Industrialization in Asia*, edited by Susan Horton. London: Routledge.

Hiroko, Takeda, Jieyu Liu, and Junko Tamashita. 2011. "Economic Restructuring and Changing Work/Family Life: The Cases of Japan and China." In *New Welfare States in East Asia: Global Challenges and Restructuring*, edited by Gyu-Jin Hwang. Cheltenham, U.K.: Edward Elgar.

Honig, Emily. 2000. "Iron Girls Revisited: Gender and the Politics of Work in the Cultural Revolution." In *Re-Drawing the Boundaries of Work, Households, and Gender*, edited by Barbara Gutwisle and Gail Henderson. Berkeley: University of California Press.

Horton, Susan. 1996. "Women and Industrialization in Asia: Overview." In *Women and Industrialization in Asia*, edited by in Susan Horton. London: Routledge.

Huffman, Matt L., Philip Cohen, and Jessica Pearlman. 2010. "Engendering Change: Organizational Dynamics and Workplace Gender Segregation, 1975–2005." *Administrative Science Quarterly* 55 (2): 255–77.

Hughes, James, and Margaret Maurer-Fazio. 2002. "The Effects of Market Liberalization on the Relative Earnings of Chinese Women." *Journal of Comparative Economics* 30 (4): 709–31.

Hughes, James, Margaret Maurer-Fazio, and Dandan Zhang. 2005. "Economic Reform and Changing Patterns of Labor Force Participation in Urban and Rural China." William Davidson Institute Working Paper 787, University of Michigan, Ann Arbor.

———. 2007. "An Ocean Formed from One Hundred Rivers: The Effects of Ethnicity,

Gender, Marriage, and Location on Labor Force Participation in Urban China." *Feminist Economics* 13 (3–4): 125–53.

Hundley, Greg. 2001. "Why Women Earn Less than Men in Self-Employment." *Journal of Labor Research* 22 (4): 817–29.

IFC (International Finance Corporation). 2010. "Economic Opportunities for Women in the Pacific." International Financial Corporation, Washington, DC.

———. 2011. *Women's Entrepreneurship in the Phillippines*, Vol. 1, Washington, DC: IFC.

IFC (International Finance Corporation) and MPDF (Mekong Private Sector Development Facility). 2006. "Women Business Owners in Vietnam: A National Survey." IFC Gender Entrepreneurship Markets Initiative and MPDF, Washington, DC.

IFC (International Finance Corporation)/NORC Indonesia. 2010. "Indonesia Small and Medium Enterprise Banking Survey." International Financial Corporation, Washington DC; NORC, Chicago. http://www.norc.org/Research/Projects/Pages/ifc-indonesia-small-and-medium-enterprise-banking-survey.aspx.

ILO (International Labour Organization). 2010. "Women in Labor Markets: Measuring Progress and Identifying Challenges." ILO, Geneva.

ILO Key Indicators of the Labour Market (KILM) database. International Labour Office. Korea Labor Force Survey and Population Census. http://kilm.ilo.org/kilmnet/.

Indonesia Family Life Survey. 2007/2008. RAND Family Life Surveys, IFLS-4. http://www.rand.org/labor/FLS/IFLS/ifls4.html.

Juhn, Chinhui, and Simon Potter. 2006. "Changes in Labor Force Participation in the United States." *Journal of Economic Perspectives* 20 (3): 27–46.

Juhn, Chinhui, and Manuelita Ureta. 2003. "Employment of Married Women and Economic Development: Evidence from Latin American Countries." Society of Labor Economists Meeting, Toronto.

Karlan, Dean, and Martin Valdivia. 2011. "Teaching Entrepreneurship: Impact of Business Training on Microfinance Clients and Institutions." *Review of Economics and Statistics* 93 (2): 510–27.

Keitel, Robert S. 2009. "Innovation in Borderless Distance Learning of English." Paper presented at the 13th UNESCO-APEID International Conference World Bank–KERIS High-Level Seminar on ICT in Education.

Kim, Elim. 2005. "Korean Women's Activities for Legislation to Guarantee Gender Equality in Employment. A Symposium: Legal Feminism in Korea." *Journal of Korean Law* 5 (2): 49–65.

Kim, Sung won, Vanessa Fong, Hirokazu Yoshikawa, Niobe Way, Xinyin Chen, Huihua Deng, and Zuhong Lu. 2010. "Income, Work Preferences and Gender Roles among Parents of Infants in Urban China: A Mixed Method Study from Nanjing." *China Quarterly* 204: 939–59.

Klapper, Leora F., and Simon C. Parker. 2010. "Gender and the Business Environment for New Firm Creation." *World Bank Research Observer* 26 (2): 237–57.

Klasen, Stephan, and Francesca Lamanna. 2009. "The Impact of Gender Inequality in Education and Employment on Economic Growth: New Evidence for a Panel of Countries." *Feminist Economics* 15 (3): 91–132.

Korenman, Sanders, and David Neumark. 1992. "Marriage, Motherhood, and Wages." *Journal of Human Resources* 27 (2): 233–55.

Kushnir, Khrystyna, Melina Laura Mirmulstein, and Rita Ramalho. 2010. "Micro, Small and Medium Enterprises Around the World: How Many Are There, and What Affects the Count?" MSME Country Indicators, World Bank/IFC, Washington, DC.

LABORSTA Internet (database). International Labour Organization, Geneva, Switzerland. http://laborsta.ilo.org/.

Lam, Alice. 1992. "The Japanese Equal Employment Opportunity Law: Its Effects on Personnel Management and Policies and Women's Attitudes." MPRA Working Paper 11559, University Library of Munich, Germany.

Lee, Bun Song, Soomyung Jang, and Jayanta Sarkar. 2008. "Women's Labor Force Participation and Marriage: The Case of Korea." *Journal of Asian Economics* 19 (2): 138–54.

Li, Shi, and Jin Song. 2011. "Changes in Gender Wage Gap in Urban China During 1996–2007." CIBC Working Paper 2011-20, University of Western Ontario, Department of Economics, London, Ontario.

Liu, Dongxiao, and Elizabeth Heger Boyle. 2001. "Making the Case: The Women's Convention and Gender Discrimination in Japan." *International Journal of Comparative Sociology* 42: 389–404.

LSB (Lao Statistics Bureau). Lao Expenditure and Consumption Survey. Vientane, Lao PDR. http://www.nsc.gov.la/index.php?option=com_content&view=article&id=50&Itemid=73&lang=.

Mammen, Kristin, and Christina Paxson. 2000. "Women's Work and Economic Development." *Journal of Economic Perspectives* 14 (4): 141–64.

Maurer-Fazio, Margaret, Rachel Connelly, Lan Chen, and Lixin Tang. 2011. "Childcare, Eldercare, and Labor Force Participation of Married Women in Urban China, 1982–2000." *Journal of Human Resources* 46 (2): 261–94.

Meng, Xin 1996. "The Economic Position of Women in Asia" *Asian-Pacific Economic Literature* 10 (1): 23–41.

Miller, Robbie Louise. 1998. "Women's Job Hunting in the 'Ice Age': Frozen Opportunities in Japan." *Wisconsin Journal of Law, Gender, and Society* 13 (2): 228.

———. 2003. "The Quiet Revolution: Japanese Women Working Around the Law." *Harvard Journal of Law and Gender* 26: 163–215.

Mincer, Jacob, and Solomon Polachek. 1974. "Family Investment in Human Capital: Earnings of Women." *Journal of Political Economy* 82 (2): S76–S108.

Morrison, Andrew, Dhushyanth Raju, and Nistha Sinha. 2007. "Gender Equality, Poverty, and Economic Growth." Policy Research Working Paper 4349. World Bank, Washington, DC.

Mu, Ren, and Dominique van de Walle. 2009. "Left Behind to Farm? Women's Labor Re-Allocation in Rural China." Policy Research Working Paper 5107, World Bank, Washington, DC.

Murthi, Mamta, Anne-Catherine Guio, and Jean Drèze. 1995. "Mortality, Fertility, and Gender Bias in India: A District-Level Analysis." *Population and Development Review* 21 (4): 745–82.

Nakavachara, Voraprapa. 2010. "Superior Female Education: Explaining the Gender Earnings Gap Trend in Thailand." *Journal of Asian Economics* 21 (2): 198–218.

NIS (National Institute of Statistics) [Cambodia]. Cambodia Socio-Economic Survey (CSES). National Institute of Statistics, Phnom Penh, Cambodia. http://www.nis.gov.kh/index.php/social-statistics/cses.

Ñopo, Hugo, Nancy Daza, and Johanna Ramos. 2011. "Gender Earnings Gaps in the World." Background paper for the *World Development Report 2012*, World Bank, Washington, DC.

NSCB (National Statistical Coordination Board) [Philippines]. 2006. Family Income and Expenditures Survey (FIES) 2006. National Statistical Coordination Board, Makati City, Philippines. http://www.nscb.gov.ph/fies/default.asp.

NSD (National Statistics Directorate) [Timor-Leste]. 2007. *Timor-Leste Survey of Living Standards*. Dili, Timor-Leste: Ministry of Planning and Finance, National Statistics Directorate. http://dne.mof.gov.tl/TLSLS/AboutTLSLS/index.htm.

NSO Mongolia (National Statistical Office of Mongolia). Living Standards Measurement Survey (LSMS). National Statistical Office, Ulaanbaatar, Mongolia. http://www.nso.mn.

NSO Thailand (National Statistical Office Thailand). Thailand Socio-Economic Survey (SES). Bangkok, Thailand: National Statistical Office. http://web.nso.go.th/tnso.htm.

Ogawa, Keiichi, and Masuma Akter. 2007. "Female Labor Force Participation in Indonesia." *Journal of International Cooperation Studies* 14 (3): 71–108.

O'Neill, June. 2003. "The Gender Gap in Wages, Circa 2000." *American Economic Review* May.

Pastore, Fransesco 2009. "The Gender Gap in Early Career in Mongolia." Discussion Paper No. 4480, Institute for the Study of Labor (IZA), Bonn, Germany.

Penn World Table 7.0. Alan Heston, Robert Summers and Bettina Aten, Penn World Table Version 7.0, Center for International Comparisons of Production, Income and Prices at the University of Pennsylvania, May 2011.

Peterman, Amber, Julie Behrman, and Agnes Quisumbing. 2010. "A Review of Empirical Evidence on Gender Differences in Nonland Agricultural Inputs, Technology, and Service in Developing Countries." IFPRI Discussion Paper 975, International Food Policy Research Institute, Washington, DC.

Pierre, G. 2011. "Recent Labor Market Performance in Vietnam through a Gender Lens." Background paper for the "Vietnam Country Gender Assessment." World Bank, Washington, DC. http://documents.worldbank.org/curated/en/2011/11/15470188/vietnam-country-gender-assessment.Vietnam Gender Assessment.

Pörtner, Claus C. 2009. "Children's Time Allocation, Heterogeneity and Simultaneous Decisions." Working Paper UWEC-2009-15, University of Washington, Department of Economics, Seattle.

Psacharopoulos, George, and Zafiris Tzannatos. 1989. "Female Labor Force Participation: An International Perspective." *World Bank Research Observer* 4 (2): 187–201.

Quisumbing, Agnes R. 1995. "Gender Differences in Agricultural Productivity." FCND Discussion Paper 5, International Food Policy Research Institute, Washington, DC.

———. 1998. "Women in Agricultural Systems." In *Women in the Third World: An Encyclopedia of Contemporary Issues*, edited by N. Stromquist, 261–72. New York: Garland.

Quisumbing, Agnes R., Jonna P. Estudillo, and Keijiro Otsuka. 2004. *Land and Schooling: Transferring Wealth Across Generations.* Baltimore, MD: Johns Hopkins University Press.

Rand, John, and Finn Tarp. 2011. "Does Gender Influence the Provision of Fringe Benefits? Evidence from Vietnamese SMEs." *Feminist Economics* 17 (1): 59–87.

Rawski, Thomas, and Robert W. Mead. 1998. "On the Trail of China's Phantom Farmers." *World Development* 26 (5): 767–81.

Sabarwal, Shwetlena, and Katherine Terrell. 2008. "Does Gender Matter for Firm Performance? Evidence from Eastern Europe and Central Asia." Policy Research Working Paper 4705, World Bank, Washington, DC.

Sabarwal, Shwetlena, Katherine Terrell, and Elena Bardasi. 2011. "How Do Female Entrepreneurs Perform? Evidence from Three Developing Regions." *Small Business Economics* 37 (4): 417–41.

Sakellariou, Chris. 2011. "Determinants of the Gender Wage Gap and Female Labor Force Participation in EAP." Paper commissioned for *Toward Gender Equality in East Asia and the Pacific: A Companion to the* World Development Report. Washington, DC: World Bank.

Sasaki, Masaru. 2002. "The Causal Effect of Family Structure on Labor Force Participation among Japanese Married Women." *Journal of Human Resources* 37 (2): 429–40.

Sasser, Alicia C. 2005. "Gender Differences in Physician Pay: Tradeoffs Between Career and Family." *Journal of Human Resources* 40 (2): 477–504.

Simavi, Sevi, Clare Manuel, and Mark Blackden. 2010. *Gender Dimensions of Investment Climate Reform: A Guide for Policy Makers and Practitioners.* Washington, DC: World Bank.

Sinha, J. N. 1967. "Dynamics of Female Participation in Economic Activity in a Developing Economy." In *Proceedings of the World Population Conference, Belgrade, 1965; Vol. 4. Migration, Urbanization, Economic Development,* 336–37. New York: United Nations.

Slootmaekers-Miranda, Veerle. 2011. "Cooking, Caring and Volunteering: Unpaid Work around the World." Social, Employment and Migration Working Papers, Organisation for Economic Co-operation and Development, Paris.

Stivens, Maila. 2006. "Family Values and Islamic Revival: Gender, Rights and the State Moral Project." *Women's Studies International Forum* 29 (4): 354–67.

SUSI (Survei Usaha Terintegrasi). 2002. Integrated Survey of Cottage and Small-Scale Firms. Badan Pusat Statistik, Jakarta, Indonesia.

Tam, Henry. 2011. "U-Shaped Female Labor Force Participation with Economic Development: Some Panel Data Evidence." *Economic Letters* 110 (2): 140–42.

Tansel, Aysit. 2001. "Economic Development and Female Labor Force Participation in Turkey: Time-Series Evidence and Cross-Province Estimates." ERC Working Paper 0105, Middle East Technical University, Economic Research Center, Ankara, Turkey.

NSO Thailand (National Statistical Office Thailand). Thailand Socio-Economic Survey (SES). Bangkok, Thailand. http://web.nso.go.th/tnso.htm.

Thomas, Duncan. 1990. "Intra-household Resource Allocation: An Inferential Approach." *Journal of Human Resources* 25 (4): 635–64.

Thomas, Duncan, and John Strauss. 1997. "Health and Wages: Evidence on Men and Women in Urban Brazil." *Journal of Econometrics* 77 (1): 159–85.

Timberlake, Sharon. 2005. "Social Capital and Gender in the Workplace." *Journal of Management Development* 24 (1): 34–44.

Tzannatos, Zafris. 1999. "Women and Labor Market Changes in the Global Economy: Growth Helps, Inequalities Hurt and Public Policy Matters." *World Development* 27 (3): 551–69.

Udry, Christopher. 1996. "Gender, Agricultural Production, and the Theory of the Household." *Journal of Political Economy* 104 (5): 1010–46.

UNIFEM, World Bank, ADB, UNDP, and DFID/UK (United Nations Development Fund for Women, World Bank, Asian Development Bank, United Nations Development Programme, and Department for International Development of the United Kingdom). 2004. "A Fair Share for Women: Cambodia Gender Assessment." UNIFEM, WB, ADB, UNDP, and DFID/UK, Phnom Penh, Cambodia.

VCCI (Vietnam Chamber of Commerce and Industry). 2006. "Targeted Policies That Support Women's Entrepreneurship Can Boost Vietnam's Economic Growth." *Business Issues Bulletin* 13 (16).

Vijverberg, W., and J. Haughton. 2004. "Household Enterprises in Vietnam: Survival, Growth, and Living Standards." In *Economic Growth, Poverty, and Household Welfare in Vietnam*, edited by David Dollar and Paul Glewwe. Washington, DC: World Bank.

Weinberg, Bruce. 2000. "Computer Use and the Demand for Female Workers." *Industrial and Labor Relations Review* 53 (January): 290–308.

White, Sally. 2006. "Gender and the Family." In *Voices of Islam in Southeast Asia*, edited by Greg Fealy and Virginia Hooker, 273–352. Singapore: Institute of Southeast Asian Studies.

Women, Business and the Law (database). World Bank, Washington, DC. http://wbl.worldbank.org/data.

Wood, Robert, Mary Corcoran, and Paul Courant. 1993. "Pay Differences Among the Highly Paid: Male-Female Earnings Gap in Law Salaries." *Journal of Labor Economics* 11 (3): 417–44.

World Bank. 2010a. "Indonesia Jobs Report: Towards Better Jobs and Security for All." World Bank, Washington, DC.

———. 2010b. Indonesia—Reconstruction of Aceh Land Administration System Project. World Bank, Washington, DC.

———. 2011a. Defining Gender in the XXI Century: Conversations with Men and Women Around the World (qualitative assessment dataset). World Bank, Washington, DC. http://go.worldbank.org/CD8RN24BP0.

———. 2011b. "Female Entrepreneurship in Mongolia." Background paper for the Mongolia Gender Assessment. World Bank, Washington, DC.

———. 2011c. "Gender Disparities in Mongolian Labor Markets and Policy Suggestions." World Bank, Washington, DC.

———. 2011d. "Gender Equality and Health." Indonesia Policy Brief 2, World Bank, Washington, DC.

———. 2011e. "Vietnam Country Gender Assessment." World Bank, Washington, DC.

———. 2011f. "Women, Business and the Law in East Asia and Pacific Region." A background paper commissioned for *Toward Gender Equality in East Asia and the Pacific: A Companion to the World Development Report.* Washington, DC: World Bank.

———. 2011g. World Development Report 2012: *Gender Equality and Development.* Washington, DC: World Bank.

———. 2012. *The Elderly and Old Age Support in Rural China: Challenges and Prospects.* Washington, DC: World Bank.

World Bank and ADB (Asian Development Bank). 2005. "Mongolia, Country Gender Assessment." World Bank: Washington DC.

World Development Indicators (WDI) database. World DataBank, World Bank, Washington, DC. http://databank.worldbank.org.

Yu, Wei-hsin. 2002. "Jobs for Mothers: Married Women's Labor Force Reentry and Part-time, Temporary Employment in Japan." *Sociological Forum* 17 (3): 493–523.

Yuan, Ni, and Sarah Cook. 2010. "Gender Patterns of Informal Employment in Urban China." In *Gender Equality and China's Economic and Social Transformation: Informal Employment and Care Provision*, edited by Xiao-Yuan Dong and Sarah Cook. Beijing: Economic Science Press.

Zuo, Jiping, and Yanjie Bian. 2001. "Gendered Resources, Division of Housework and Perceived Firness: A Case in Urban China." *Journal of Marriage and Family* 63: 1122–33.

Agency: Voice and Influence within the Home and in Society | 4

In much of the world, women have a more limited voice and influence than men in decision making in their homes, their communities, and society. Women are also more likely to be victims of gender-based violence. The inability of women to voice and act on their preferences negatively affects their own welfare and is detrimental to development.

This chapter analyzes women's agency in East Asian and Pacific countries. The chapter defines *agency* as the ability of individuals or groups to give voice to and act on their preferences and to influence outcomes that affect them and others in society. Agency is affected by and also affects individuals' ownership of and control over endowments and their access to economic opportunities (Kabeer 1999). The discussion of agency in this chapter also includes the ability of countries to ensure the safety and security of women in their homes and in society, because gender-based violence and trafficking of women reflects the extreme deprivation of women's agency in society.

Within a household or partnership, one's relative power affects the strength of one's voice and influence in household decisions, such as how to spend or invest family resources. Similarly, at the community or societal level, the relative power of individuals and groups affects their ability to act on their preferences and influence outcomes in the economic, social, and political domains. The relative power of different members of society, which often differs systematically by gender, reflects a complex combination of one's personal characteristics, prevailing social norms, and the broader legal and institutional environments.

Agency—important for gender equality and development

The ability to act on one's preferences, regardless of one's gender, and to translate those preferences into desired outcomes is a development objective in its own right. As discussed in chapter 1, development not only involves raising incomes or reducing poverty, but also involves a process of expanding freedoms and choices available to all people (Sen 1999). Agency is a measure of a person's well-being, reflecting both the ability to achieve as well as actual achievements (Sen 1992).

Women's agency enhances development. When women are free to make choices, that freedom positively affects all levels of society. Increasing women's voice and influence in the home has been found to improve children's education, health, and welfare (Duflo 2003; Fiszbein and Schady 2009; Haddad, Hoddinott, and Alderman 1997; Thomas 1995), as discussed in chapter 2. Increasing women's representation in firm ownership and management, and on corporate boards, also increases gender equality within firms and increases the provision of nonwage benefits to workers (Cohen and Huffman 2007; Ely 1995; Hultin and Szulkin 2003; Rand and Tarp 2011). Increasing women's representation in elected office not only ensures that decisions are more representative of the voting population, but also can lead to increased provision of public goods, better natural resource management, and increased reporting of crimes against women (Agarwal 2009; Ban and Rao 2009; Beaman et al. 2012; Chattopadhyay and Duflo 2004). Global evidence shows that violence against women has lasting negative effects on economic development in addition to causing significant social, psychological, and physical harm to those who experience and witness it (Morrison and Orlando 2004; Morrison, Ellsberg, and Bott 2007). Reducing gender-based violence thus results in healthier workers and higher economic productivity, with dynamic benefits across generations.

Economic growth and development can, in turn, contribute to strengthening women's agency in some areas. As discussed in earlier chapters, growth and development result in better education and health outcomes for women—and better human capital outcomes for women contribute directly to stronger voice and influence, whether in the home, in the economy, or in society. Economic development, as measured by gross domestic product (GDP) per capita, is also associated with higher levels of civic activism, including on issues related to gender equality (figure 4.1). Civic activism is a measure of "collective agency"; that is, it is the space for both male and female citizens to express their voice in the public

sphere.[1] To the extent that development is accompanied by stronger legal and judicial systems, more-developed societies provide women (and men) with better access to justice, which strengthens their voice and protects them against the extreme deprivation of agency.

Growth and development alone are not enough to enhance women's agency in all its dimensions. As shown in previous chapters, women have made positive strides toward gender equality in education and health, yet gender gaps remain in access to assets and economic opportunity. Increasing women's ability to earn and accumulate assets is critical for strengthening their voice and influence in society and making them less vulnerable to domestic violence and other types of abuse. Moreover, the relationship between economic development and women's political representation, an important pathway toward agency in society, is unclear (figure 4.2). Although development can contribute to strengthening women's agency in some dimensions, data show that improvements in a number of other areas are not automatic. Thus, governments need to develop policies that actively raise women's agency if they are to induce meaningful changes toward gender equality in agency.

This chapter aims to strengthen understanding of gender and agency in East Asian and Pacific countries, and to lay the foundation for identifying policy priorities to strengthen women's voice and influence. The analysis in the chapter focuses on agency in the following three domains:

- Agency in the household and in individual decisions is examined through household decision making, control of resources, and reproductive decisions.
- Agency in the public sphere is examined through women's participation and representation in the private sector, civil society, politics, and public institutions.
- Safety and security in expressing one's agency are examined through the prevalence of gender-based violence, an extreme deprivation of agency.

The form of agency that is most frequently measured is the decision-making power of men and women (Kabeer 1999; Mason 2005; McElroy 1990). Agency may be more explicitly measured by examining women's mobility in the public domain, their participation in public action, and the incidence of gender-based violence (Kabeer 1999). Some researchers have assessed gender differences in bargaining power within a household by examining the extent to which people's choices change when factors affecting their bargaining power, such as education, relative earnings, or asset holdings, change (Duflo 2003; Quisumbing and Maluccio 2003; Thomas 1990, 1992).

Agency at the household level is difficult to measure, since negotiations for decision-making power often occur within private spaces in the household. Furthermore, since indicators of agency are of relatively new interest to the international development community, many have not yet been measured over time. Where possible, the chapter will present information on how women's agency has evolved over time in the region. When data on different dimensions of agency are not available over time, however, the chapter will present the most recent evidence available—both qualitative and quantitative.

The remainder of the chapter is structured as follows. The next section analyzes the state of agency in the region. The third section analyzes the factors that influence agency, and the fourth section identifies key policy priorities for promoting gender equality in voice and influence in the region. These directions for policy are discussed in further detail in chapter 6.

Despite the geographic proximity between East Asian countries and the Pacific Islands, their development experiences and paths toward gender equality have been different, especially with respect to women's voice and influence. For this reason, this chapter distinguishes, where possible, between the progress made and the challenges faced by East Asian and Pacific countries in the different domains of agency.

FIGURE 4.1 There is a positive relationship between economic development and civic activism

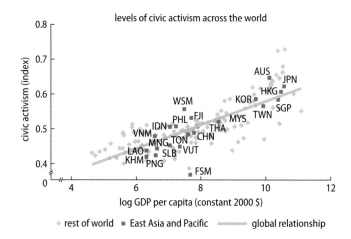

Source: World Development Indicators (WDI) database; Indices of Social Development (ISD) database.
Note: GDP = gross domestic product.

FIGURE 4.2 There is no clear relationship between economic development and women's representation in parliament

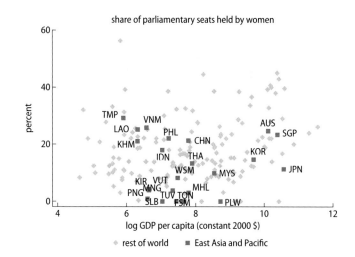

Source: WDI database; ISD database.
Note: GDP = gross domestic product.

The state of women's agency in East Asia and the Pacific

The East Asia and Pacific region has experienced uneven gains in women's agency over the past two decades. Although women now have more household decision-making power, more influence and voice in the public sphere, and new laws to protect their choices and interests in society, progress has been uneven across countries, and many challenges still remain. Women in East Asia had the highest representation in national parliaments in the developing world in the 1990s; yet, the share of women in parliaments has reached a plateau and fallen behind other regions in the past decade. Women have little representation among parliamentarians in the Pacific. Women's participation in the private sector has increased, but women remain a minority among firm owners and on corporate boards. The sections that follow examine evidence on women's voice and influence in the home and in the public domain, as well as evidence on gender-based violence in the region.

Agency in the household and in personal decisions

Women's household decision-making power in the region is relatively high. Women from all wealth quintiles in East Asia and the Pacific are more likely to have control over large purchases and over decisions to visit family and relatives than women in other developing regions, and they are among the most likely to have control over their own earnings. However, a higher share of women in East Asia than in the Pacific control their own income and have a say in other household decisions, reflecting a great intraregional variation.

Women in East Asia, in particular, have high levels of autonomy. According to Demographic and Health Survey (DHS) data, nearly 70 percent of married women ages 15 to 49 in Cambodia and Indonesia report that they control their own earnings; 31 and 28 percent, respectively, report joint control of their earnings with their husbands. Only 1 percent of women in Cambodia, and 3 percent of women in Indonesia, report that their husbands decide how their earnings are used (figure 4.3). Women in Cambodia and Timor-Leste report high levels of control not only over their own income, but also over their husband's income (DHS data, various years).[2]

Women in the Pacific have relatively less control over their own earnings. Between 13 and 15 percent of women in Marshall Islands, Samoa, and Tuvalu report that their husbands have control over their wives' cash earnings (figure 4.3). Studies in different countries in the Pacific also find high levels of financial control by men in the household. In Kiribati, for example, 19 percent of women report that their partners do not allow them to make any financial decisions for household expenditures, and 12 percent of women report that they cannot exercise control of their own incomes because their partners take it away. When intimate partner violence occurs, women have an even weaker voice over household financial decisions. Twenty-three percent of women who experience intimate partner violence report that their partners do not allow them to make any financial decisions related to household expenditure, and 22 percent report having no control of their own income.

Most women in the region have the freedom to make other household decisions, whether related to their own health care, household purchases, or visits to family and relatives. Women in the Pacific have slightly less ability to make decisions on their own health care and household purchases, on average, than women in East Asia. In the Philippines, 94 percent of women make their own health care decisions solely or jointly with their spouse, whereas only 70 percent of women in the Marshall Islands do so (figure 4.4). Similar patterns occur in how much say wives have in decisions regarding visits to her family and relatives. However, in some areas in East Asia, women also have relatively little decision-making power with regard to certain choices. For instance, 18 percent of women surveyed in Indonesia report that men have the final say in making decisions on household purchases. Women in Tuvalu and the Marshall Islands report having the lowest control over their own health care decisions and household purchases (figure 4.4; DHS data, various years).

Where DHS data are not available, country surveys from Kiribati, the Solomon Islands, and Vanuatu suggests that women have relatively low levels of autonomy with respect to household decision making. This data indicates a high prevalence of controlling behavior by husbands and male partners over household, financial, and mobility decisions. Of partnered women, 58 percent and 69 percent reported the experience of some sort of controlling behavior by their partners in the Solomon Islands and Vanuatu, respectively. This sort of behavior includes preventing her from seeing family, wanting to know where she is at all times, forbidding contact with other men, and controlling access to health care. Controlling behaviors by husbands are often correlated with a lack of agency in other dimensions, as well as with personal characteristics that hinder agency, such as poor endowments and economic opportunities, discussed in more detail below. In both the Solomon Islands and Vanuatu, women with little to no education are more likely to have partners who exhibit controlling behaviors than women with higher educational attainment. Women who have experienced intimate partner violence are also significantly more likely to experience controlling behaviors than women who have not experienced violence (SPC and NSO 2009; VWC and NSO 2011). In that sense, the factors that inhibit women's agency are likely to be mutually reinforcing.

Globally, women living in wealthier households are likely to have more decision-making power than women living in poorer households. This pattern bears out in Pacific countries, but less so in East Asia. Evidence from countries in other regions of the world illustrates that women in wealthier households have a wider set of choices and face fewer financial constraints. As a result, they have more freedom to make purchasing decisions and rarely have to forgo human capital investments in their children or to ration their access to goods and services. Available data for East Asia suggest, however, that women in wealthier households do not have substantially more control over decisions than women in households with lower income-levels (figure 4.5). Conversely, data

FIGURE 4.3 **Who decides how wives' cash earnings are used varies across the region**

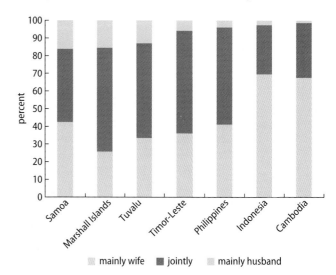

person who decides how the wife's cash earnings are used (%)

mainly wife   jointly   mainly husband

Source: DHS data, various years.
Note: Among currently married women ages 15–49 who receive cash earnings for employment.

FIGURE 4.4 **A majority of wives control decisions regarding their own health care and household purchases**

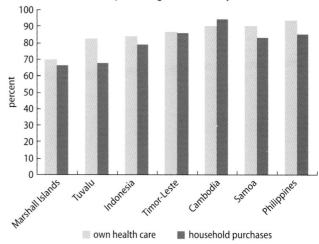

percentage of currently married women ages 15–49 who control health care and purchasing decisions solely or with husband

own health care   household purchases

Source: DHS data, various years.

from countries in the Pacific show a different picture. For instance, in Vanuatu, 50 percent of women in households at the bottom of the income distribution have to ask permission to visit family or others, while 41 percent of women in wealthier households have to do so (VWC and NSO 2011).

Women have made advances in their ability to make reproductive decisions in nearly all countries in the region. Reductions in fertility rates and fertility gaps—defined as unwanted fertility—are observed across most countries in East Asia, suggesting that women have gained greater control over their reproductive decisions. In many East Asian and Pacific countries, the use of contraceptive methods—whether modern or traditional—has become increasingly more common. For example, in Cambodia, the percentage of married women using modern methods of family planning increased from 19 percent to 35 percent between 2000 and 2010. The percentage of married women using traditional methods increased from 5 percent to 16 percent during the same period (NIS and DGH 2011).

Most countries in both East Asia and the Pacific have seen decreases in the number of births per woman (figure 4.6). In 2009,

China, Thailand, and Vietnam had the lowest fertility rate in the region, at less than two births per woman. The Pacific Islands (barring Fiji, at 2.7 births per woman) have the highest fertility rates in the region, at over 3.5 births per woman. While all countries in the region have experienced a decrease in female fertility rates, Timor-Leste experienced a spike of 7.8 births per woman in 2003 before declining to 5.7 births per woman in 2009–10. Despite progress, fertility remains the highest among the poorest Timorese, at 7.3 births per woman, compared to 4.2 births among the wealthiest. Women with no education have significantly more children than women with more than a secondary education: 6.1 versus 2.9 children per woman. The high fertility observed in Timor-Leste can be partially attributed to weakened health systems due to prolonged political conflict (World Bank 2011d).

Countries with the highest fertility rates also have the highest percentages of adolescent pregnancies. As can be seen in figure 4.6 and figure 4.7, the countries in the region with the highest fertility rates also have the highest percentage of pregnant teenagers ages 15 to 19. Among young women in the Pacific in this age range, a large number of pregnancies are unwanted or mistimed (figure 4.8). In the Marshall Islands, Solomon Islands, and Nauru, more than half of young women report having had an unwanted pregnancy (Kennedy et al. 2011).

In some countries where fertility rates have decreased, women still have more children than they desire. In the past 50 years in the Philippines, the average number of children per woman decreased from 7.0 children to 3.3 children (Costello and Casterline 2002). However, the fertility gap in the Philippines is still high, with an average number of children per mother of 3.3 and a desired number of children per mother of only 2.4 (NSO [Philippines] and ICF Macro 2009). This may be more the result of norms, or weak service delivery than spousal control. In fact, 65 percent of women who do not use contraceptives mention health concerns, fear of side effects, and inconvenience of use; 18 percent cite

FIGURE 4.5   Women in East Asia have greater control over decisions than in other regions
(percent)

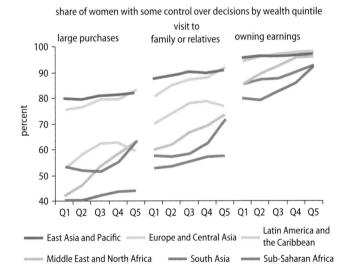

share of women with some control over decisions by wealth quintile

Source: World Bank 2011b.
Note: Country coverage for East Asia and Pacific includes Cambodia, Indonesia, and the Philippines.

opposition by women or their partner or religious factors; 23 percent cite issues of access, distance, or cost. Only 2 percent cite lack of knowledge of method or source (NSO [Philippines] and ICF Macro 2009). Women who experience violence also seem more likely to have an unplanned pregnancy. In Kiribati, for example, 22 percent of women who experienced domestic violence had an unplanned pregnancy, compared to 12 percent of those who did not experience domestic violence (SPC, Ministry of Internal and Social Affairs, and Statistics Division 2010).

Advancement toward gender equality in agency in the household in one dimension may not translate into progress in other dimensions. A study of 21 communities in Thailand found that women are relatively free to make fertility-related decisions and participate in the labor market, but they have only moderate levels of freedom of movement and high levels of fear about disagreeing with their husbands on household matters (Mason and Smith 2003). Other studies show even more striking differences in levels of agency across dimensions in countries in the Pacific. For instance, data collected for the international Social Institutions and Gender Index indicate that women in Papua New Guinea face a relatively high level of agency in some public dimensions, such as freedom of movement and freedom of dress, while they experience very low levels of agency in their households, for example, with respect to violence (SIGI, 2010 data).

Agency in the public domain

Women in East Asia have made dynamic strides in participation and influence in the private sector, civil society, and political institutions. Current levels are still far from equal. In the Pacific, women have experienced little change in many public domains and continue to face large challenges.

The private sector

Women's participation and leadership in the private sector has made progress in East Asia. In Hong Kong SAR, China, the percentage

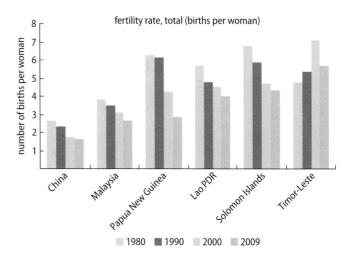

FIGURE 4.6 **Fertility rates in the region have declined over the past three decades**

fertility rate, total (births per woman)

1980 ■ 1990 ■ 2000 ■ 2009

Source: WDI database, 2011 data.

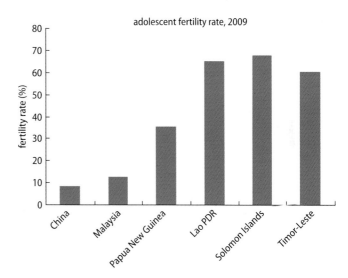

FIGURE 4.7 **Adolescent fertility is especially high in the Pacific**

adolescent fertility rate, 2009

Source: WDI database, 2009 data.
Note: Adolescent fertility rate is the number of births per 1,000 women ages 15-19.

of women on corporate boards increased from less than 5 percent to almost 9 percent over the past decade (Mahtani, Vernon, and Sealy 2009). East Asia has a high percentage of firms owned by women and firms with women in the top management compared to other developing regions. Mongolia, the Philippines, and Vietnam have among the highest levels of female participation in business

FIGURE 4.8 **The rate of unwanted and mistimed pregnancies is especially high in the Pacific**

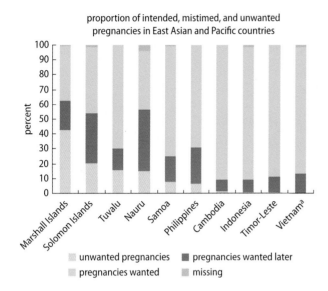

proportion of intended, mistimed, and unwanted pregnancies in East Asian and Pacific countries

legend:
- unwanted pregnancies
- pregnancies wanted
- pregnancies wanted later
- missing

Source: Demographic and Health Surveys, various years.
a. Percent distribution of births to women 15–19 in the five years preceding the survey, except for Vietnam, where data refer to three years preceding the survey.

ownership (Enterprise Surveys, 2006–09 data).

Still, within companies, women are less likely to be found in management and decision-making positions. The share of female directors ranges from 10 percent in the Philippines and 7 percent in China and Thailand to 5 percent in Indonesia and under 2 percent in the Republic of Korea and Japan (CWDI 2010). Although 67 percent of all publicly traded companies in Hong Kong SAR, China, have at least one woman on the board, only 15 percent of those companies have more than one (Mahtani, Vernon, and Sealy 2009). Globally, only one country in the world, Norway, has succeeded in having more than 30 percent female representation on corporate boards, while only one-third of countries have female board representation of over 10 percent (CWDI 2010).

Despite the presence of women as firm owners in several countries in the region, women make up a much lower share of female top managers and so are less likely to influence firm decisions. Most countries in the region have firms with female participation in

ownership. However, owning an asset might be significantly different from having the principal voice in managing and running an enterprise. In the five countries in the region for which data are available—Indonesia, Malaysia, Mongolia, the Philippines, and Vietnam—more firms have female participation in ownership than female top managers (figure 4.9).

Civil society and grassroots movements
East Asian women have seen some improvements in the strength of their voice and their ability to actively participate in civil society and grassroots movements. Data show a positive relationship between the strength of civil society and gender equality. Civil society groups and women's organizations and movements, in particular, have been important in creating broader space for women to have voice and influence in society. The size and nature of civil society have not been static over the past two decades. Figure 4.10 shows that even though wealthier countries such as Australia and Japan still have a more thriving civil society than poorer countries, over the past two decades there has been an increase in the presence of civil society in poorer countries. As a whole, civil society has been strengthened in the region, which has in part been influenced by increasingly amicable relationships between civil society and governments.[3]

Although most civil society organizations (CSOs) are focused on broad development issues—poverty reduction, education, and health—more organizations now focus on a wider range of gender issues than two decades ago. Of those that currently focus on gender, some operate at the local level while others operate at the national level. In Indonesia, the civil society organization PEKKA (Women-Headed Household Empowerment Program) was created to address the needs of widows who were victims of conflict in Aceh, to improve their access to legal and financial assistance, and to improve their overall welfare. The program provides training for village paralegals who focus on domestic violence and family law, and also holds district

FIGURE 4.9　**Many firms have female participation in ownership, but fewer have a female top manager**

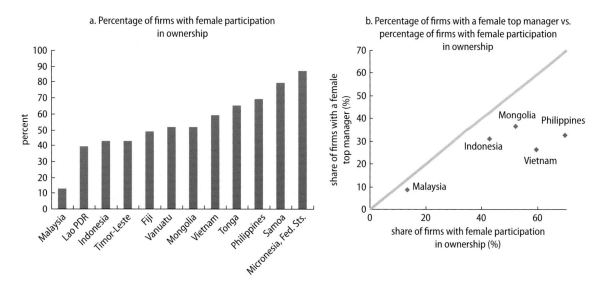

Source: World Bank estimates using Enterprise Surveys database, 2006–09 data.

forums to bring together judges, prosecutors, police, nongovernmental organizations, and government officials to raise awareness of gender issues (PEKKA 2012; World Bank 2011b). In Papua New Guinea, the number of CSOs targeting their assistance to women and enhancing their access to services has expanded. For instance, some provide services such as microcredit and savings to economically empower women. Many Mother's Groups have flourished throughout this period to address local women's needs such as nutrition and health services. Through their services, they enable women to engage actively in their children's health care (Imai and Eklund 2008). Examples of gender-focused activist groups include Gender and Development for Cambodia, the National Council of Women in Thailand (under the patronage of the Queen), and the Liberal Women's Brain Pool in Mongolia. Others that operate at the international level include Save the Children, Coordination of Action Research on AIDS, Mobility Asia (CARAM-Asia), and End Child Prostitution in Asian Tourism (ECPAT). CSOs interested in tackling policy issues such as eliminating human trafficking, increasing political representation

FIGURE 4.10　**Civic activism has grown in the low- and middle-income countries in the region**

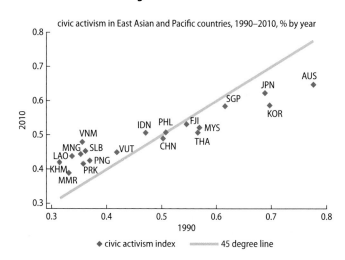

Source: ISD database.

of women, and changing laws to decrease the discrimination of women have also emerged throughout the region.

Pacific countries have seen some progress in the growth of gender-focused CSOs. Civil society continues to be a vital feature of development in the Pacific Island countries, especially in delivering services (Swain 2000). The

Pacific region has seen an increase of CSOs at the country level as well as multi-island network organizations in the past few decades. Even though only a few organizations focus solely on the promotion of gender equality, an increasing number of organizations are recognizing the importance of taking a more active role in promoting gender equality by giving voice to women.

Politics

For countries in the region that have experienced growth in female political representation, progress has been slow, and in most countries, levels remain below what is commonly perceived to be an acceptable threshold for women's voice in the political domain: 30 percent of electoral seats at the national and local levels (Agarwal 2010a, 2010b). When women do enter politics or public administration, they tend to be in lower-ranked positions and sectors and positions perceived as "female," such as government ministries of women, health, education, or social welfare, and not some more traditionally male ministries, such as finance or planning, where government resource allocations are typically made (World Bank 2011b).

The share of women in national parliaments varies tremendously across the region. Data on female representation in national parliaments provide insight into the level of agency in political decision making. As shown in figure 4.11, many East Asian countries, including Cambodia, Lao PDR, the Philippines, and Singapore, have increased the share of parliamentary seats held by women since 2000. In contrast, Mongolia has experienced a steep decline in female representation during the same period, from about 10 percent in 2000 to less than 4 percent in 2010 (figure 4.11). Figure 4.12 illustrates the variation across the region with respect to parliamentary seats held by women, with Timor-Leste, Lao PDR, and Vietnam having the highest percentage of seats held by women and the Pacific Islands being on the lower end of the spectrum.

Women's voice and participation in politics in the Pacific remains largely absent. Women in Pacific countries have made little progress in terms of political representation in parliament over the past decade. Currently, women represent about 2.5 percent of political leadership roles in the Pacific, and the Federated States of Micronesia, Nauru, Palau, and the Solomon Islands did not have female representation in parliament as of December 2011 (PARLINE database).

Several countries have elected or appointed female heads of state in recent years. Australia, Indonesia, Korea, New Zealand, the Philippines, and, most recently, Thailand have or have had a woman heading their government since 2000. At the ministerial level, Palau and Samoa have the highest percentage of women ministers (at 25 percent and 23 percent, respectively) in the East Asia and Pacific region. These percentages are similar to those found in Australia (23 percent) and slightly below those of New Zealand (29 percent). As of 2010, Nauru, the Solomon Islands, Tuvalu, and Vanuatu had no women ministers appointed to a ministerial position (PARLINE database).

For some countries, the level of female representation at the national level is not necessarily reflected at the local level, and vice versa. In Mongolia, women make up 4 percent of parliament's members at the national level, but they represent 22 percent at the subnational level (UNDP 2010). Overall, local government–level participation in Cambodia was 15 percent in 2007, similar to the national estimate, but elected women at the commune level and people's level was only 11 percent and only 3 percent at the people's chief level (Labani and others 2009).

Protection from gender-based violence and loss of freedom

Violence and the threat of violence deprive all people of their freedom and undermine their well-being. This section discusses the persistence of domestic violence against women, the intergenerational impacts of violence against children, and rising concerns about human trafficking.

Gender-based violence

Countries in the East Asia and Pacific region have among the highest numbers of trafficked people in the world, and the Pacific has the highest incidence of violence against women in the world. Gender-based violence is internationally recognized as "any act of gender-based violence that results in, or is likely to result in, physical, sexual or psychological harm or suffering to women, including threats of such acts, coercion or arbitrary deprivation of liberty, whether occurring in public or in private life" (UN 1993). Violence against women can take the form of physical, sexual, and psychological harm and can take place within the private or the public sphere, or can be in the form of the human trafficking of women and girls, among others. Violence against women is historically underresearched and underreported because of a variety of factors, including the sensitivity of the topic and concerns for the safety of respondents. Despite the lack of easily comparable international statistics over time, the increasing availability of studies shows that violence against women is a major concern for many countries in the region, and particularly in the Pacific.

FIGURE 4.11 **Women's representation in parliament has grown in much of East Asia but has experienced little change in the Pacific Islands**

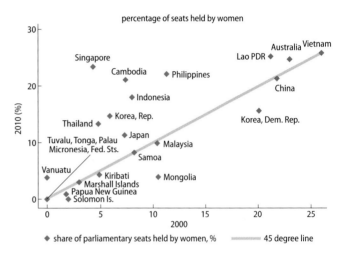

Source: PARLINE database, 2012 data.

FIGURE 4.12 **Women's representation in parliament is low, especially in the Pacific**

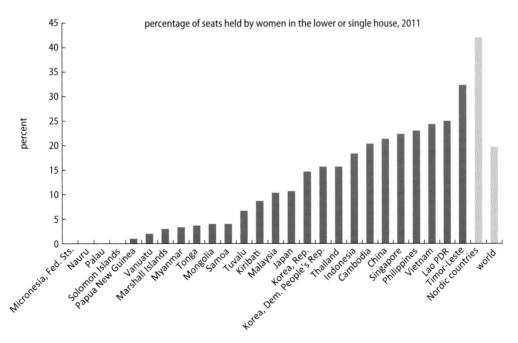

Source: PARLINE database, 2012 data.

Consistent with international patterns, women in the region are at far greater risk from violence by an intimate partner or somebody they know than from violence by other people. In East Asia and the Pacific, the prevalence of violence by an intimate partner has a wide range (figure 4.13). Some areas in East Asia, including the provinces of Bangkok and Nakhonsawan, Thailand, have a high incidence rate (44 percent). Evidence from the Pacific Islands suggests that violence is endemic. As shown in figure 4.13, 68 percent of ever-married women ages 15 to 49 in Kiribati, 64 percent in the Solomon Islands, and 60 percent in Vanuatu have experienced physical and/or sexual violence by an intimate partner (SPC and NSO 2009; SPC, Ministry of Internal and Social Affairs, and Statistics Division 2010; VWC and NSO 2011). Although no nationally representative data exist for Papua New Guinea, recent regional studies suggest that violence is just as prevalent (Ganster-Breidler 2010; Lewis et al. 2008).

Sexual violence at the hands of an intimate partner is a significant concern in much of the Pacific. In most of East Asia, barring Thailand, physical violence against women far exceeds

sexual violence. In the Pacific, however, both physical and sexual violence are extremely high. In the Solomon Islands, sexual violence by an intimate partner is more prevalent than physical violence. Sexual violence by nonpartners is also prevalent, as is childhood sexual abuse (SPC and NSO 2009).

Emotional and psychological violence, as well as harassment, are also common in the region. In 2005, over 30 percent of the region's women in the labor force reported having experienced some form of gender-based harassment—verbal, physical, or sexual (Haspels et al. 2001; UN 2006). Psychological and emotional violence is also pervasive and debilitating, but such violence is often underreported because it is frequently viewed as normal behavior.[4] In Vietnam, over half of ever-married women reported lifetime emotional abuse from their spouse, with 25 percent reporting abuse in the past 12 months (World Bank 2011a). In Cambodia, the Philippines, and Vietnam, the prevalence of emotional violence far exceeds that of physical and/or sexual violence by an intimate partner. In Kiribati, the Solomon Islands, and Vanuatu, the high prevalence of emotional violence by an intimate partner matches the high incidence of physical and/or sexual violence (World Bank 2011a; SPC and NSO 2009; SPC, Ministry of Internal and Social Affairs, and Statistics Division 2010; VWC 2011). Conversely, women in Timor-Leste experience a high prevalence of physical violence, yet very low levels of emotional and sexual violence (NSD [Timor-Leste], Ministry of Finance, and ICF Macro 2010).

Violence against children—Intergenerational impacts

In some societies in the Pacific there is still have violence against children. One study found that sexual violence has increased, in part because of low access to services and poverty faced by families and the rising prevalence of the logging, mining, and fishing industries, which employ large numbers of single men who seek the services of young women (UNIFEM 2010). The prevalence of sexually related violence against girls under

FIGURE 4.13   Violence against women is high in the region

percentage of ever-partnered women aged 15–49 who have experienced sexual and/or physical intimate-partner violence

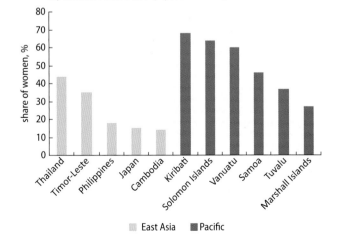

Source: DHS reports, various years.
Note: Data for Thailand are for Bangkok and Nakhonsawan provinces only; data for Japan are for Yokohama only.

the age of 15 in Vanuatu is 30 percent; and for most of these children, assaults are repetitive and carried out by family members (54 and 55 percent, respectively) (VWC and NSO 2011). In Vanuatu, the figure was 40 percent. Child prostitution in Fiji, Papua New Guinea, and the Solomon Islands is an organized venture with formal and informal brothels in urban centers (UNIFEM 2010).

Data also show that effects of violence are transmitted across generations. Domestic violence, and its acceptance, continues to be passed from parent to child (Fehringer and Hindin 2009). In Vanuatu, an estimated 57 percent of children whose mothers experienced violence either saw or heard the violence when it happened. Not only did these children witness the violent episodes, but many (17 percent) also experienced a beating at the same time (VWC and NSO 2011). As shown in figure 4.15, children in Kiribati who witnessed violence between parents while growing up are more likely to experience violence themselves as adults than those who did not witness violence between their parents (SPC, Ministry of Internal and Social Affairs, and Statistics Division 2010). Daughters of violent men have also been shown to be more likely to endure violence from their husbands later in life. In Timor-Leste, women whose fathers beat their mothers are more than twice as likely to experience emotional, physical, or sexual violence from an intimate partner than women whose fathers or mothers were not abusive, at 59 percent versus 24 percent (NSD, Ministry of Finance, and ICF Macro 2010).

The intergenerational consequences of violence also affect child development. In the Solomon Islands, one study shows that children who witness violence are more likely to grow up very timid or withdrawn, to repeat a year or more in school, and, in many cases, to be aggressive and likely to run away from home at an early age (SPC 2009).

Human trafficking

Human trafficking is also a growing concern throughout the region. An increase in female migration in the past decade has

FIGURE 4.14 The incidence of sexual intimate-partner violence is significantly higher in the Pacific

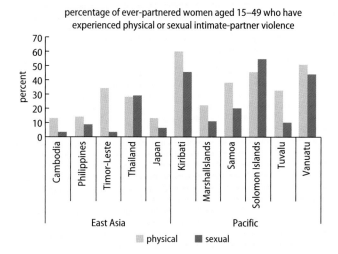

Source: DHS reports, various years.
Note: Data for Thailand are for Bangkok and Nakhonsawan provinces only; data for Japan are for Yokohama only.

FIGURE 4.15 Gender-based violence can have inter-generational consequences

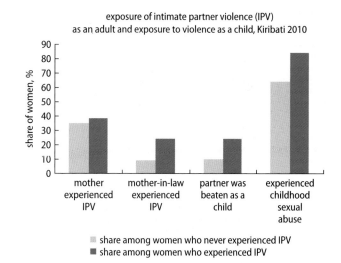

Source: SPC, Ministry of Internal and Social Affairs, and Statistics Division 2010.

increased economic opportunities as well as increased the risk of being trafficked (ILO 2009). Human trafficking encompasses forced labor, sex trafficking, bonded labor, debt bondage among migrant laborers, involuntary domestic servitude, forced child labor, sale of children (including bride price), child conscription to be soldiers, and child sex trafficking (U.S. Department of State 2011). Although estimates are difficult to generate, the International Labour Organization estimates that Asian and Pacific countries account for over half of all trafficked victims worldwide, at an estimated 1.36 million, most of whom are women and girls (ILO 2008). Studies show that women and girls are the main victims of sexual exploitation in East Asia and the Pacific. Women are also more likely than men to be trafficked for economic exploitation. Although women and children seem to be at higher risk for exploitation, men are also trafficked within and from the region. For instance, in Thailand, women make up the majority of those who are trafficked for commercial sexual exploitation; however, men, women, and children from neighboring countries are also trafficked for labor purposes, for example in fisheries and fishing related industries and domestic work (ILO 2008; UNFPA 2006, U.S. Department of State 2011).

Human trafficking is pervasive throughout the region; however, the nature of the industry varies. The Greater Mekong region and Indonesia are two main hubs for human trafficking. Thailand is both a destination and source country. The country serves as a transit hub to other Asian countries, Australia, the United States, and Western Europe (UNFPA 2006). In two southern counties in Yunnan, China, most women and children are trafficked for forced marriage or adoption. Rural men are willing to pay substantial sums for a trafficked bride who can bear children and extend the family line. Families will pay traffickers for infant boys, whom they will adopt as their own (ILO 2005). Although internal bride trafficking is frequently reported in China, systematic empirical research does not exist. In countries such as Japan and Korea, the majority of human trafficking reported is in the sex trafficking industry. In Cambodia, one study found a strong link between migration from domestic work into commercial sexual exploitation (51 percent of commercially sexually exploited women and girls were previously

domestic workers). The same study found that the majority (89 percent) of child domestic workers are female (E. Brown 2007).

Explaining progress and pending challenges

Women's ability to exercise voice in their homes and in society has experienced progress in some areas and challenges in others across East Asia and the Pacific. A complex combination of factors that affect progress includes socioeconomic characteristics, social norms and practices, and a country's legal environment, including women's access to justice. The strength of civil society is an indicator of voice in itself, but it also functions as a means by which individuals can collectively influence decisions that affect gender inequality. Because the extent of women's agency is driven by a combination of factors, this section discusses several influencing factors in turn.

Individual characteristics

Women's agency is affected in a fundamental way by their endowments and economic opportunities. High educational attainment, good health, economic assets, and their own earnings can enable women to influence their circumstances in accordance with their preferences.

As a woman's education level increases, it expands her knowledge and opportunities and improves her ability to translate her preferences into desired outcomes. Increases in education are positively correlated with women's increased bargaining power (Cochrane 1979; Jejeebhoy 1995). Education creates opportunities for women to connect with the world by removing barriers to knowledge and information. Higher academic achievement can also facilitate better economic opportunities and decrease economic dependency on others. In Indonesia, women who obtain an educational level of secondary school or higher are more likely than less-educated women to participate in decisions involving their own health care, make household purchases, and engage more regularly in social activities (DHS 2002–03 data).

Higher education can also facilitate women's entry into politics, whereas little or no formal education can greatly hinder women's participation in politics. In Rudong County in China, women involved in local government had a much higher level of education than average local female villagers; about 68 percent of them had reached senior high school education levels or above (Wang and Dai 2010). Where women have low levels of education—overall, as well as relative to men—they are even less likely to participate in politics. As a result, their voice, which often reflects different preferences than men's, is even less likely to be heard (UNDP 2010). In some contexts, the legal environment can exacerbate this situation. In Cambodia, indigenous women's participation in local politics remains low because they lack the education level and language skills necessary to be active participants. The law in Cambodia mandates that a person must speak, read, and write Khmer in order to run for political office (Maffii and Hong 2010).

Women who are more educated are less likely to experience violence in their household and hold higher perceptions of self-worth with regard to violence against women. Gender-based violence is more prevalent among illiterate women, partly because women are more likely to be economically dependent on their husbands and feel they have to endure an abusive relationship to survive. This situation is evidenced in some countries in the Pacific where bride price, young age of marriage, and lack of education for women perpetuate violence; leaving the marriage would involve having to repay the bride price, and the women commonly have no resources to do so (UNIFEM 2010, SPC and NSO 2009; SPC, Ministry of Internal and Social Affairs [Kiribati], and Statistics Division 2010; VWC and NSO 2011). In Samoa, 48 percent of women with vocational or higher education agree that a husband is justified in hitting or beating his wife, compared with the 69 percent of women with primary or lower levels of education who believe

wife beating is legitimate (MOH [Samoa], Bureau of Statistics, and ICF Macro 2010).

More educated women have greater economic opportunity, which in turn increases their agency. Women with more education are more equipped to take on skilled occupations and do better in the labor market, on average, than those with little or no education. Moreover, gender gaps in earnings tend to be smaller as females move up the skills ladder. Furthermore, education and skills training is invaluable to entrepreneurship. Low levels of skills and inability to access information have been suggested as hindering women in the Solomon Islands from becoming viable entrepreneurs (Haque and Greig 2011).

Increases in women's educational attainment and relevant training contribute to increasing women's participation in leadership roles in the private sector. However, despite advancements in educational attainment in much of the world, qualitative evidence suggests that the women continue to face gender-specific obstacles in the workplace. Female board members in Hong Kong SAR, China, for instance, argue that in order for companies to appoint qualified—well educated and trained—women to their corporate boards, there needs to be support for women moving up the career ladder to become board members (Mahtani, Vernon, and Sealy 2009).

Women's assets and own income can also increase women's agency within and outside the household in multiple ways. For example, power within the household has been attributed to individual control over economic resources such as individual incomes, assets, and wealth (Quisumbing and Maluccio 2003; Rammohan and Johar 2009; Thomas and Chen 1994). Evidence from the literature on intrahousehold resource allocation shows that increases in women's share of earned income in the household result in an allocation of resources that better reflects the preferences of women, and subsequently leads to improvements in women's overall status in the household and society (Ashraf, Karlan, and Yin 2006; Duflo 2003; Rangel 2005). Evidence also suggests that a woman's ownership and control of her own assets and income are associated with a decreased risk of intimate partner violence, at least in the medium term (ICRW 2006; Panda and Agarwal 2005; Pronyk et al. 2006; Swaminathan, Walker, and Rugadya 2008). A woman's income can also positively affect the accumulation of assets, which significantly affects her ability to leave an abusive partner, to cope with shocks, and to invest and expand her earnings and economic opportunities (World Bank 2011b). Women's own health has been found to improve women's relative positions in households, even when households experience adverse shocks (Aizer forthcoming; Dercon and Krishnan 2000).

Economic empowerment in the form of employment can also increase agency by increasing a woman's physical mobility. The ability to move freely means greater ability of women to expand their knowledge, broaden their networks, and create outlets to exercise collective agency through engagements in unions, professional associations, women's groups, and other types of civil society organizations. Civil society has worked to create an environment that enables women to better exercise their agency as well as a vehicle through which women can exercise their collective agency to change rules, regulations, laws, and social norms that positively affect gender equality.

Improvements in women's socioeconomic characteristics may not be sufficient to increase their voice and influence in society, however. Progress in women's economic position can threaten preexisting social norms with regard to women's role in the household and society and lead to undesirable outcomes, at least in the short run, including increased violence against women (Hjort and Villanger 2011; Panda and Agarwal 2005). For instance, in Timor-Leste, the proportion of women who have ever experienced physical violence from an intimate partner is highest among women with more than secondary education (46 percent), women belonging to households in the highest wealth quintile (45 percent), and women who are employed for cash (43 percent) (NSD, Ministry of Finance, and ICF Macro 2010). Evidence suggests that increases

in violence may be transitional phenomena, as men and women adjust to new roles. This transition can be aided by a supportive legal and regulatory environment.

Male attributes, attitudes, and behaviors also correlate with gender-based violence. Studies in the Pacific Islands of Kiribati, the Solomon Islands, and Vanuatu found that the most significant risk factors associated with women experiencing physical and/or sexual violence in their lifetime are the characteristics of a woman's husband or partner. A woman is significantly more likely to experience violence if her husband or partner controls her behavior, drinks alcohol or home brews, has affairs with other women, is violent with other men, or is unemployed (SPC and NSO 2009; SPC, Ministry of Internal and Social Affairs, and Statistics Division 2010; VWC and NSO 2011). In the Philippines, husbands from the lowest wealth quintile and with the least educational attainment are the most likely to physically abuse their wives (Philippine Commission on Women 2012). In Vanuatu, the Solomon Islands and Kiribati, a husband's lack of employment and high alcohol consumption and drug use are correlated with a higher likelihood of physical and/or sexual intimate-partner violence (SPC and NSO 2009; SPC, Ministry of Internal and Social Affairs [Kiribati], and Statistics Division 2010; VWC and NSO 2011).

Social norms and practices

Social norms and practices provide powerful undercurrents that influence preferences, values, and the social behavior that govern gender relations and outcomes. Social norms that perpetuate gender inequality in voice and influence can range from explicit to subtle. They can encourage behavior that on the surface seems to reflect choice but in fact can constrain what is possible for women, whether in the home or in society (Kabeer 1999).

Social norms regarding women's traditional role in the household and in society affect her social bargaining power. In most countries in the region and in the world, housework, child rearing, and care of the elderly are normally considered the responsibility of a woman, while men are considered the main financial provider of the family. Qualitative evidence from Fiji and Papua New Guinea shows that girls are expected to work around the house from a very early age, while boys have fewer household responsibilities and are expected to become wage earners (World Bank 2011e). As discussed in chapter 3, these social norms influence the other factors that affect agency, such as earnings capacity and asset accumulation, as well as constrain the choices an individual can make. In other words, traditional social norms may strongly discourage women from pursuing occupations traditionally dominated by men, as well as make it difficult for men to take on traditionally female roles, thus affecting the agency of both genders.

Female expectations in the household often mean that women have less time—and thus fewer networks—to exercise influence outside the home. Other societal expectations, such as having larger families, can increase women's time spent in the household and minimize other activities in society (Freeman 1997; Pritchett 1994). Focus group discussions in Fiji suggest that although men and women may agree in principle on the idea of equal economic opportunity for both genders, in practice, men are the more feasible breadwinners, since women have household responsibilities that family and society expect them to attend to first. Female youth in the community of Baulevu stated that men in the household can hold formal jobs more easily than women because women first have the responsibility to care for and feed the children as well as tend to other household needs before leaving for work (World Bank 2011e).

Gender norms and societal expectations about women's ability to participate in the public sphere hinder participation. Data from the World Values Survey show that many men and women from across the region believe that men make better leaders than women (figure 4.16). Women's participation in voting has sometimes been compromised by their lack of information. One study shows that in rural areas of China, women ask their

male family members to write the ballot for them at an election, or they take the opinion of their male family members. In that region of China, many people still think of women as less capable (*disuzhi*), and norms dictate that they should confine their activities to the domestic settings. This belief, largely based on perceptions that have been transmitted across generations and people, leads people to discourage women from voting or participating in public office (Wang and Dai 2010). Low female representation and participation in politics (as well as in leadership positions in the private sector) can reinforce each other, so that women are less likely to enter politics and other leadership positions because of social beliefs that men are better leaders than women, and women's absence prevents women from demonstrating their ability to lead (World Bank 2011c).

In most countries and in the Pacific Islands in particular, violence against women is perceived as acceptable or justifiable by both men and women. In the Solomon Islands, 73 percent of women believe that a husband is justified in beating his wife under certain

FIGURE 4.16 Many men and women in the region believe that men make better political leaders than women

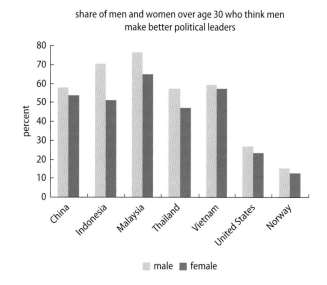

Source: World Values Survey, 2006 and 2007 data.

circumstances.[5] In Vietnam, 64 percent of women ages 15 to 49 accept violent treatment from husbands as normal (Vietnam Multiple Indicator Cluster Survey, 2006 data). Men also perpetuate these social norms: 81 percent of men in Timor-Leste believe that beating wives is socially acceptable and justifiable under certain circumstances (NSD, Ministry of Finance, and ICF Macro 2010). These attitudes and acts of violence may be intensified during times of conflict. Violence against women is believed to have reached its height in some areas of the Papua New Guinea Highlands during tribal fighting in 1995–96. During that time, gang rape of women was often considered to be a normal aspect of intervillage conflict (Brouwer, Harris, and Tanaka 1998).

Long-standing cultural practices, such as bride price, affect women's perceptions of their roles within marriage, and therefore their voice. In Vanuatu, about 81 percent of marriages involve a bride price paid to the family of the bride. Over half of women ages 15 to 49 in this country believe that, if a bride price is paid, a wife becomes the husband's property (VWC and NSO 2011). In both Vanuatu and the Solomon Islands, bride price is significantly associated with intimate partner violence (SPC and NSO 2009; VWC and NSO 2011). In East Asia, paying for a bride is a tradition across China, Indonesia, Myanmar, and Thailand (Anderson 2007). Although the practice has become less important in Indonesia, in rural China, where the tradition is still practiced, the groom pays for the rights to the woman's labor and reproductive capabilities (Boomgaard 2003). Similar to what is found in the Solomon Islands and Vanuatu, many young girls in China grow up with the belief that they will eventually become a man's property with little control over their lives (P. Brown 2003; Zhang 1999).

Social expectations of men's roles in society can influence men's actions and harm women's safety and overall agency. The incidence of intimate partner violence is often associated with societal expectations of men's roles in the home and in society, and can be exacerbated by male alcoholism and financial hardship (Heise 2011). In the region, men are subject to a substantial amount of social pressure to be the main provider in the household. Therefore if a man's wife defies traditional social norms and starts earning a higher income than her husband, for instance, this situation may challenge the role of the man as provider of the family, creating stress and resistance.

Deep-seated attitudes that fuel discrimination and enable gender inequality continue to be perpetuated in society by being taught to children. Recent studies show that social attitudes held by both men and women fail to foster an environment conducive to having equality between men and women. In Kiribati and Vanuatu, 56 percent and 50 percent of women, respectively, believe that a good wife must obey her husband at all times. Women in these countries also believe a man should show his wife that he is the boss (61 percent and 40 percent, respectively) (SPC and NSO 2009; SPC, Ministry of Internal and Social Affairs, and Statistics Division 2010). Children grow up absorbing these attitudes and behaviors, accepting them as norms they must follow, and girls internalize their subordinate role in the household and lesser status in society (Bourdieu 1977; Kabeer 1999; see also chapter 2).

The legal and institutional environment

The legal setting and access to justice form the environment in which men and women can voice and act on their preferences. Whether women and men are equally supported under the law and whether their rights are protected in practice are critical to women's ability to have voice and influence in society directly. This environment also affects voice by affecting the channels through which women build their access to resources and economic opportunity, which, in turn, affects their voice. The law and access to justice are shaped by, but can also shape, the norms that affect women's agency in society.

Equally as important is women's equal access to the judiciary system, which can be influenced by social norms or socioeconomic characteristics. For example, a legal court may be a day and a half trip away from home, yet social norms discourage women from sleeping outside the household. Or higher illiteracy rates of elderly women hinder their ability to know their rights. Countries in East Asia and the Pacific have made varying degrees of progress in guaranteeing and enforcing equal rights for men and women. As discussed in the previous section and in chapter 2, several countries in the region have plural legal environments, in which the interaction of customary or religious law and statutory law means that women's legal stature can vary substantially across ethnic groups, even within a country.

International conventions, national laws, and institutions
Nearly all countries in the East Asia and Pacific region have acceded to and ratified international commitments to reduce gender-based discrimination and promote agency for women as laid out by the Convention on the Elimination of All Forms of Discrimination Against Women (CEDAW), with a few exceptions. The ratification of CEDAW indicates a public commitment by governments to abide by a set of internationally recognized standards regarding gender equality. As of the end of 2011, only six countries in the world have not ratified the CEDAW agreement, including Palau and Tonga in the Pacific (CEDAW 2012).

The signing of CEDAW has served as an instrument to open up space for people to argue for legal and institutional reform to promote women's agency. In several East Asian countries, legal and institutional reforms came about after the ratification of CEDAW, at the behest of CSOs and government agencies arguing for the fulfillment of their countries' international commitments. For instance, individuals, CSOs, and government agencies tasked with women's issues in Cambodia, Indonesia, the Philippines, Thailand, and Vietnam used CEDAW conventions and frameworks as the basis for promoting women's participation in local government and for examining whether existing laws on violence and discrimination

against women were aligned with the convention. The organizations then used this information to advocate for an overhaul of existing laws or propose new ones to rectify omissions (UNIFEM 2010).

Beyond CEDAW, many countries in the region have put in place domestic laws to support the advancement of women and gender equality (UNDP 2010). In the past 10 years, gender equality laws such Vietnam's Gender Equality Law (2006), the People's Republic of China Law on the Protection of Rights and Interests of Women (amended in 2005), Lao PDR's Law on the Development and Protection of Women (2004), and the Philippines' Magna Carta of Women (2009) have been adopted with the aim of providing a more comprehensive approach to addressing gender equality. Most countries in the region have also adopted domestic violence legislation over the same period, including Cambodia, China, Indonesia, Korea, Lao PDR, Malaysia, Mongolia, Papua New Guinea, the Philippines, Thailand, Timor-Leste, and Vietnam. Papua New Guinea, for example, reformed its criminal code in 2002 to dramatically change its sexual assault regime by introducing grading of offenses according to the gravity of harm and eliminating marital immunity. Still, many countries (mostly Pacific Island states) lack adequate legislation for gender equality (UNDP 2010).

Most countries in the region have equal rights under inheritance laws, as discussed in chapter 2. Evidence from India suggests that women who own a house or land significantly reduce their risk of marital violence (Agarwal and Panda 2007). The majority of countries in East Asia no longer differentiate by gender in statutory law. Cambodia, China, Lao PDR, Mongolia, Thailand, and Vietnam do not have plural legal systems and have legislation for property and inheritance rights that have no discrimination against women. However, plural legal systems exist in Indonesia, Malaysia, the Philippines, and Singapore, and these laws discriminate against women in inheritance (see chapter 2 for more details). In the Pacific, Kiribati and Tuvalu have unequal statutory legislation, and although equal inheritance laws exist in Fiji, Papua New Guinea, Samoa, the Solomon Islands, and Vanuatu, customary law practices regarding land rights are recognized by the constitution and may lawfully discriminate against women (Jivan and Forster 2007). Although many areas of the Pacific region have traditionally had matrilineal land ownership—where land has historically been transmitted through mothers' lines—in practice, men most often make decisions regarding land management (Stege et al. 2008).

In regard to gender-based violence, gaps in the law still remain across the region (UNDP 2010). As illustrated in table 4.1, several countries in the region continue to have legal gaps in the protection of women against gender-based violence. Many of the countries where violence against women is the most prevalent do not have legislation against it, including Kiribati, the Marshall Islands, Papua New Guinea, Samoa, the Solomon Islands, Timor-Leste, and Tuvalu.

Important gaps also occur in laws protecting against human trafficking, despite efforts by many countries to pass specific legislative provisions. Between 2005 and 2008 about 10 countries in Asia and the Pacific (including South Asia) introduced new laws or modified old ones (UNODC 2009). However, some countries still lack comprehensive laws to protect men and women vulnerable to trafficking. Thailand and Vietnam, for example, do not have provisions on the exploitation of humans. Some progress has been made on this front in Indonesia, Malaysia, and the Philippines, where mandates against exploitation of women were introduced in the recent past (2007, 2007, and 2003, respectively). In Lao PDR, trafficked humans are treated as victims who have legal immunity from criminal prosecution for prostitution, and they are provided specific services to reunite them with their families, namely legal, medical, and counseling services, all imparted by the Lao Women's Union (U.S. Department of State 2008).

TABLE 4.1   Legislation against gender-based violence

Region/ Country	Domestic violence	Sexual assault/rape	Sexual harassment at workplace	CEDAW 1979 (c)	CEDAW optional protocol 1999 (d)
Pacific					
Fiji	Yes	Yes	No	1995	Not signed/Not ratified
Kiribati	No	Yes	No	2004	Not signed/Not ratified
Palau	NA	NA	NA	NA	Not signed/Not ratified
Papua New Guinea	No	Yes	No	1995	Not signed/Not ratified
Marshall Islands	No	Yes	No	2006	Not signed/Not ratified
Micronesia, Fed. Sts.	No	No	No	2004	Not signed/Not ratified
Samoa	No	Yes	Yes	1992	Not signed/Not ratified
Solomon Islands	No	Yes	No	2002	2002
Tonga	NA	NA	NA	Not signed/Not ratified	Not signed/Not ratified
Tuvalu	No	Yes	No	1999	Not signed/Not ratified
Vanuatu	Yes	Yes	No	1995	2007
East Asia					
Cambodia	Yes	Yes	Yes	1992	Signed
China	Yes	Yes	NA	1980	Not signed/Not ratified
Indonesia	Yes	Yes	No	1984	Signed
Lao PDR	Yes	Yes	NA	1981	Not signed/Not ratified
Malaysia	Yes	Yes	No	1995	Not signed/Not ratified
Mongolia	Yes	Yes	No	1981	2002
Philippines	Yes	Yes	Yes	1981	2003
Thailand	Yes	Yes	Yes	1985	2000
Timor-Leste	No	Yes	Yes	2003	2003
Vietnam	Yes	Yes	No	1982	Not signed/Not ratified

Source: UNDP 2010.

Even when countries have appropriate legislation in place, women may be unprotected by the legal system because the laws remain largely unenforced. A recent study highlighted that officers in the Fiji Police Force Sexual Offences Unit, which was set up in 1995, have unwelcoming attitudes when dealing with female victims (UNFPA 2010). The same is true in some areas in Cambodia, where many law enforcement officials are either unaware of the existence of the 2005 *Law on the Prevention of Domestic Violence and the Protection of Victims*, or continue to believe that domestic violence is an internal family problem (CAMBOW 2007). Even in cases where the police or other formal institutions condemn these acts, they are unable to pursue them further because of inadequate training to respond to the reports or fear of reprisal from the perpetrators, especially in cases in which they are people of influence (U.S. Department of State 2011). In Vanuatu, after a long period of lobbying by various CSOs and by the Vanuatu Women's Center, in 2008, the government passed the Family Protection Act, which focused on advancing women's rights. Implementation and enforcement of the law did not occur in more remote areas, leaving women unprotected and living under the previous legal (or traditional) system (AusAID 2008).

Furthermore, the interaction of customary practice and statutory law means that women's legal stature can vary substantially across ethnic groups, even within a country. Citizens in East Asian and Pacific countries often face a plurality of legal systems within a single country. Statutory laws interact—and often compete—with customary (and sometimes religious) laws and practices. In the Pacific Islands, for example, virtually all constitutions state that the constitution is the supreme law, but they simultaneously recognize customary law (UNDP 2010). As a result, inheritance practices vary substantially across the region, from matrilineal systems in most of Micronesia and in parts of Melanesia to patrilineal systems in most other Pacific countries. The interaction between customary and statutory law can even result in discriminatory inheritance practices that vary substantially across subpopulations within a single country (box 4.1).

States that adopt a commitment to revise and develop laws and regulations that promote gender equality create more conducive environments for women to exercise their agency. As women's labor force participation has continued to increase, especially in less traditional occupations, some governments have been proactive and adjusted their protection and antidiscrimination laws. In the Philippines, for example, 85 percent of an estimated 2.5 million domestic workers are women. Despite their number, until recently domestic workers were excluded from enjoying the full range of rights guaranteed to other women in traditional occupations (in industry) under labor law. A bill is presently under way to address the gap and to guarantee women's right to decent working conditions and protection from abuse, trafficking, and exploitation (ADB et al. 2008).[6]

Over the past few decades, several countries in the region have put in place temporary special measures to promote female participation in political leadership. Gender quotas set a fixed goal for having women in decision-making positions, with the aim of at least a critical minority of 30 to 40 percent (Agarwal 2010a, 2010b). In the political sphere, quotas can be reserved seats (constitutional and legislative), legal candidate quotas (constitutional and legislative), and political party quotas (voluntary). Although reservation of seats regulates the number of women elected, the latter two types of quotas set a minimum for the share of women on the candidate lists, either as a legal requirement or as a measure written into the statutes of individual political parties. In East Asia, countries including China, Indonesia, Korea, the Philippines, Thailand, and Timor-Leste had reservation systems for the single or lower house, for the upper house, or at the subnational level (The Quota Project, 2012 data). Recently, the revised Pacific Platform for Action 2005–2015 called for governments in the Pacific subregion to put in place affirmative action measures to enhance women's participation in politics.

BOX 4.1   Gender and land tenure in a plural legal environment: The case of the Solomon Islands

Land tenure in the Solomon Islands is characterized by multiple, overlapping arenas, norms, and institutions emanating from customary practice, the state, and Christianity. The intersection of customary and state legal systems allows only a small number of individuals, predominantly men, to exert control over customary land. This situation has occurred to the detriment of female landowners, who have often found themselves excluded from both decision-making processes and the distribution of financial benefits from the use of land.

In Guadalcanal Island, customs dictate that women be excluded from discussions about land-related issues that occur in formal and state arenas such as courts or land acquisition hearings. A male child or brother is usually appointed as spokesperson for land-related issues. Women often have less education and, hence, are considered less able to understand the state legal system and manage land transactions. Moreover, some inhabitants of Guadalcanal state that custom dictates that women "no save tok" (cannot/must not talk) about land, and that they must stand behind men when speaking about land in the public arena.

Furthermore, women are limited in their ability to hold land titles. Although people commonly assert that "women are the real landowners," land and court records generally show names of a small number of male leaders as landowners. Though the state legal system requires that the titleholders consult with other landowners before dealing in the land, they often fail to do so, limiting women's roles and participation.

The state legal system also tends to recognize only a small number of individuals with the customary authority to speak about land inside a public arena, thereby turning the customary "right to speak" into effective ownership. A review of court records for West Guadalcanal suggests that the witnesses and parties to a dispute are predominantly senior male leaders. In addition to these individuals having greater authority to speak about the land, these rights are compounded by the court system being based on Western legal principles and an adversarial system. Land disputes are sometimes compared to warfare and are matters for men. Women and children are often advised to stay away from meetings regarding disputes. Hence, as hearings are generally conducted by male chiefs, clerks or judges are also likely to act as a further impediment to women's involvement.

Development of mechanisms for ensuring transparency and accountability is essential for sustained peace and security. Land programming would benefit from being gender sensitive and from paying attention to differential impacts on state legal frameworks and their implementation.

Source: Monson 2010.

Other countries, though not mandating quotas, have taken a strong stance on female representation in government as well as in the private sector. Lao PDR's National Socio-Economic Development Plan (NSEDP) 2011–2015 has stipulated a target to increase the female staff of high-ranking positions in government, political parties, and civil society organizations to at least 15 percent, and to increase women members in the National Assembly to more than 30 percent. Other goals include increasing the number and percentage of women in the paid workforce and in professional careers (Lao PDR Ministry of Planning and Investment 2011). China's Outline for the Development of Chinese Women 2011–2020 states that women's participation should be a 30 percent female-to-male ratio in village-level government committees and a 50 percent ratio in neighbor committees in urban areas, with at least 10 percent of village committee heads being women.

Access to justice
Providing access to the justice system—by reducing financial costs, bringing services to remote areas, and helping overcome social and psychological constraints—is critical to helping women exercise their agency. Legal reform should not be limited to fixing the laws; it should also extend to ensuring access to the legal system. Women in poorer

areas can be particularly disadvantaged in this regard because, compared with men, they may have lower education levels, travel outside their communities less often, have fewer resources to pay for the service, and face higher levels of discrimination once they reach places such as police stations and courts. Moreover, when they do have physical access to the justice system, they often face the risk of being treated unequally. In Indonesia, the legal system is vastly underutilized by subsets of women because courts and police stations are not financially or physically accessible. A recent study in Indonesia found, for example, that most female heads of household were unable to access courts to obtain divorce certificates. The average cost of a case filed with the religious courts is about US$90, or four times the monthly per capita income for those living below the poverty line. In civil court, the cost of divorce with a lawyer is US$1,100, or 52 times the monthly per capita income (Sumner and Lindsey 2011).

Institutional reforms and opportunities for women

Strengthening the capacity of institutions to enforce the law can improve the probability of the law being adopted and used. Once legal reform comes about, institutional capacity needs to be reinforced to prevent hindering implementation, access, and enforcement of the law. Moreover, institutions' role as implementers of the law can be hindered further by prevailing social norms. Evidence from many countries shows that government institutions charged with implementing laws related to preventing discrimination against women and enforcing gender equality initiatives lack the human capacity, financial resources, and influence to address gender issues. For

TABLE 4.2 **Political affirmative action in East Asia and the Pacific**

Country	Year	Quota	Quota type	Results
China	2007	Legislated quotas for the single or lower house	Reserved seats, 22% women	637 of 2,987 seats (21%) held by women in 2008 election
Indonesia	2008	Legislated quotas for the single or lower house	Quota of one in three candidates on a political party list to be women	101 of 560 seats (18%) held by women in 2009 election
Korea, Rep.	2004	Legislated quotas for the single or lower house; legislated quotas at the subnational level; voluntary quotas adopted by political parties	Political party quota of 50% women on candidate lists for proportional representation in elections; recommendation for political parties to include 30% women candidates	44 of 299 seats (15%) held by women in 2008 election
Philippines	1991	Legislated quotas at the subnational level; voluntary quotas adopted by political parties (Philippine Democratic Socialist Party, 25% quota for women)	Women to be one of three sectoral representatives that sits in every municipal, city, and provincial legislative council	62 of 280 seats (22%) held by women in 2010 election
Thailand	n.a.	Voluntary quotas adopted by political parties.	The Democratic Party has a target of 30% women candidates for election	79 of 500 of seats (16%) held by women in the 2011 elections
Timor-Leste	2006 and 2011	Legislated quotas for the single or lower house	Modification of the Electoral Law to require that 33% of the candidates of each party be female (from 25%)	18 of 65 seats (28%) held by women in 2007 election

Source: The Quota Project, 2012 data.

instance, in a report to the CEDAW Committee in 2009, the government of Papua New Guinea identified the lack of structures for promoting gender equality as a challenge to carrying out the tasks outlined in international commitments. Similarly, the Lao PDR National Strategy for the Advancement of Women in 2006 and Vietnam National Strategy for the Advancement of Women in 2010 cited low capacity of their national institutions as a problem, and asked for help to strengthen those institutions—through training and more resources—to make them more effective. To address the constraints in institutional capacity, countries such as the Philippines and Vietnam worked on improving their institutional structures and reinforcing mandates already in place to promote women's agency and subsequently promote gender equality (box 4.2).

Several countries use advocacy organizations that are part of government to proactively address women's issues. The Lao Women's Union and the Vietnam Women's Union are two examples of organizations formed to promote information sharing among women, educate women at all levels of society, and promote women's active participation in society. They mobilize women to be active participants in their community (and in political party–related activities) and work to protect women's rights and interests.

Collective agency: Women's organizations and the space for civic action

Civil society functions both as an indicator of voice and participation and as a lever to create the conditions that enable women to change the legal environment, public priorities, and social norms. It also can facilitate dialogue and cooperation between parties that can influence the other determinants of agency. Collective agency through civil society does not necessarily need to focus on women-specific issues, but instead, it provides a forum in which men and women can exercise voice on any issues they care about.

Civil society organizations, social movements, and civic activists have helped expand women's agency by magnifying their voice in public forums and strengthening women's influence. Having a collective voice, through mass movements, is critical

---

BOX 4.2   Strengthening state mechanisms for gender equality

Gender equality laws in the region also focus on the improvement of existing gender equality structures. The Philippine Magna Carta for Women, for example, strengthened the mandate of the Philippine Commission on Women as the primary policy-making and coordinating body on women and gender equality concerns, and emphasized the placement of qualified women in all government departments and their attached agencies. These measures include bringing local government units, government-owned and -controlled corporations, and other government instrumentalities into discussions of how best to ensure a gender balance in their workforce. The Magna Carta for Women strengthened the gender and development focal points by increasing resources and support. The government will evaluate the increased allocation based on the law's influence in

making 95 percent of the budget gender-responsive. The law will be subject to an annual audit by the Commission on Audit.

Vietnam's Law on Gender Equality emphasized state responsibility to promote gender equality by identifying a state management agency to unify efforts by all of the government's gender equality entities as well as responsibilities of agencies to mainstream gender in their work. Moreover, Articles 20–22 of the law outline a process for integrating gender into legislation. A designated National Assembly Committee for Social Affairs, together with other parliamentary committees, shall verify the integration of gender equality issues in the draft law and ordinances by drafting agencies.

Source: UNDP 2010.

in the process of influencing policy, culture, and social environments. Grassroots movements have been instrumental in changing attitudes and behaviors that maintain gender inequalities, such as those that limit women's participation in politics, by promoting new ideas and actively sharing information in the mainstream. In Indonesia, in 1997, a group of women got together to protest policies in response to the crisis that made them unable to afford powdered milk for their children. The group, known as Voice of Concerned Mothers (*Suara Ibu Pedulu*), was the first group led by women in the country since the beginning of Suharto's New Order (Robinson and Bessell 2002). These protests flourished and evolved into a collective voice representing mothers throughout the country. Its success prompted wider protests against the government, which contributed not only to ending the Suharto regime but also to creating new space for women to participate actively in the post-Suharto period (Rinaldo 2002).

CSOs and citizen movements have often facilitated dialogue and cooperation between ordinary women and those in public office. In Cambodia, CSOs helped promote awareness and political education among women through organizations such as the Women's Media Centre of Cambodia. Party-affiliated organizations, semigovernmental unions, as well as autonomous groups have served as incubators for female leaders to run for office or become appointees in key government positions. These gender grassroots movements worked effectively in promoting the participation of women in local politics through information dissemination campaigns. They concentrated much of their efforts throughout Cambodia producing written and graphic awareness materials—explaining women's rights, responsibilities, and the process for participating—and encouraging women to play an active role in politics. Currently, an increasing number of women organize forums at provincial levels to champion gender-related causes and to pressure politicians into action (Singh 2009).

In matters related to increasing women's role in policy making, intense lobbying efforts by gender-focused CSOs and civic activists have prompted changes in social behavior and engaged the commitment of politicians. Institutional factors such as parliamentary frameworks, coupled with societal norms, continue to hinder women's entrance onto the political stage (UNDP 2010). For instance, the Center for Asia-Pacific Women in Politics—based in the Philippines with subregional offices in Fiji, Korea, Mongolia, and Nepal—is an active advocate for increasing the roles of women in politics. This and other similar CSOs actively lobby for expanding the capacity of women in decision-making roles and working toward gender equity in representation. In the Pacific, patterns showed decreasing participation of women in politics, prompting the group Women's Rights and Advocacy in the Pacific (WRAP) to recommit their lobbying efforts to promote women in public office and political leadership (WRAP 2011).

Policy approaches to promote gender equality in agency

The analysis thus far has focused on the state of agency in the East Asia and Pacific region and identified the factors that advance or constrain gender equality in agency. Knowing those factors can help to identify specific policies and programs to strengthen women's voice and influence in practice. What follows is an initial discussion, based on the evidence presented, about policy priorities for promoting gender equality in agency. Measures to increase women's endowments and economic opportunity, such as those described in previous chapters, contribute to strengthening their voice within the household and in society. Educated women in good health, with assets and income, are better able to act on their preferences and influence outcomes that affect themselves and others in society. In addition, several other policy approaches can directly promote women's agency.

This section addresses the following factors: (a) supporting initiatives to transform gender norms and practices, (b) strengthening the legal and institutional environment, (c) increasing women's access to justice, (d) enabling women's participation in politics and policy making, (e) pursuing a multipronged approach to reducing gender-based violence, and (f) creating space for women's collective agency.

Supporting initiatives to transform gender norms and practices

Social norms are not static, and several factors can influence them. The preferences that men and women express at the individual and household level as well as in the public sphere are in part determined by their socially constructed gender roles. Individual experiences as well as large-scale political and economic processes are capable of bringing about dramatic, and often rapid, social change. China's 1949 Communist Revolution, for example, had the effect of reducing gender inequalities to an unprecedented extent (Bian, Logan, and Shu 2000; Whyte and Parish 1984; Wolf 1985). Moments of social change and structural transformation create opportunities for people to rethink their roles and choices. In East Asia, the process of rapid urbanization that is currently under way brings with it the possibility of newly defined roles for men and women. As more women participate in remunerative activities, rigid social norms are challenged. The education system can be a vital source to change gender inequality by promoting social norms from a young age. The integration of gender equality principles into the school and professional curricula can tackle the value system of children early on and challenge discriminatory social norms. In Indonesia, a recent project evaluated the textbooks used in various classes and found that they contained gender-biased material; the researchers identified messages that ignored or condoned sexual harassment, gender-based violence, and gender-based stereotypes (Utomo et al. 2009). Evidence

of the positive effects of changing the curriculum is available for adults in Thailand, where gender sensitivity was integrated into the curriculum in the Chulalangkorn medical school. Evaluation of the program showed that respondents were more aware of gender issues and tended to apply gender concepts and concerns to their work and personal lives (WHO GWH 2007).

Provisions of information through television programming can also play a critical role in changing social norms, especially with respect to fertility and gender-based violence. Evidence shows that people can be prompted to rethink gender roles in society when they are exposed to new information and experiences that challenge existing norms. In Brazil, despite strong traditional norms in favor of having many children, increased exposure to the opposite behavior by popular women in soap operas led to a measurable decline in fertility (Chong and La Ferrara 2009; La Ferrara, Chong, and Duryea 2008). In India, increased exposure to television contributed to decreased acceptance of wife beating, lower fertility rates, and noticeable shifts away from son preference (Jenson and Oster 2008).

Policies that promote women's voice and participation in public settings may have positive impacts for future generations. Recent evidence from India shows that the use of political reservation policies improved not only how people view female politicians but also how they view their own children and their future opportunities, as well as how children view their own ambitions (Beaman et al. 2012).

Strengthening the legal and institutional environment

As discussed earlier, although countries have made some advances, they must continue to improve the legal protections of women to ensure that they have equal rights under the law. Apart from legal reforms, institutions must have the capacity to enforce the law and provide adequate services. This section presents some of the policies and programs that countries should continue to pursue. These strategies have contributed to improving the environment for women by enabling them to exercise agency in their own household and community, and have helped to decrease gender-based violence in countries and communities.

Actively participate in international treaties that promote gender equality
The ratification of international treaties such as CEDAW and the Beijing Declaration and Platform for Action signals governments' commitment to gender equality. Through active participation, governments help policy makers, CSOs, individuals, and development agencies strengthen their position, enabling them to call for further reforms of the civil and criminal laws to make them consistent with international standards.

After ratifying international conventions against the discrimination of women, the government's next step is to review the laws and the way institutions function to ensure that they actively promote equality between the genders under the law, actively promote nondiscrimination based on individual characteristics, and legally empower the state to eliminate all discrimination based on gender. In cases in which overall legal reform is not possible, governments should identify priority areas. For instance, in contexts where women's agency in the household is weak, reforms can focus on marriage, divorce, maintenance laws, and the protection of women from gender-based violence. Governments should also commit to undertaking regular assessments to make sure the laws are upheld, fill legal gaps, and monitor progress toward gender equality. In countries where plural systems of law coexist, governments should continuously assess customary practices to ascertain whether they curtail women's agency and develop strategies to address them.

Strengthen the capacities of institutions to enforce the law
The judicial system fails women when it is reluctant to pursue crimes against them and

refuses to uphold judgments in their favor largely on the basis of their gender. Governments must make financial and personnel investments to ensure that public sector personnel have the administrative capacity to enforce the law and are able to follow gender equality principles. Police forces in several countries in the region—including Indonesia, Malaysia, and Thailand—have been criticized for being too passive in investigating trafficking and enforcing antitrafficking laws (U.S. Department of State 2011). Gender training regarding gender-related matters, including human trafficking, should be mandatory for all law enforcement personnel, including judges, lawyers, police officers, mediators, and social workers.

Increasing women's access to justice

Countries should take steps to make the judicial system more accessible to women so that women can exercise their agency in the courts when needed. For instance, mobile courts, such as those in rural areas of China and Indonesia, provide a solution to the problem of accessibility and security for women who wish to exercise their rights in the legal system but lack transportation. Technology can help extend basic legal services; for instance, basic legal transactions can use telephone hotlines and websites. For women with few economic resources, governments can waive or subsidize the costs of legal aid to ensure access to the judicial system.

Enabling women's participation in politics and policy making

Countries may adopt affirmative action policies if the context requires it. In both the private and public sectors, voluntary or mandatory affirmative action policies in many countries have increased the representation of women throughout institutions, from entry-level positions to managerial posts. In many countries, private sector companies actively pursue these policies on their own; in other countries, public sector institutions take the lead in promoting gender-based temporary special measures to signal to the rest of the labor market. In policies, the range of affirmative action mechanisms also varies. Quotas, for example, can be in the form of constitutional changes to reserve a certain number of government posts for women, both through legislative and (formal or informal) political party quotas. Quotas can be used informally or be formally mandated at the subnational or national levels (Dahlerup 2006). Countries should evaluate the suitability of affirmative action measures well before embarking on their implementation. Affirmative action measures can open doors for women in politics and public office and enable them to move into positions of power. Affirmative action measures can also help to transform people's views about the efficacy of female political leaders by increasing the number of women participating in electoral politics. However, under such measures, the perceptions that women are less qualified may persist and, in some cases women may hesitate to take such positions because of concern they'll be perceived as less capable.

Pursuing a multipronged approach to reducing gender-based violence

Reducing gender-based violence requires action on a number of fronts: efforts to increase women's voice within the household; enactment and enforcement of appropriate legislation and strengthening of women's access to justice; provision of adequate support services for victims of violence; and use of the media to provide information on women's rights, to increase social awareness and to shift social norms with respect to violence. The Fiji Women's Crisis Centre, for example, provides crisis counseling as well as legal, medical, and other support to women and children who have experienced gender-based violence. The crisis centre encourages male advocacy through programs that train men from police, military, community and religious agencies to more easily recognize and prevent gender-based

violence in their communities (Fiji Women's Crisis Centre 2012).

Creating space for women's collective agency

Collective action has drawn private life into the public arena, identifying and addressing gender bias in statutory, religious, and customary law (UNRISD 2005). It has also reduced the hold of social norms that block greater gender equality. During the debate in Cambodia, which led to passage of the 2005 Law on Prevention of Domestic Violence and Protection of Victims, the draft law was denounced for being antagonistic to Khmer culture. Parliamentarians criticized it for "providing women with too many freedoms and rights, which will cause them to be so happy with their freedom that they do not respect ancient Cambodian customs. . . . A cake cannot be bigger than the cake pan (as cited in Frieson 2011)." The Cambodia Committee of Women, a coalition of 32 nongovernmental organizations, persistently lobbied the government and the Ministry of Women's Affairs to secure the legislation's passage (World Bank 2011c) .

Notes

1. Civic activism is defined as the set of practices among citizens which demand greater involvement and scrutiny of public decisions and outcomes and is often used as a proxy for agency. Examples of civic activism are memberships in civil society organizations, petitions, protests, and peaceful demonstrations.

2. Demographic and health survey data for the countries studied in this chapter are available through MEASURE DHS, ICF International, Calverton, MD. http://www.measuredhs.com/. NSD [Timor-Leste], Ministry of Finance, and ICF Macro 2010; NIS and DGH [Cambodia] 2011; BPS (Statistics Indonesia) and ORC Macro 2003.

3. How individual civil society organizations (CSOs) interact with government varies greatly within countries; distinct paths evolved in the relationship over time among more developed countries in East Asia. In Korea, prior to 1990,

the relationship between the government and civil society was not close in part because their goals did not seem to be aligned. After 1990, the relationship changed because the government sought to find common ground with CSOs (Kim 1998). The role of civil society in Japan also continues to evolve; it had a limited role in the 1990s because of a cumbersome regulatory framework that made entry of new CSOs and operations of existing ones difficult (Amemiya 1999; Yamaoka 1999). A new legal framework was put forth in 2006, which prompted some positive changes, especially in the relationship between the government and CSOs (Lowry 2008).

4. Psychological and emotional violence is defined by acts or threats of acts, such as shouting, controlling, intimidating, humiliating, and threatening the victim. This type of violence may include coercive tactics.

5. Circumstances include if she does not do her housework to his satisfaction, if she disobeys him, if she refuses to have sex with him, if she asks him whether he has other girlfriends, if he suspects that she is unfaithful, and if he finds out that she has been unfaithful.

6. The Kasambahay bill is still awaiting Senate approval as of mid-June 2012. See http://www.pia.gov.ph/news/index.php?article=231339654008.

References

ADB (Asian Development Bank), CIDA (Canadian International Development Agency), European Commission, National Commission on the Role of Filipino Women (NCRFW), United Nations Children's Fund (UNICEF), United Nations Development Fund for Women (UNIFEM), and United Nations Population Fund (UNFPA). 2008. *Paradox and Promise in the Philippines: A Joint Country Gender Assessment.* http://www.adb.org/documents/paradox-and-promise-philippines-joint-country-gender-assessment.

Agarwal, Bina. 2009. "Gender and Forest Conservation: The Impact of Women's Participation in Community Forest Governance." Ecological Economics 68 (11): 2785–99.

———. 2010a. "Does Women's Proportional Strength Affect Their Participation? Governing Local Forests in South Asia." *World Development* 38 (1): 98–112.

———. 2010b. *Gender and Green Governance: The Political Economy of Women's Presence within and beyond Community Forestry.* New York: Oxford University Press.

Agarwal, Bina, and Pradeep Panda. 2007. "Toward Freedom from Domestic Violence: The Neglected Obvious." *Journal of Human Development* 8 (3): 359–88.

Aizer, Anna. Forthcoming. "The Gender Wage Gap and Domestic Violence." *American Economic Review.*

Amemiya, Takako. 1999. "Japan." In *Philanthropy and Law in Asia: A Comparative Study of the Nonprofit Legal Systems in Ten Asia Pacific Societies,* edited by Thomas Silk, 131–62. San Francisco: Jossey-Bass.

Anderson, Siwan. 2007. "The Economics of Dowry and Brideprice." *Journal of Economic Perspectives* 21 (4): 151–74.

Ashraf, Nava, Dean Karlan, and Wedley Yin. 2006. "Tying Odysseus to the Mast: Evidence from a Commitment Savings Product in the Philippines." *Quarterly Journal of Economics* 121 (2): 635–72.

AusAID. 2008. "Vanuatu Country Report." http://www.ausaid.gov.au/Publications/Documents/ResVAW_vanuatu.pdf.

Ban, Radu, and Vijayendra Rao. 2009. "Is Deliberation Equitable? Evidence from Transcripts of Village Meetings in South India." Policy Research Working Paper 4928, World Bank, Washington, DC.

Beaman, Lori, Esther Duflo, Rohini Pande, and Petia Topalova. 2012. "Female Leadership Raises Aspirations and Educational Attainment for Girls: A Policy Experiment in India." *Science* 335: 582–86.

Bian, Yanjie, John R. Logan, and Xiaolong Shu. 2000. "Economic Reform and the Gender Wage Gap in China." In *Re-Drawing Boundaries: Work, Household, and Gender in China,* edited by Barbara Entwisle and Gail E. Henderson, 111–33. Berkeley: University of California Press.

Boomgaard, Peter. 2003. "Bridewealth and Birth Control: Low Fertility in the Indonesian Archipelago, 1500–1900." *Population and Development Review* 29 (2): 197–214.

Bourdieu, P. 1977. *Outline of a Theory of Practice.* Cambridge: Cambridge University Press.

BPS (Statistics Indonesia) and ORC Macro. 2003. *Indonesia Demographic and Health Survey 2002–2003.* Calverton, MD: BPS and ORC Macro.

Brouwer, Elizabeth C., Bruce M. Harris, and Sonomi Tanaka, eds. 1998. *Gender Analysis in Papua New Guinea*. Washington, DC: World Bank.

Brown, Eleanor. 2007. *Out of Sight, Out of Mind? Child Domestic Workers and Patterns of Trafficking in Cambodia*. Geneva, Switzerland: International Organization for Migration.

Brown, Philip. 2003. "Dowry and Intrahousehold Bargaining: Evidence from China." Working Paper 608, William Davidson Institute, University of Michigan, Ann Arbor.

CAMBOW (Cambodian Committee of Women) 2007. *Violence Against Women: How Cambodian Laws Discriminate Against Women*. Phnom Penh, Cambodia.

CEDAW (Convention on the Elimination of All Forms of Discrimination Against Women). 2012. http://www.un.org/womenwatch/daw/cedaw/.

Chattopadhyay, Raghabendra, and Esther Duflo. 2004. "Women as Policy Makers: Evidence from a Randomized Policy Experiment in India." *Econometrica* 72 (5): 1409–43.

Chong, Alberto, and Eliana La Ferrara. 2009. "Television and Divorce: Evidence from Brazilian Novelas." *Journal of the European Economic Association* 7 (2–3): 458–68.

Cochrane, S. H. 1979. *Fertility and Education. What Do We Really Know?* Baltimore: Johns Hopkins University Press.

Cohen, Philip, and Matt Huffman. 2007. "Working for the Woman: Female Managers and the Gender Wage Gap." *American Sociological Review* 72: 681–704.

Costello, Marilou, and John B. Casterline. 2002. "Fertility Decline in the Philippines: Current Status, Future Prospects." Paper prepared for the Expert Group Meeting on Completing the Fertility Transition, Population Division, United Nations, New York, March 11–14.

CWDI (Corporate Women Directors International). 2010. "CWDI/IFC 2010 Report: Accelerating Board Diversity." CWDI and International Finance Corporation (IFC), Washington, DC.

Dahlerup, Drude. 2006. "Introduction." In *Women, Quotas and Politics*, edited by Drude Dahlerup, 3–31. London and New York: Routledge.

Dercon, Stefan, and Pramila Krishnan. 2000. "In Sickness and in Health: Risk Sharing within Households in Rural Ethiopia." *Journal of Political Economy* 108 (4): 688–727.

DHS (Demographic and Health Surveys). MEASURE DHS, ICF International, Calverton, MD. http://www.measuredhs.com/.

Duflo, Esther. 2003. "Grandmothers and Granddaughters: Old-Age Pensions and Intrahousehold Allocation in South Africa." *World Bank Economic Review* 17 (1): 1–25.

Ely, Robin J. 1995. "The Power in Demography: Women's Social Constructions of Gender Identity at Work." *Academy of Management Journal* 38: 589–634.

Enterprise Surveys (database). World Bank, Washington, DC. http://www.enterprisesurveys.org.

Fehringer, Jaya, and Michelle J. Hindin. 2009. "Like Parent, Like Child: Intergenerational Transmission of Partner Violence in Cebu, Philippines." *Journal of Adolescent Health* 44 (4): 363–71.

Fiji Women's Crisis Centre. 2012. "Male Advocacy on Women's Human Rights." http://www.fijiwomen.com/index.php?option=com_content&view=article&id=94&Itemid=115.

Fiszbein, Ariel, and Norbert Schady. 2009. "Conditional Cash Transfers: Reducing Present and Future Poverty." Policy Research Series, World Bank, Washington, DC.

Freeman, Richard. 1997. "Working for Nothing: The Supply of Volunteer Labor." *Journal of Labor Economics* 15 (1): S140–66.

Frieson, Kate 2011. "Cambodia Case Study: Evolution toward Gender Equality." Background paper for the *World Development Report 2012*, World Bank, Washington, DC.

Ganster-Breidler, Margaret. 2010. "Gender-Based Violence and the Impact on Women's Health and Well-being in Papua New Guinea." *DWU Research Journal* 13.

Haddad, Lawrence, John Hoddinott, and Harold Alderman, eds. 1997. *Intrahousehold Resource Allocation in Developing Countries: Models, Methods, and Policy*. Baltimore: International Food Policy Research Institute and Johns Hopkins University Press.

Haque, Tobias, and Froniga Greig. 2011. "Increasing the Participation of Women Entrepreneurs in the Solomon Islands Aid Economy." World Bank, Washington DC. http://documents.worldbank.org/curated/en/2011/01/13430122/increasing-participation-women-entrepreneurs-solomon-islands-aid-economy.

Haspels, Nelien, Zaitun Mohamed Kasim, Constance Thomas, and Deirdre McCann.

2001. "Action against Sexual Harassment at Work in Asia and the Pacific." Technical report prepared for the ILO/Japan Regional Tripartite Seminar on Action against Sexual Harassment at Work in Asia and the Pacific, Penang, Malaysia, October 2–4.

Heise, Lori. 2011. "What Works to Prevent Partner Violence? An Evidence Overview." Department for International Development (DFID) Working Paper, DFID, London.

Hjort, J., and E. Villanger. 2011. "Backlash: Female Employment and Domestic Violence." Technical report, preliminary working paper. World Bank, Washington, DC.

Hultin, Mia, and Ryszard Szulkin. 2003. "Wages and Unequal Access to Organizational Power: An Empirical Test of Gender Discrimination." *Administrative Science Quarterly* 44: 453–72.

ICRW (International Center for Research on Women). 2006. *Property Ownership and Inheritance Rights of Women for Social Protection: The South Asia Experience.* Washington, DC: ICRW.

ILO (International Labour Organization). 2005. "The Mekong Challenge. Analysis Report of the Baseline Survey for the TICW Project Phase II in Yunnan Province." The Mekong Sub-regional Project to Combat Trafficking in Children and Women, International Labour Office, Bangkok.

———. 2008. "Women's Entrepreneurship Development in Aceh:Gender and Entrepreneurship Together (GET Ahead) Training Implementation: Impact Assessment." ILO Jakarta Office, Jakarta, Indonesia.

———. 2009. "Operational Indicators of Trafficking in Human Beings: Results from a Delphi survey Implemented by the ILO and the European Commission." ILO, Geneva, Switzerland.

Imai, Katsushi, and Per A. Eklund. 2008. "Women's Organizations and Social Capital to Reduce Prevalence of Child Malnutrition in Papua New Guinea." *Oxford Development Studies* 36 (2): 209–33.

ISD (Indices of Social Development) database. International Institute of Social Studies in The Hague. http://www.indsocdev.org/home.html.

Jejeebhoy, Shireen. 1995. *Women's Education, Autonomy, Reproductive Behavior: Experiences from Developing Countries.* Oxford: Clarendon Press.

Jensen, Robert, and Emily Oster. 2009. "The Power of TV: Cable Television and Women's Status in India." *Quarterly Journal of Economics* 124 (3): 1057–94.

Jivan, Vedna, and Christine Forster. 2007. *Translating CEDAW into Law: CEDAW Legislative Compliance in Nine Pacific Countries.* Suva, Fiji: UNDP Pacific Centre and UNIFEM Pacific Regional Office.

Kabeer, Naila. 1999. "Resources, Agency, Achievements: Reflections on the Measurement of Women's Empowerment." *Development and Change* 30 (3): 435–64.

Kennedy, Elissa, Natalie Gray, Peter Azzopardi, and Mike Creati. 2011. "Adolescent Fertility and Family Planning in East Asia and the Pacific: A Review of DHS Reports." *Reproductive Health* 8: 11. doi:10.1186/1742-4755-8-11.

Kim, H. 1998. *Study on the Civil Society and Non-Governmental Organizations in Korea.* Seoul: Yonsei University, Dong-Su Yon Ku, Institute of East and West Studies.

Labani, Sepideh, Carla Zabaleta Kaehler, and Paula de Dios Ruiz. 2009. "Gender Analysis of Women's Political Participation in 7 South-East Asian Countries: Bangladesh, Cambodia, the Philippines, Indonesia, Sri Lanka, East Timor and Vietnam." Enjambra Contra la Explotación Sexual. http://www.bantaba.ehu.es/obs/files/view/Gender_analysis_of_women's_political_participation.pdf?revision_id=79226&package_id=79202

La Ferrara, Eliana, Alberto Chong, and Suzanne Duryea. 2008. "Soap Operas and Fertility: Evidence from Brazil." RES Working Paper 4573, Inter-American Development Bank, Washington, DC.

Lao PDR Ministry of Planning and Investment. 2011. "Seventh Five-Year National Socio-Economic Development Plan 2011–2015." http://www.wpro.who.int/countries/lao/LAO20112015.pdf.

Lewis, Ione, Bessie Maruia, and Sharon Walker. 2008. "Violence against Women in Papua New Guinea." *Journal of Family Studies* 14: 183–197.

Lowry, Cameron. 2008. "Civil Society Engagement in Asia: Six Country Profiles." Manuscript prepared for the Asia-Pacific Governance and Democracy Initiative, East-West Center, Washington, DC.

Maffii, Margherita, and Sineath Hong. 2010. "Political Participation of Indigenous Women in Cambodia." *Asien* 114–15 (April): 16–32.

Mahtani, Shalini, Kate Vernon, and Ruth Sealy. 2009. *Women on Boards: Hang Seng Index 2009*. Community Business and Cranfield School of Management. http://www.communitybusiness.org/images/cb/publications/2009/WOB.pdf.

Mason, Karen Oppenheim. 2005. "Measuring Women's Empowerment: Learning from Cross-National Research." In *Measuring Empowerment: Cross-Disciplinary Perspectives*, edited by D. Narayan, 89–102. Washington, DC: World Bank.

Mason, Karen Oppenheim, and Herbert L. Smith. 2003. "Women's Empowerment and Social Context: Results from Five Asian Countries." Paper presented at the workshop on "Measuring Empowerment: Cross-Disciplinary Perspectives," World Bank, Washington, DC, February 4–5.

McElroy, Marjorie B. 1990. The Empirical Content of Nash-Bargained Household Behavior. *Journal of Human Resources* 25(4): 559–83.

MOH (Ministry of Health) [Samoa], Samoa Bureau of Statistics, and ICF Macro. 2010. *Samoa Demographic and Health Survey 2009*. Apia, Samoa: Ministry of Health, Samoa. Available from the MEASURE DHS website http://www.measuredhs.com/publications/publication-fr240-dhs-final-reports.cfm.

Monson, Rebecca. 2010. "Women, State Law and Land in Peri-Urban Settlements on Guadalcanal, Solomon Islands." Briefing note. World Bank, Washington, DC.

Morrison, Andrew, Mary Ellsberg, and Sarah Bott. 2007. "Addressing Gender-Based Violence: A Critical Review of Interventions." *World Bank Observer* 22 (1): 25–51.

Morrison, Andrew, and Maria Beatriz Orlando. 2004. *The Costs and Impacts of Gender-Based Violence in Developing Countries: Methodological Considerations and New Evidence*. Washington, DC: World Bank.

NIS (National Institute of Statistics) [Cambodia Ministry of Planning], and DGH (Directorate General for Health) [Ministry of Health]. 2011. *Cambodia: Demographic and Health Survey 2010—Final Report*. Calverton, MD: MEASURE DHS.

NSD (National Statistics Directorate) [Timor-Leste], Ministry of Finance [Timor-Leste], and ICF Macro. 2010. *Timor-Leste Demographic and Health Survey 2009–10*. Dili, Timor-Leste: NSD and ICF Macro.

NSO (National Statistics Office) [Philippines] and ICF Macro. 2009. *Philippines Demographic and Health Survey 2008*. Calverton, MD: NSO and ICF Macro.

Panda, Pradeep, and Bina Agarwal. 2005. "Marital Violence, Human Development and Women's Property Status in India." *World Development* 33 (5): 823–50.

PARLINE database on national parliaments. Inter-Parliamentary Union. http://www.ipu.org/wmn-e/world.htm.

PEKKA (Perempuan Kepala Keluarga [Women Headed Household Empowerment Program]). Accessed 2012. http://www.pekka.or.id/8/index.php.

Philippine Commission on Women. 2012. http://pcw.gov.ph/statistics/201205/statistics-violence-against-filipino-women.

Pritchett, Lant H. 1994. "Desired Fertility and the Impact of Population Policies." *Population and Development Review* 20 (1): 1–55.

Pronyk, Paul, James Hargreaves, Julia Kim, Linda Morison, Godfrey Phetla, Charlotte Watts, Joanna Busza, and John Porter. 2006. "Effect of a Structural Intervention for the Prevention of Intimate-Partner Violence and HIV in Rural South Africa: A Cluster Randomized Trial." *Lancet* 2368 (9551): 1973–83.

Quisumbing, Agnes R., and J. Maluccio. 2003. "Resources at Marriage and Intra-household Allocation: Evidence from Bangladesh, Ethiopia, Indonesia, and South Africa." *Oxford Bulletin of Economics and Statistics* 65: 283–327.

The Quota Project (database). International IDEA, Stockholm University, and Inter-Parliamentary Union. http://www.quotaproject.org/.

Rammohan, Anu, and Meliyanni Johar. 2009. "The Determinants of Married Women's Autonomy in Indonesia." *Feminist Economics* 15 (4): 31–55.

Rangel, Marcos. A. 2006. "Alimony Rights and Intrahousehold Allocation of Resources: Evidence from Brazil." *Economic Journal* 116 (July), 627–58.

Rand, John, and Finn Tarp. 2011. "Does Gender Influence the Provision of Fringe Benefits? Evidence from Vietnamese SMEs." *Feminist Economics* 17 (1): 59–87.

Rinaldo, Rachel. 2002. "Ironic Legacy: The New Order and Indonesian Women's Groups."*Outskirts* 10. http://www.chloe.uwa.edu.au/outskirts/archive/volume10/rinaldo.

Robinson, Kathy, and Sharon Bessell. 2002. "Introduction to the Issues." In *Women in Indonesia: Gender, Equity and Development*, edited by K. Robinson and S. Bessell, 1–12. Singapore: Institute of Southeast Asian Studies.

Sen, Amartya. 1992. "Missing Women." *British Medical Journal* 304: 587–88.

———. 1999. *Development as Freedom*. Oxford, U.K.: Oxford University Press.

SIGI (Social Institutions and Gender Index). Poverty Reduction and Social Development Unit, OECD Development Centre. http://genderindex.org/.

Singh, Umakant. 2009. "Women and Men United to End Violence against Women and Girls: Four Years Impact Evaluation Report 2005–2009." Women's Media Centre of Cambodia, Phnom Penh, Cambodia.

SPC (Secretariat of the Pacific Community), Ministry of Internal and Social Affairs [Republic of Kiribati], and Statistics Division [Republic of Kiribati]. 2010. *Kiribati Family Health and Support Study: A Study on Violence against Women and Children*. Nouméa, New Caledonia: SPC.

SPC (Secretariat of the Pacific Community) and NSO (National Statistics Office) [Solomon Islands]. 2009. *Solomon Islands Family Health and Safety Study: A Study on Violence against Women and Children*. Nouméa, New Caledonia: SPC.

Stege, Kristina, Ruth Maetala, Anna Naupa, and Joel Simo. 2008. "Land and Women: The Matrilineal Factor: The Cases of the Republic of Marshall Islands, Solomon Islands and Vanuatu." Pacific Islands Forum Secretariat, Suva, Fiji.

Sumner, Cate, and Tim Lindsey. 2011. "Courting Reform: Indonesia's Islamic Courts and Justice for the Poor." *International Journal for Court Administration* (December).

Swain, P. 2000. "Development, Globalization, Civil Society and Non-Governmental Organizations in the Pacific." Background paper prepared for the Stakeholders Workshop on NGO Capacity Building, Port Vila, Vanuatu, November 6–8.

Swaminathan, Hema, Cherryl Walker, and Margaret A. Rugadya, eds. 2008. *Women's Property Rights, HIV and AIDS and Domestic Violence: Research Findings from Two Districts in South Africa and Uganda*. Cape Town, South Africa: HSRC Press.

Thomas, Duncan. 1990. "Intra-household Resource Allocation: An Inferential Approach." *Journal of Human Resources* 25 (4): 635–64.

———. 1992. "The Distribution of Income and Expenditures within the Household." *Annales d'Economie et de Statistique* 29: 109–36.

———. 1995. "Like Father, Like Son, Like Mother, Like Daughter, Parental Resources and Child Height." Papers 95-01, RAND Reprint Series.

Thomas, Duncan, and Chien-Liang Chen. 1994. "Income Shares and Shares of Income." Labor and Population Working Paper 94-08, RAND Corporation, Santa Monica, California.

UN (United Nations). 1993. "Declaration on the Elimination of Violence against Women," December 20, http://www.un.org/documents/ga/res/48/a48r104.htm.

———. 2006. "In-Depth Study of All Forms of Violence against Women: Report of the Secretary-General." A/61/122/Add.1, July 6. UN General Assembly, New York.

UNDP (United Nations Development Programme). 2005. *Community Capacity Enhancement Strategy Note*. New York: UNDP. http://www.undp.org/content/undp/en/home/librarypage/hiv-aids/community-capacity-enhancement-strategy-note/.

———. 2010. *Asia-Pacific Human Development Report: Power, Voice and Rights: A Turning Point for Gender Equality in Asia and the Pacific*. Colombo, Sri Lanka: Macmillan.

UNFPA (United Nations Population Fund). 2006. *State of the World Population 2006: A Passage to Hope: Women and International Migration*. http://www.unfpa.org/swp/2006/english/introduction.html.

———. 2010. "An Assessment of the State of Violence Against Women in Fiji." Suva as cited in CEDAW/C/FIJI/2-4 p. 50, Jan 28, 2010.

UNIFEM (United Nations Development Fund for Women). 2010. "Ending Violence Against Women and Girls: Evidence, Data and Knowledge in the Pacific Island Countries. Literature Review and Annotated Bibliography." UNIFEM Pacific Regional Office, Suva, Fiji.

UNODC (United Nations Office on Drugs and Crime). 2009. "International Framework for Action to Implement the Trafficking in Persons Protocol," United Nations, New York.

UNRISD (United Nations Research Institute for Social Development). 2005. "Women Mobilizing to Reshape Democracy." In

Gender Equality: Striving for Justice in an Unequal World, edited by UNRISD, 167–80. Geneva: United Nations.

U.S. Department of State. 2008. *Laos Trafficking in Persons Report 2008.* Washington, DC: U.S. Department of State.

———. 2011. *Trafficking in Persons Report 2011.* Washington, DC: U.S. Department of State.

Utomo, Iwu, Peter McDonald, Terence Hull, Ida Rosyidah, Tati Hattimah, Nurul Idrus, Saparinah Sadli, and Jamhari Makruj. 2009. "Gender Depiction in Indonesian School Text Books: Progress or Deterioration." Australian Demographic and Social Research Institute, Australian National University, Canberra.

Vietnam: Multiple Indicator Cluster Survey. 2006. World Bank, Washington, DC. http://microdata .worldbank.org/index.php/catalog/31.

VWC (Vanuatu Women's Center) and NSO (National Statistics Office). 2011. *Vanuatu National Survey on Women's Lives and Family Relationships.* Port Vila, Vanuatu: VWC.

Wang, Zhengxu, and Weina Dai. 2010. "Women's Participation in Rural China's Self-Governance: Institutional, Socioeconomic, and Cultural Factors in a Jiangsu Country." Discussion Paper 69, China Policy Institute, University of Nottingham, U.K.

WHO GWH (World Health Organization, Department of Gender, Women and Health). 2007. "Integrating Gender into the Curricula for Health Professionals." Meeting Report, December 4–6. WHO, Geneva.

Whyte, Martin King, and William L. Parish. 1984. *Urban Life in Contemporary China.* Chicago: University of Chicago Press.

Wolf, Margery. 1985. *Revolution Postponed: Women in Contemporary China.* Stanford, California: Stanford University Press.

World Bank. 2011a. "Vietnam Country Gender Assessment." World Bank, Washington, DC. http://documents.worldbank.org/curated/ en/2011/11/15470188/vietnam-country-gender-assessment.

———. 2011b. Defining Gender in the XXI Century: Conversations with Men and Women Around the World (qualitative assessment dataset). World Bank, Washington, DC. http:// go.worldbank.org/CD8RN24BP0.

———. 2011c. *World Development Report 2012: Gender Equality and Development.* Washington, DC: World Bank.

———. 2011d. "Reproductive Health at a Glance: Timor-Leste." World Bank, Washington, DC.

WRAP (Women's Rights and Advocacy in the Pacific). 2011. "Promoting Gender Equity in the Pacific: Recommendations for Pacific Island Forum Leaders," WRAP Reports, WRAP, Wellington, New Zealand.

Yamaoka, Yoshinori. 1999. "Japan." In *Philanthropy and Law in Asia,* edited by Thomas Silk, 163–98. San Francisco: Jossey-Bass.

Zhang, Heather. 1999. "Female Migration and Urban Labour Markets in Tianjin." *Development and Change* 30 (1): 21–41.

Gender and the Region's Emerging Development Challenges | 5

The world is more integrated now than it has ever been in its history, bringing regions together economically, socially, and culturally. The East Asia and Pacific region is one of the most dynamic in the world and has been at the forefront of the trend toward greater global integration. Over the course of two decades, many countries in the region have changed the structure of their economies and have gradually opened to greater inflows and outflows of people, goods, and physical capital. Technological advances have changed production processes in all sectors, from better market access in agriculture to mechanization in manufacturing. The region has seen shifts in the geographic patterns of settlement and work as the center of economic activities has moved from rural to urban areas and as international and intranational migration has increased. Rapid declines in fertility and mortality are anticipated to change the region's demographic profile—in some countries, the proportion of the population over age 65 is already larger than the proportion that are 15 and under.

This chapter examines how the following five key trends are affecting the men and women of the East Asia and Pacific region:

(a) globalization and integration, (b) migration, (c) urbanization, (d) aging populations, and (e) information and communication technology (ICT). Each of these trends involves a number of gender-related challenges but also presents new and encouraging prospects for women. Policy makers in the region will have to manage these challenges while simultaneously fostering new opportunities:

- *Globalization.* The trend toward greater global economic integration, which has been particularly marked in the East Asia and Pacific region, has the potential to increase economic opportunities for women and to narrow the gender wage gap. However, globalization also raises the likelihood that shocks will be transmitted from country to country through integrated markets. Many studies, including those from the recent global financial crisis, have found that these shocks have a gender-differentiated effect along a wide range of outcomes, including human capital investment, labor force participation, earnings, and mortality.
- *Migration.* The economic growth experienced in the region over the past three decades has spurred significant

180   TOWARD GENDER EQUALITY IN EAST ASIA AND THE PACIFIC

movements of people within and across countries in search of better economic opportunities. Women make up nearly half of these migrants in the East Asia and Pacific region. Migration can provide women with more and better economic opportunities, but it can also put them at greater risk for exploitation, abuse, and human trafficking.

- *Urbanization.* Unprecedented increases in migration to urban areas have taken place in recent decades as men and women move in search of jobs. Urbanization can transform women's lives through improved access to infrastructure, education, and health services and a wider range of economic opportunities. Women's ability to take advantage of these opportunities depends on whether gender-sensitive infrastructure and service delivery have been considered and whether they are able to balance their home and work lives.

- *Aging.* The population of the East Asia and Pacific region is aging rapidly, in many countries at a faster pace than in the rest of the world. Women, both the elderly and the young, are likely to face particular challenges as a result of this demographic change. Although women tend to live longer than men, they have fewer economic assets and resources and are less likely to be covered by formal social security systems. They are also more likely to be caregivers to the elderly than men.

- *Information and communication technology (ICT).* Advances in information and communication technologies are revolutionizing the ways in which men and women in the region are exposed to ideas, share knowledge, and network. ICTs can empower women by opening new economic opportunities, breaking down information barriers, changing social norms, and enabling collective action. Evidence suggests that women in the region may still have less access to information technology than men, reducing their ability to harness their transformative potential.

Globalization

Globalization presents opportunities as well as challenges for reducing gender inequality. The opening up of domestic economies to specialization spurred by international trade may increase economic opportunities and women's empowerment, but this trend also brings with it an increased risk that shocks will be transmitted from country to country through increasingly integrated markets.

Growth in trade and capital flows in East Asia and the Pacific

Developing countries all over the world are experiencing structural change at a rapid pace. Since 1987, the share of both male and female employment in manufacturing and services in developing countries has grown faster than in developed countries, reflecting changes in the global distribution of production and labor (World Bank 2011b). These trends have been mirrored in the East Asia and Pacific region (chapter 3).

Behind these changes in economic structure lie the powerful global forces of economic and social integration and economic reforms. In many countries in the region, reforms during the 1990s and early 2000s laid the foundation for institutional change and rapid growth. The export of goods and services as a proportion of gross domestic product (GDP) rose from 17 percent in 1980 to 43 percent in 2008. However, these flows are volatile. The recent financial crisis has caused exports to fall to 35 percent of GDP in the region. Within the region, the ratio of exports to GDP varies widely. For example, in Vietnam, this ratio has increased hugely over time, from 7 percent in 1986 to over 70 percent in the late 2000s, while in Indonesia, the ratio has remained between 20 and 40 percent over the past three decades.[1]

Globalization and gender

These trade and capital flows have generated new employment and income-generation opportunities within the region, particularly for women.[2] However, these flows have also been associated with greater economic volatility, because global or regional macroeconomic shocks—such as the Asian financial crisis of 1997–98, the food and fuel crisis of 2004 and 2008, and the global financial crisis—can spread more rapidly throughout the world. This section explores the ways in which globalization has changed the lives of men and women in the region.

Globalization and increased economic opportunities for women

Women have played an important role in the expansion of export-oriented manufacturing during the process of economic development and structural transformation (Standing 1999; Wood 1991). Increases in trade and mobile capital have expanded job opportunities in the East Asia and Pacific region, particularly for women who make up a large proportion of the labor force in the export-oriented manufacturing sector in the region and around the world (World Bank 2011b).

Export-oriented firms are more likely to hire women than non-export-oriented firms, both across the world and within the region (figure 5.1). The East Asia and Pacific region is second only to Europe and Central Asia in the proportion of female workers working in the export-oriented sector. Within East Asia, the fraction of female workers is substantially higher in export-oriented firms than in non-export-oriented firms, with Indonesia as a key exception in this regard (figure 5.1b).

Female workers predominate in certain export-oriented sectors, such as the garment sector. These sectors have grown substantially and quickly in recent years. Figure 5.2 depicts employment in Vietnam's textile and apparel sectors, which grew significantly between 2000 and 2008. Women are more likely to work in these sectors than men in Cambodia, Indonesia, the Lao People's Democratic Republic, and Vietnam, although overall garments and textiles account for only a

small fraction of the female workforce. Among female workers ages 15–34, who are the most likely to be employed in the manufacturing sector, employment in the garment industry is substantial in some countries in the region, from approximately 8 percent of 15- to 34-year-old females in Indonesia to over 15 percent in Vietnam (figure 5.3). Employment patterns have shifted over time

FIGURE 5.1   **Women are more likely to work in export-oriented firms than in non-export-oriented firms**

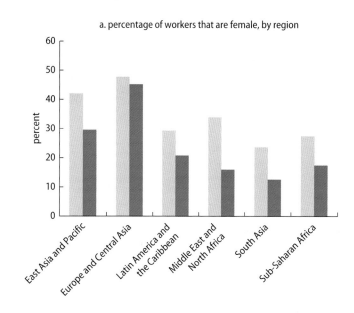

a. percentage of workers that are female, by region

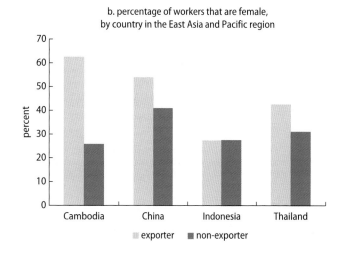

b. percentage of workers that are female, by country in the East Asia and Pacific region

☐ exporter   ■ non-exporter

Source: Enterprise Surveys database, 2002–05 data.
Note: The data are not available for examining similar patterns in Pacific countries.

FIGURE 5.2   Employment in textile and apparel in Vietnam grew substantially between 2000 and 2008

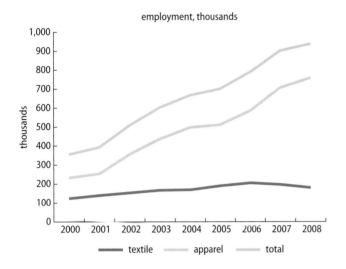

employment, thousands

Sources: Frederick and Staritz 2011.

FIGURE 5.3   Women predominate in the garment sector in four East Asian countries

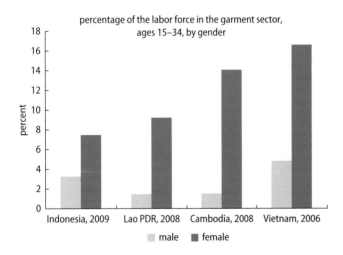

percentage of the labor force in the garment sector, ages 15–34, by gender

Sources: World Bank staff estimates using Cambodia Socioeconomic Survey (CSES) (NIS Cambodia), 2008 data; Indonesia National Labor Force Surveys (SAKERNAS) (BPS Indonesia), 2009 data; Lao Expenditure and Consumption Survey (LECS) (LSB Lao PDR), 2008 data; VHLSS (GSO Vietnam), 2006 data.

in the electronics industry. In earlier waves of export-orientated growth in Japan, the Republic of Korea and Taiwan, China, the labor force in electronics was predominantly female. Analysis of recent household data

from the East Asia and Pacific region suggests that with the exception of Indonesia, electronics production does not appear to be an overwhelmingly female area of work.

The experience of developed countries in the region shows that the demand for female labor in export-oriented industries declines as countries transition from labor-intensive to capital-intensive manufacturing. In Korea, women constituted 70 percent of workers in export processing zones in 1990 compared to 42 percent of employees in manufacturing overall (Kusago and Tzannatos 1998), a substantial rise over the 28 percent of women employed in manufacturing in 1972 (Seguino 1997). However, the dominance of females in the export-oriented sector has declined since heavy chemical and manufacturing industries, such as steel, cars, and shipbuilding, began to increasingly replace light manufacturing industries (Kong 2007). In Taiwan, China, a concerted shift toward capital- or technology-intensive exports in the late 1970s and early 1980s was accompanied by a steady decline in the share of female wage workers in the manufacturing sector until the mid-1990s (Berik 2000).

Although women are extensively employed in the sectors that are expanding as a result of increased trade opportunities in many East Asian and Pacific countries, as sectors become more technology, skill, and capital intensive, employers may ultimately replace female workers with more highly skilled men (Gamberoni and Reis 2011; Tejani and Milberg 2010). For example, in Malaysia, only 40 percent of the workers in the Special Economic Zones are now female, down from 60 percent two decades ago (IFC 2008). Several possible explanations for this include (a) differences in the level and content of men's and women's education; (b) discrimination and gender segregation in higher skilled jobs; (c) tight female labor markets that can force female wages to rise; and (d) the view that men deserve more secure employment and are less likely to leave paid work to fulfill domestic responsibilities" (Seguino and Grown 2006, 8; Berik forthcoming; Gamberoni and Reis 2011; Tejani and Milberg 2010).

The wages of men and women respond differently to trade liberalization and export-oriented growth, depending on (a) the type of growth seen, (b) a country's comparative advantage, and (c) institutional and policy-related factors (Berik, van der Meulen Rodgers, and Zveglich 2004; Black and Brainerd 2002; Oostendorp 2009; Seguino 2000).[3] During periods of export-oriented growth, gender wage gaps narrowed in Korea while they widened in Taiwan, China (Seguino 2000). The difference in patterns is attributed to increasing capital mobility in Taiwan, China, which moved capital out of labor- and female-intensive industries, whereas regulations in Korea limited the movement of capital (Seguino 2000). Similar evidence from Bangladesh indicates that a movement up the value chain in the Bangladeshi garment industry had negative repercussions for female workers, who are less likely to work in the more skill-intensive occupations (Frederick and Staritz 2011). In Taiwan, China, and in Korea, the wage gap between men and women widened in sectors with strong trade competition between 1980 and 1999 (Berik, van der Meulen Rodgers, and Zveglich 2004). In Taiwan, China, greater export orientation adversely affected both men's and women's wages, but it reduced gender wage inequality because male employees faced a greater wage penalty than female workers did (Berik 2000).

The expansion of labor market opportunities as a result of trade liberalization may also encourage investment in education among young women. For example, the expansion of call centers and other economic opportunities for women linked to globalization have increased female educational investment in India (Munshi and Rosenzweig 2004; Oster and Millet 2010; Shastry 2010). The evidence suggests that the relationship between investment in education and labor market opportunities depends on the skill bias of export-oriented growth and that growth in low-skill intensive opportunities may in fact reduce educational investment. Growth in the unskilled

manufacturing sector in Mexico increased the school dropout rate (Atkin 2010).

Gender-related challenges of globalization

Despite expanding economic opportunities for women, globalization poses challenges in the form of increased exposure to externally driven shocks and in the type and quality of work conducted by women in the export-oriented sector. Externally driven shocks have had a negative effect on women in a number of outcomes related to health, education, and employment.

Men and women work in different occupations and industries, so they tend to be affected differently by economic shocks. In addition, because crises vary in the sectors they affect and in their propagation mechanisms, the employment and wage effects on men and women also vary across crises. For example, in the Philippines, the food and fuel crisis of 2008 affected employment and wages in different ways than the financial crisis of 2009 (Menon and van der Meulen Rodgers forthcoming). Following the food and fuel crisis, more unskilled workers lost their jobs than skilled workers, whereas skilled workers suffered more than unskilled workers during the financial crisis. Women's likelihood of employment dropped more than that of men during both crises but, although most of the decline for men came from falling wage employment, women lost work in both wage employment and self-employment. In Cambodia, analysis showed that the financial crisis resulted in labor market churning, in which high job destruction was followed by an even larger creation of low-quality jobs. Women accounted for the greatest share of job losses but also the largest share of jobs created (Bruni et al. 2012).

Across the world, as well as in the East Asia and Pacific region, both women's employment and the total number of hours that women work tend to increase during periods of economic crisis (Frankenburg, Smith, and Thomas 2000; Lim 2000; Tork and Mason 2009).[5] Using data from 63 developing and transition countries, Bhalotra and Umana-Aponte (2010) found that a 10 percent drop in a country's GDP is associated with a 0.74 percentage point increase in women's work participation.[4] Those authors also found that, in the East Asia and Pacific region, the employment of rural women and married women with a child under the age of five was more sensitive to cyclical variation than other women, possibly because they tend to be closer to their subsistence requirements.

Women's employment in the East Asia and Pacific region may be disproportionately affected by crises because of perceptions that they should leave job opportunities for male workers. In Korea, more women than men dropped out of the labor force and became "discouraged" workers during the 1997–98 crisis (Kim and Voos 2007). The increase in dropouts was concentrated among young, single women and outweighed the increased labor force participation of married women who entered the labor market to maintain their family's income (Kang 1999, cited in Kim and Voos 2007). The majority of employers targeted women workers for voluntary resignation, especially if those women were from double-income families or were married. Anecdotal evidence suggests that women in some sectors in Korea, such as banking, were forced to resign from their permanent positions and were then rehired as temporary employees (Kim and Voos 2007). The disproportionate dismissal of female employees during the crisis occurred partly because of employers' perception that hiring female workers cost them more than hiring male workers (Kong 2007) and partly because employers viewed men as the main family breadwinners and hence believed that women should step aside (Kim and Park 2006).

Female health has been found to be more susceptible to shocks than male health, both in the region and elsewhere (Strauss and Thomas 2008). Females are more likely to suffer from physical and mental health deteriorations during periods of crisis (Dercon and Krishnan 2000; Frankenberg et al. 2008; Friedman and Thomas 2007). Data from 59 developing countries show a large negative association between per capita GDP and infant mortality, with a 1 percent decline in per capita GDP being associated with a 10 to 15 percent increase in infant mortality on average. However, the effect on females is approximately twice the effect on males—a 1 percent decline in per capita GDP increases the mortality rates of boys by 0.27 per thousand born, whereas it raises that of girls by 0.53 per thousand. The quality and quantity of women's and girls' diets are also more likely to have been disproportionately affected by the food and financial crisis than those of men and boys (Jones et al. 2009). In addition, tensions in households increase during periods of economic stress. During the financial crisis of 2009, both men and women reported an increase in the number of arguments between husbands and wives related to their limited financial resources, which sometimes led to violence (Turk and Mason 2009).[6]

Evidence from macroeconomic crises suggests that, although children's school enrollment declines in response to these shocks, the gender differences are small relative to their overall impacts (Sabarwal, Sinha, and Buvinic 2010). In Indonesia and the Philippines, the 1997–98 crisis was associated with declines in children's school enrollment or increases in child labor (Frankenberg, Beard, and Saputra 1999; Lim 2000; Thomas et al. 2004). In the Philippines, the enrollment of female children declined slightly at the elementary level in 1998–99, whereas male enrollment increased substantially during that year (Lim 2000). At the same time, child labor among boys ages 10 to 14 increased in both urban and rural areas, and labor force participation rates of girls increased only marginally. In Indonesia, real education expenditures as well as the share of the household budget spent on schooling declined between 1997 and 1998 (Thomas et al. 2004). Households spent more on the education of young men (ages 15–19) than young women and mobilized the money to do so by reducing their expenditures on the education of younger male and female children (ages 10–14) and of older females (ages 15–19).

Finally, globalization poses challenges in the type and quality of work conducted by women in the export-oriented sector. Gender segregation continues to exist. Women are less likely to be employed in managerial or professional positions and are more likely to be in part-time or informal subcontracted jobs (Seguino 2000).[7] Older female workers also face barriers in these industries. For example, in Shenzhen, the largest of the Special Economic Zones (SEZs) in China, female workers under 25 years of age make up 90 percent of workers in garment and electronics plants (Summerfield 1995, cited in Davin 2004). This bias may be because firms in these zones often use single-sex dormitories to accommodate migrant workers, a situation that is incompatible with marriage and family formation (Davin 2004; Ngai 2004).

Many studies have suggested that employment in Export Processing Zones (EPZs) is characterized by unsafe working conditions and the suppression of labor rights; however, studies do not make clear whether workers in these zones are worse off than their rural and urban counterparts (Murayama and Yokata 2008). Governments set the legal framework for employment practices in SEZs, which may have the same labor standards as those in the country as a whole or may be unique to the SEZs. For example, in the Philippines, all national laws officially apply in EPZs. In practice, however, the Philippines Economic Zone Authority, often allows firms in the EPZ to circumvent these laws (McKay 2006).

Implications for policy related to globalization

The expansion of employment opportunities in the East Asia and Pacific region due to globalization has been associated with greater female access to income and rising empowerment, particularly among young rural female migrants. However, challenges remain for women working in these sectors. As in the rest of the economy, women are less likely to be employed in managerial

positions and more likely to be employed in temporary positions or in the informal sector than men. Furthermore, experience from developed countries in the region suggests that, as economies develop and industries move up the value chain, female employment in the export-oriented sector may decline as so-called male industries emerge.

Policies that reduce gender gaps in economic opportunities, and in particular in labor force participation and gender based occupational and industrial segregation, are necessary to ensure that women are able to fully benefit from the economic opportunities brought about by greater globalization. Since women are disproportionately employed in the export-orientated sector and in special economic zones, there may be scope for policy makers to promote the creation of zone-level social services that are sensitive to women's needs.

Greater economic integration also entails risks, notably an increased exposure to employment shocks, which will have gender-differentiated impacts. To address these risks, policy makers need to design social protection programs that adequately account for the different risks faced by female and male workers. Recognizing the gender-differentiated impacts of labor market programs and policies enacted to mitigate crises will lead to better policy making. Evidence from the region suggests that gender blind policies enacted in response to shocks may have gender differentiated impacts.

Migration

The movement of people, both on a national and international scale, has increased across the world in the past two decades because of improvements in transportation technology and infrastructure (World Bank 2008). Worldwide, women account for almost half of all international migrants, rising from below 47 percent of migrants in 1960 to 49 percent in 2010 (UNPD 2008). Studies have noted a shift in the motive for migration among female migrants. Previously, women predominantly migrated for marriage or as dependents of spouses or other male family members who worked abroad. Increasingly, women are migrating independently to improve their economic opportunities and to fulfill the role of primary breadwinner for their families (Sørensen 2005).

Migration and gender

The economic growth experienced in the East Asia and Pacific region over the past three decades has spurred a significant intraregional movement of people. Evidence of these flows is highly visible within the region— from the 200 million people travelling across China to be with their families for the Lunar New Year to the substantial cross-border migration flows seen in the Greater Mekong Subregion (World Bank 2008). Economic booms in East Asian cities have created labor force shortages, which have been met by a growing supply of migrant labor as people have moved to urban areas to improve the well-being of their families and communities. The flow of migrants to several cities in the region has become increasingly female, and women comprise a significant proportion of

the international migration from the region as a whole (Guzmán 2006; UN 2001, cited in Hugo 2003).

This section examines the following in the context of the East Asia and Pacific region: (a) emerging migration trends and gender, (b) the socioeconomic characteristics of male and female migrants, (c) factors influencing female migration, (d) the mixed impact of migration on women's status, and (e) the mixed impact of migration on those who stay behind.

Gender-differentiated migration trends in the East Asia and Pacific region

The often informal and undocumented nature of female migration in the region has resulted in significant information gaps, but three salient facts have emerged in recent years. First, the numbers of women migrating are increasing. At the turn of the 21st century, the number of female migrants in the region is estimated to have surpassed that of male migrants (Lee 2005). In Fiji and Tonga in 2005, approximately 50 percent of migrants were female (World Bank 2006b). The bulk of female migrants in the region come from Indonesia and the Philippines. Thai women are underrepresented among legal migrants but dominate among irregular migrants or those presumed to be trafficked. Thailand has also been an important destination for irregular migrant women, such as Myanmar and Lao women working as domestic workers (Piper 2009). More than half of the Lao migrants and close to half of the Cambodian migrants to Thailand are women (World Bank 2006c).

Second, the proportion of women in the migrant population is increasing. A study in China covering six provinces found that, between 1995 and 2000, the migration rates for women increased twice as quickly as those for men (de Brauw et al. 2002). The annual proportion of women migrating, of Indonesian workers, reached approximately 80 percent in 2007, an increase from about 70 percent in 2000 (figure 5.4). In Vietnam, women constituted 57 percent of individuals who had migrated internally or who were

FIGURE 5.4 The share of international female migrants has increased over time in Indonesia and Vietnam

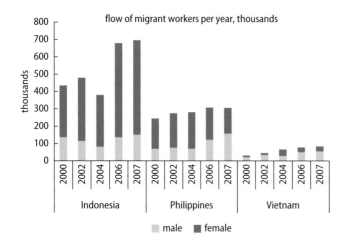

Source: World Bank forthcoming.
Note: Figure is based on data from the Indonesia Ministry of Manpower, the Philippines Overseas Employment Administration, and the Vietnam Ministry of Labor, Invalids and Social Affairs.

international migrants in the previous five years (Pierre 2011).

Third, female migrants who used to migrate out of the region are increasingly finding opportunities to migrate within the region. Although the most frequent destination for female migrants from Indonesia is Saudi Arabia, more and more of these women are finding employment in other Asian countries such as Hong Kong SAR, China; Malaysia; Singapore; and Taiwan, China. This trend differs from those of Indonesian male migrants. Although the majority of male migrants find work in Malaysia, many are now going to Saudi Arabia (Nguyen and Purnamasari 2011).

Socioeconomic factors characterizing male and female migrants

The socioeconomic factors that characterize male and female migrants in the region appear to be similar. Migrants, both male and female, are young and relatively low-skilled but are not necessarily among the least educated in their country of origin (Jampaklay et al. 2009; World Bank 2006c; World Bank, forthcoming).[8] The exception to this trend is the Philippines, where both male and female migrants tend to be older and highly educated (Cabegin and Alba 2011). High-skilled female migration differs from low-skilled migration: the destination countries for this type of migration are usually high-income developed countries in North America and Western Europe. Female migrants participate in all types of migration. They can be found among both internal and international migrants and within temporary and permanent flows; they can be regular or irregular migrants.

Factors influencing female migration

A combination of factors have influenced the feminization of migrant flows in the region. The perceived comparative advantage of women in growing industries has led to an increase in female labor demand and to the expansion of labor opportunities for women in both the formal and informal sectors in urban areas. Women are more likely than men to work in export-oriented industries, where employers perceive them to have a comparative advantage.[9] China's Pearl River Delta region is one of the most popular destinations for internal rural migrants. Young rural migrant women constitute 65 to 70 percent of the region's labor force because they are favored by the region's transnational clothing, textile, toy, electronic, and other labor-intensive manufacturing and processing firms (Gaetano and Jacka 2004).

A rapid increase in demand for domestic workers has occurred as a result of urbanization and an expansion of the middle class in many countries in the world. Women migrate to work abroad as domestic workers in response to gender-specific labor demand. Female migrant workers from East Asian and Pacific countries are also employed as domestic workers in other regions, including the Middle East, North America, and Western Europe. Approximately 1 million migrant domestic workers are in Saudi Arabia, with the majority coming from Indonesia, the Philippines, and Sri Lanka. In 2003, an estimated 200,000 migrant domestic workers were found in Hong Kong SAR, China, and 155,000 were found in Malaysia. In Singapore, one of every seven households employs a domestic worker, the majority of whom are migrants (UNFPA 2006).

More women are moving independently for educational and employment reasons. When migration is not restricted, many women choose to migrate to obtain new skills, increase their income, and broaden their education. In Vietnam, approximately 14 percent of male and female migrants migrate to attend school; since females are more likely to migrate, they constitute a larger fraction of those migrating for schooling (Pierre 2011). Women are overrepresented in the brain drain, which may be a consequence of their unequal access to labor markets in developing countries. Econometric estimates show that emigration of highly skilled women is higher the poorer the country of origin (Dumont et al. 2007). Although this can be beneficial to the individual in terms of opportunities, the female brain drain can be detrimental to the economic growth of source countries.

The informal sectors of cities have many opportunities for women as well as men, promoting rural to urban migration. The informal labor sector in Cambodia has absorbed female migrants from rural areas in work such as street peddling, manual labor, domestic service, and garbage collection (UNIFEM 2005). Male migrants conduct different, generally brawn-intensive, types of work, including construction, mining, fishing, and logging.

In recent decades, the sex and entertainment industries in Asian cities and in cross-border areas have grown substantially. Almost all women involved in these industries are rural-to-urban migrants with circular, or temporary, patterns of migration (Hugo 2003). Global tourism, which has been promoted as a national policy in countries such as Indonesia, Malaysia, the Philippines, and Thailand for decades, is cited as the primary reason for the rapid growth of the sex and entertainment industry (Hugo 2005). Cambodia and Lao PDR also now face this social challenge.

The mixed impact of migration on women's position in households and communities

Migration can increase a woman's empowerment, economic opportunity, knowledge, and skills, as well as increase her participation as an active member of society. Through these channels, a woman can contribute to economic growth, increase the well-being of her family, and have an intergenerational effect by acting as a role model for other women and girls.[10] However, female migrants are more likely to be at risk of exploitation, abuse, and human trafficking than male migrants.

Many women become empowered when they move from rural to urban areas and away from familial and rural community social controls. Increased autonomy, access to information, and status can empower women to create new identities for themselves. Rural migrant women in Chinese cities experience more autonomy and independence than they did at home (Zhang 1999). Migrant women who return home may act as role models to other women by transferring or demonstrating newly acquired skills, ideas, attitudes, and knowledge (Hugo 2005; UNFPA 2006). Migrants employed as domestic workers in Chiang Mai and Mae Sot, Thailand, expressed the desire to establish businesses based on the new skills that they've learnt when they return home (Punpuing et al. 2005). The women of rural Anhui and Sichuan, China, who participated in circular migration returned home having adopted the urban norm of desiring only one child; they also experienced lower rates of domestic violence than women who did not migrate. These women also believe that women should be able to choose a marriage partner and that divorce is acceptable (Connelly et al. 2010).

Empowerment may arise in the form of participation in collective organizations, including nongovernmental organizations and labor associations that lobby for gender equality. In the past 15 years, domestic workers in Hong Kong SAR, China, have been active in organizing and participating in political protests that concern not only local migrant workers' rights but also global, transnational, and human rights (Constable 2009).

Migration provides women with increased economic opportunities, which can improve the standard of living of the migrants and their families. For many female rural-to-urban migrants in China, their migration is the first time they earn wages and choose how to spend them (Connelly et al. 2010). In Cambodia, the porous border to Thailand allows Khmer female migrants to sell home and farm products in Thailand's markets at higher prices, where the per capita GDP is 12 times that of Cambodia. Khmer women also engage in paid and formal employment as shop assistants, fruit sellers, and sweet sellers for mostly Thai employers. This largely circular and nonpermanent migration ranges from day migration to stays of two to three weeks at a time (UNIFEM 2005). Econometric evidence that combines a survey of Indonesian maids and factory workers in Malaysia with data from the Indonesia Family and Life Survey shows that these young female migrants may gain an additional US$80 to US$130 per month compared with their earnings if they stay in Indonesia. These income gains are as high as five times their income in Indonesia (Tan and Gibson 2010).

Migrants contribute to the economic development of their destination countries through their competencies and skills. Furthermore, they contribute to the economic development of the sending countries through remittances and increased experience and knowledge upon returning. Female migrants can improve the well-being of family members at home and potentially foster economic growth. In Indonesia, the proportion of households receiving remittances from female migrants between 2000 and 2007 far exceeded the proportion receiving remittances from male migrants. Their contribution to their family back home—in remittances as a percentage of per capita consumption among recipient households—also surpassed that of male migrants from 2000 to 2007 (Nguyen and Purnamasari 2011).[11]

In destination areas, women are particularly vulnerable to exploitation because of their status as migrants and because they are women. Women are more likely to be found in occupations where they are subject to labor exploitation and health risks, which can make migration detrimental to a woman's well-being. Furthermore, women's progress in empowerment may be negated upon returning to their home community.

Gender-based labor segregation channels migrant women into occupations that may subject them to physical and psychological abuse as well as labor exploitation and human rights violations (Yamanaka and Piper 2005). Many female migrants in the region, both documented and undocumented, work as live-in maids, caregivers, entertainers, sex workers, and other service employees (Yamanaka and Piper 2003). Labor legislation generally does not cover these occupations, leaving female migrants vulnerable to exploitation.

The largest segregated occupation, and one of the most isolating, is domestic work. Indonesian domestic workers employed in Malaysia typically work 16- to 18-hour days, seven days a week, without holidays (Human Rights Watch 2004). In the Chiang Mai and Mae Sot provinces of Thailand, almost 98 percent of domestic workers worked for more than 12 hours a day (UNESCAP 2007). Over half of these domestic workers were subject to verbal abuse, 10 percent experienced physical abuse, and 14 percent stated that they experienced some form of sexual harassment (Punpuing et al. 2005).[12] In addition to these abuses, domestic workers frequently had their freedom of movement and their communication with the outside world limited by their employers.

One of the greatest vulnerabilities women and girls face is trafficking for prostitution and forced labor. A highly profitable and growing industry, human trafficking is the third most lucrative illicit business in the world after arms and drug trafficking, and it is a substantial source of organized crime revenue (ILO 2008; UNFPA 2006). The International Labour Organization (ILO) estimates that at least 2.5 million trafficking victims are currently being exploited worldwide and that another 1.2 million are trafficked annually, both across and within national borders. Asia and the Pacific regions account for over half of these trafficked victims—an estimated 1.36 million (ILO 2008).[13] Women often arrive in destination countries such as Malaysia through legal channels but without a job. These women become particularly vulnerable to trafficking because of financial hardship and limited information.

Male and female migrants often have jobs that subject them to health risks. The prevalence of female temporary migrants' work in the sex and entertainment industries increases their likelihood of acquiring HIV. In China, male migrants and non-migrants have similar rates of casual and commercial sex; however, female temporary migrants have rates of casual and commercial sex 14 and 80 times the rates for female nonmigrants, respectively (Yang and Xia 2005). Male migrants, who dominate industries such as mining and construction, face increased health risks, including risk of death, partly as a result of lax occupational safety regulations. In China, between 2001 and 2005, an average of 6,222 workers died in coal mining accidents each year; the majority of the workers were migrants (IOM 2009). Other effects of working in mines develop over time: pneumoconiosis, a lung disease caused by dust inhalation, accounts for 83 percent of all occupational disease recorded in China (Ministry of Health figures, Su 2005, as cited in IOM 2009).

Migrants' gains in female empowerment in the destination area may not have lasting impacts once women return to rural areas. Context matters in determining empowerment outcomes, as shown in many studies. Mason and Smith (2003, 2) compare autonomy measures across five Asian countries and argue that "community is a far stronger predictor of women's empowerment than are individual traits" (see the discussion in chapter 4). Evidence from China suggests that rural women migrants who return home experience a decrease in empowerment because of the existing community patriarchal inequality factors rather than individual characteristics (Connelly et al. 2010).

The mixed impact of migration on those who stay behind

Migrants' spouses, children, parents, and communities are also affected by their absence, and the evidence shows that the impact is mixed. Women and girls who are left behind may face more financial hardships, difficulties disciplining children, less access to food, as well as loneliness and isolation. Frequently, women who are left behind must engage more in income-generating activities to compensate for the income lost by the migrant relative if the latter does not send adequate remittances or remittances on a regular basis. Women left behind in rural China experience a substantial reallocation of traditional farm labor, with older women

taking on most of the added hours in farm work. This additional obligation seems to be a persistent effect and comes at the cost of fewer hours in local off-farm work, with no signs of increased decision-making responsibilities over the household's farming activities. However, men who are left behind do not experience this reallocation (Mu and van de Walle 2009).

Migration affects the employment of those left behind. In the Philippines, having a migrant in the household reduces the labor force participation and hours worked of nonmigrant relatives of both men and women, who substitute income for more leisure (Rodriguez and Tiongson 2001). In Indonesia, migration reduces the working hours of remaining household members by 33 hours per week if the migrant is male. This negative relationship is not observed for households with female migrants (Nguyen and Purnamasari 2011). Furthermore, in Indonesia, female migration may reduce the labor force participation of children ages 6–18 by 17 percentage points (Nguyen and Purnamasari 2011). This impact is not seen with male migration. Anecdotal evidence from the Pacific Islands of Fiji, Samoa, and Tonga suggests a similar phenomenon. Regular remittances from female migrants such as nurses, teachers, domestic workers, and caregivers have induced those who receive remittances to increase their leisure time by resigning from their jobs or dropping out of school. This response can be a problem because it implies a total dependence on remittances (UNESCAP 2007).

The responsiveness of investment in education tends to vary according to the sex of the migrant. In Indonesia, migration may have a slightly positive impact on school enrollment among households with male migrants. By comparison, migration does not appear to have strong effects on children's schooling among households with female migrants, perhaps because the absence of mothers often makes it harder to monitor children's activities (Nguyen and Purnamasari 2011).

Implications for policy related to migration

The number of female migrants in the East Asia and Pacific region is now estimated to have surpassed that of male migrants. Women have benefited from being more likely than men to be employed in growing export-oriented industries such as clothing, textile, toy, and electronics manufacturing, and they are more likely to provide services that are in demand, such as live-in maids or caregivers. By increasing a woman's economic opportunities and ability to generate income, migration can increase her empowerment, economic opportunities, knowledge, and skills, thus also improving the well-being of her family. It can also increase her participation as an active member of society and make her a role model for other women and girls. However, migrant women are more likely than men to work in occupations where they may be subject to labor exploitation and health risks, or, even worse, physical and psychological abuse. These occupations are generally not covered by labor legislation, leaving female migrants vulnerable to exploitation. The biggest danger women migrants face, especially those with no job and little money, is trafficking for prostitution and forced labor, a huge and growing illegal industry that exploits vulnerable women.

Improved laws, safety nets, and knowledge transfers can all help to mitigate the vulnerabilities experienced by migrant women. Some specific policy recommendations include improving the legal and social protections of female migrants, strengthening the monitoring and credibility of labor recruitment agencies, and developing and providing welfare and support services to assist female migrants. Governments in both sending and receiving countries should actively address the issue of human trafficking through prevention, protection, and prosecution. Gender-sensitive training for people involved in the migration process will increase their ability to identify and assist abused female migrants and those trafficked or at risk of being trafficked.

Urbanization

Urbanization has increased across the world and, for the first time in world history, the urban population accounts for more than 50 percent of the world's population globally (figure 5.5). At the global level, the Latin America and Caribbean region has the highest rate of urbanization in the world, which is expected to reach close to 90 percent by 2050. Since 1950, countries in the East Asia and Pacific region have experienced population shifts away from rural areas to urban centers. Individuals in developing countries, in search of better economic and social opportunities, are attracted to urban areas. With their concentration of population and economic activities, cities make a major contribution to countries' national incomes. As economic activity becomes concentrated, local welfare in urban areas may improve, although remote areas tend to lag behind until development proceeds and living standards converge (World Bank 2008).

Urbanization affects all aspects of life, from the family and community networks that people rely on to the economic activities conducted by men and women. Access

FIGURE 5.5. **The urban population now accounts for more than 50 percent of the world's population**

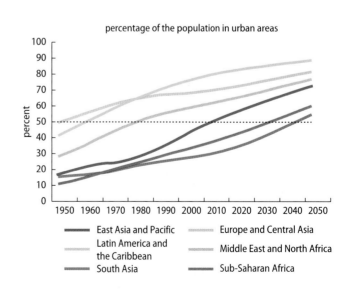

Source: UN DESA Population Division 2010.

FIGURE 5.6   **Urbanization is expected to be rapid in East Asia**

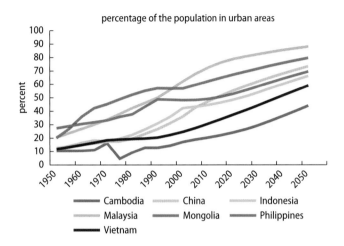

percentage of the population in urban areas

Source: UN DESA Population Division 2010.

FIGURE 5.7   **The rates of urban growth are predicted to vary substantially across countries in the Pacific**

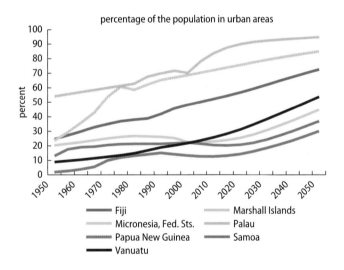

percentage of the population in urban areas

Source: UN DESA Population Division 2010.

to urban labor markets is likely to affect female participation in the formal labor market as well as the composition of productive work conducted (Moser 1993). Urbanization is also likely to alter the time use patterns of both men and women and, in particular, to reduce the time spent on housework for women. Gender differences in the time devoted to nonmarket activities are greater

in rural areas, where more limited access to water, sanitation, and energy increases female time spent on basic life-sustaining activities, as discussed in chapter 3.

This section analyzes urbanization patterns and how they affect men and women in the region. It then discusses the opportunities and challenges that men and women face in urban settings and highlights some policy priorities for tackling these challenges.

Trends in urbanization in the East Asia and Pacific region

Although the rate of urbanization differs across countries in the region, one observation remains clear: in all developing countries in the East Asia and Pacific region, the fraction of the population living in urban areas is expected to grow over the next half century. Figure 5.6 illustrates trends in the proportion of the population living in urban areas for East Asia. The growth of the urbanized population is substantial: in 1950, 12 percent of the Chinese population was living in urban areas, and by 2000 this number had risen to 36 percent and is estimated to increase to 73 percent by 2050. Urbanization in the Pacific varies substantially across countries (figure 5.7). In Palau, over 90 percent of the population is expected to be living in urban areas by 2050; in contrast, less than 30 percent are expected to do so in Papua New Guinea.

Urbanization and gender

Urban areas provide many opportunities for men and women: health care, education, and financial services are better developed and easier to access, and labor markets present a broader range of employment opportunities than in rural areas. This section discusses the opportunities and challenges of urbanization from a gender perspective in three dimensions: economic opportunities, service delivery, and agency.

Economic opportunities
Like their male counterparts, women have access to more economic opportunities in

urban areas than in rural areas. Urban labor markets offer a wide variety of occupations, from manufacturing and services to clerical activities.[14]

Limited access to child-care services may limit women's ability to take advantage of these opportunities, however. A study on gender differences in labor market behavior in Mongolia found that women spend about twice as much time on household duties as men, independent of whether they participate in the labor market, and that the number of young children decreases female labor market participation in urban areas (World Bank 2011a). The economic restructuring of the 1990s, in which Mongolia reduced state-sponsored child-care provision, may partially explain the time burden on women. Although a set of policies was passed to increase children's enrollment in kindergarten, only about half of the children ages 2 to 6 were enrolled in a kindergarten in 2007, and urban areas face serious constraints in available schools (World Bank 2011a).

Service delivery

Public services are less expensive to provide in urban areas than in rural areas, because of population density and economies of scale (UNFPA 2007). Figures 5.8 and 5.9 show the percentage of the rural and urban populations with access to improved sanitation facilities and water sources. In rural areas in Cambodia, only 15 percent of the population had access to improved sanitation in 2005, in contrast to urban areas, where 60 percent had access (figure 5.8). Similarly, in Mongolia, only 45 percent of the rural population had access to safe water sources, whereas 94 percent of the urban population did (figure 5.9).

Access to improved water and sanitation services is particularly important for women, who are often responsible for the collection, management, and use of the domestic water supply. Increasing access to improved water sources is likely to reduce women's time spent on domestic water management and to allow more time for other activities. Inadequate access to water may reduce women's

FIGURE 5.8   In most East Asian and Pacific countries, urban areas have better access to improved sanitation

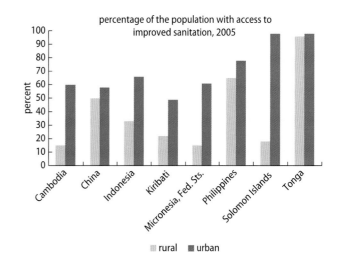

Source: World Development Indicators (WDI) database.

FIGURE 5.9   Rural areas have lower access to improved water sources than urban areas in the majority of countries in the region

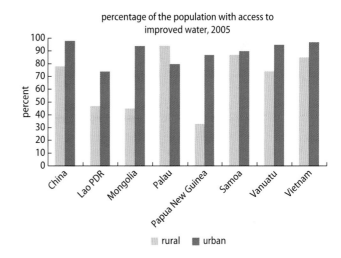

Source: WDI database.

economic opportunities by limiting access to home-based income-generating activities such as food production and sale, livestock raising, and other service-oriented business activities (Noel, Soussan, and Barron 2007). Poor sanitation can cause the spread of infectious diseases such as cholera, polio, or hepatitis. Finally, evidence from outside the region shows that improving water sources increases girls' school attendance (Koolwal and van de Walle, forthcoming).

Transportation is crucial for ensuring that urban populations are able to benefit from the social and economic potential of urban areas. Women and men have different transport needs and patterns. Women's transport needs and commuting patterns are associated with responsibilities in the household as well as income-generating activities. Studies show that women in urban areas travel frequently in off-peak times and conduct multiple stops on a single trip. In contrast, men tend to use transport at peak times for reaching work (World Bank 2010b). Women's commuting patterns also vary over their life cycle and according to their reproductive and domestic duties.[15]

In urban areas, transport systems may particularly benefit men since they are often focused on the major routes to and within the city (World Bank 2010b). Pricing structures during peak times are also likely to encourage longer trips relative to multiple short trips. Given the pricing structures of urban transport, and that men use transport for income-generating activities, transportation costs can constrain women's mobility if the intrahousehold distribution of transport expenditure is skewed toward financing trips for those earning incomes outside the household (World Bank 2010b).

Safety and cultural concerns may also limit women's access to certain modes of transport. Evidence from Cambodia suggests that female garment workers experience security concerns, principally from accidents, robberies, threats, and sexual assaults, during their commute to and from work (World Bank 2006a). Cultural concerns, such as men and women sharing a single crowded vehicle, may also constrain women's access to communal transport (World Bank 2010b).

Agency
Urbanization can contribute to changing norms and may alter women's roles within the society and households. Cities are a melting pot of people and ideas, which change traditional ways of life, structures, and norms. The density and diversity of the urban population can increase women's access to networks and to information (World Bank 2008). Evidence from Indonesia indicates that urban women are more likely to be the sole decision makers on a number of household matters than rural women (Rammohan and Johar 2009).

Some dimensions of women's agency may, however, be more restricted in urban areas. Violence in urban areas may be more pronounced because traditional and cultural norms are less likely to guide behavior and neighborhood networks are less developed (Blank 2008; UNFPA 2007). Exploitative occupations such as sex work are also more likely to be found in urban areas. For instance, in the capital of Papua New Guinea, a large fraction of unemployed young women engage in sex work to bolster their incomes (Blank 2008). Although urban children have better access to education in Cambodia, security concerns for daughters traveling to school were found to be higher among parents in urban areas than in rural areas (ADB 2004).

Implications for policy related to urbanization

Growing urbanization in the East Asia and Pacific region has presented women with increased economic opportunities and greater empowerment. However, evidence shows that women in urban areas continue to sort into sectors that yield lower incomes and that are more likely to be informal, despite the more extensive opportunities. The choices of women's economic activities are constrained in part by limited access to affordable child care. Cultural concerns in some countries in the region may also limit women's access to certain modes of transport, thus making it more difficult for them to commute. Women also face higher security risks than men do in urban areas, particularly from exploitation, sexual abuse, and assault.

Whether women will be able to take full advantage of the wide range of opportunities available to them in urban areas will depend on whether the services and infrastructure exist in these cities to enable them to do so. Thus, policy makers need to ensure that their child care, education, infrastructure, transportation, and water and sanitation policies take into account women's specific social and cultural needs. They should also adopt rigorous laws and policies to protect women in urban areas from the risk of violence and exploitation, as discussed in the migration section above.

Aging populations

The world's population is aging. By 2045, the number of people age 60 years and older will exceed the number of those under the age of 15 for the first time in history (UN 2007). In Asia, this process is estimated to be even faster; the milestone of elderly people outnumbering children will be reached five years earlier, in 2040 (UN 2007). As a group, Asians age 64 and above are expected to more than triple, from 207 million in 2000 to 857 million in 2050. This demographic change is likely to have significant repercussions for economic development, the quality of life, and the role of public policy. In addition, population aging is likely to have a gender-differentiated effect among both older and younger cohorts. Several noteworthy challenges face the East Asia and Pacific region in managing these rapid demographic changes, notably how the aging population will maintain its standard of living, the types of formal and informal supports that are needed to do so, the impact on the working-age population, and the impact on health care and other services.

Trends in population aging in the East Asia and Pacific region

The high-income economies in East Asia are experiencing rapid population aging. Most emerging countries in the region have also begun this process; dependency ratios are

already increasing in many middle income countries in East Asia and the Pacific. For example, fertility rates have dropped significantly in Mongolia and Vietnam, and dependency ratios have increased dramatically in middle-income countries, including China, Indonesia, and Thailand, and in countries in the Pacific, namely Fiji, Timor-Leste, and Tonga.[16] Dependency ratios are expected to continue increasing and, given that women live longer than men, female dependency ratios will likely exceed male dependency ratios in the future (figures 5.10).

Population aging and gender

Life expectancy at birth has improved for both men and women since the 1990s. Reduced fertility and a decreased risk of maternal mortality have contributed to the improvements in life expectancy of women in many parts of the world (World Bank 2011b). Although the gender gap in life

FIGURE 5.10   The old age dependency ratio is expected to increase for both men and women in the next two decades; the female ratio will exceed the male ratio in the future

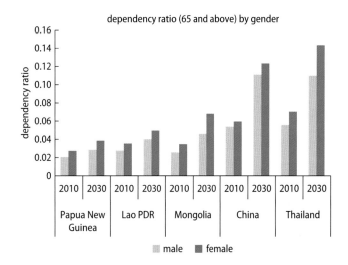

Source: HNPStats (Health Nutrition and Population Statistics) database, Population Projection, World Bank.
Note: Old age dependency ratio of males and females is defined as the ratio of the male and female population 65 years of age and above over the working-age population (15–64 years of age) of both genders.

FIGURE 5.11 **The gender gap in life expectancy at birth is lower in the East Asia and Pacific region than in many other regions**

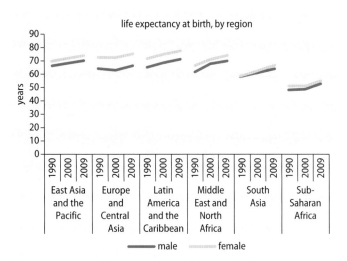

Source: WDI database.

FIGURE 5.12 **Life expectancy at birth in the region has improved for both men and women since 1990, although gender gaps have widened in some countries**

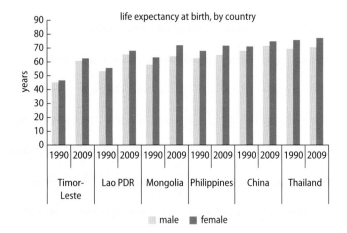

Source: WDI database.

expectancy is lower in the East Asia and Pacific region than in many other regions (figure 5.11), it has widened for many countries in the past two decades. In some countries, the widening gender gap in life expectancy is substantial, for example, in Mongolia (figure 5.12).

Older women have different access to resources than older men, in part because of

their different life histories (Hooyman 1999; INSTRAW 1999). Even as gender gaps continue to close for today's youth, the gender gaps of the past are embodied in today's adults through less schooling, different work experiences, lower rates of pension coverage, and less control over assets, among other things. Women's social arrangements within the family and the community are different, and they tend to outlive their spouses. Therefore, a careful examination of the implications of an aging society must consider the perspectives of both women and men.

The debate is open as to whether men or women are more likely to suffer from being less vested in the labor market and in formal social security mechanisms such as pensions, since women are more likely to be vested in familial relations. Mothers' closer relationships with their children might lead to larger intrafamily transfers later in life (Aboderin 2004). Their abilities in home production might outlast men's abilities in the workforce, and psychologically they may be better able to handle old age, both because it poses a smaller disruption to their previous roles and because they tend to establish a broader and deeper array of friendships than men.

Gender differences in consumption poverty and housing quality are small among the elderly in many East Asian countries (table 5.1) (Friedman et al. 2003; Knodel 2009; Knodel and Chayovan 2008; Masud, Haron, and Gikonyo 2008). In countries where data at the individual level are available, elderly men systematically report higher levels of individual income than elderly women (Masud, Haron, and Gikonyo 2008; Ofstedal, Reidy, and Knodel 2004). This is consistent with women's lower participation in the workforce: women are less likely to earn income from work or to draw pensions during old age. However, differences in individual incomes do not appear to translate into significant differences in measures of material well-being at a household level; that is, elderly women are not more likely than elderly men to live in poor households.

Marital status, notably widowhood, tends to play a greater role than gender in

TABLE 5.1 **Evidence shows no systematic gender differences in consumption poverty among the elderly, regardless of family status**

Percentage of elderly individuals (65 and over) living under US$1.25 per day

Country	Marital status	Male	Female	Total
Indonesia	*Married*	**29.3**	**33.0**	**30.4**
	Widowed	**24.7**	**28.0**	**27.4**
Cambodia	*Married*	**8.2**	8.5	8.3
	Widowed	**17.6**	9.8	11.1
Lao PDR	*Married*	27.3	27.9	27.5
	Widowed	26.4	28.8	28.3
Philippines	*Married*	13.9	13.2	**13.6**
	Widowed	12.4	11.9	**12.0**
Thailand	*Married*	0.1	0.1	**0.1**
	Widowed	0.1	0.2	**0.2**
Timor-Leste	*Married*	25.9	18.9	24.1
	Widowed	27.9	23.1	24.6
Vietnam	*Married*	11.6	14.9	12.9
	Widowed	8.6	14.9	13.8

Source: World Bank estimates based on East Asia and Pacific household poverty monitoring database.
Note: Bolded numbers are statistically significant differences between male- and female-headed households and between widowed and married households at the 10 percent level.

FIGURE 5.13 **Women ages 65 and above are far more likely to be widowed than men**

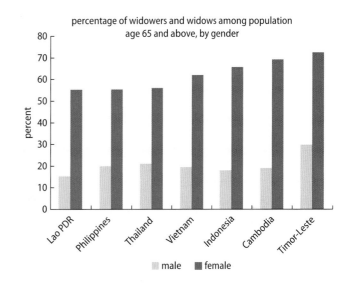

Source: World Bank staff estimates using CSES (NIS Cambodia), 2008 data; SUSENAS (BPS Indonesia), 2009 data; LECS (LSB Lao PDR), 2008 data; Mongolia LSMS (NSO Mongolia), 2007–08 data; Philippines Family Income and Expenditures Survey (NSCB Philippines), 2006 data; Thailand Household Socio-Economic Survey (NSO Thailand), 2009 data; Timor-Leste Survey of Living Standards (SLS) (NSD Timor-Leste) 2007, data; and VHLSS (GSO Vietnam), 2008 data.

determining the well-being of elderly men and women. Women are far more likely than men to be widows (figure 5.13). In Cambodia, the percentage of women who have lost a spouse is more than triple that of men, and women are also less likely to have living children than their male counterparts (Knodel 2009). In Thailand, 80 percent of elderly men are living with their spouses, compared to only about half of elderly women (Knodel and Chayovan 2008). In Indonesia, Cambodia and Timor Leste, the differences are dramatic. In Indonesia and Cambodia, approximately 65 percent of elderly women are widowed compared to approximately 20 percent of men; in Timor-Leste, 73 percent of elderly women are widowed compared to 30 percent of men.

Marital status is the most important correlate of household circumstances among the elderly. Across the region, unmarried elderly are more likely than married elderly to live with their children and receive support from their children, either directly through transfers or through their living arrangements (Friedman et al. 2003; Knodel and Chayovan

2008; Knodel and Zimmer 2009; Ofstedal, Reidy, and Knodel 2004). However, no systematic evidence shows that widows and widowers are consistently likely to be poorer than married elderly individuals (table 5.1), although they are more likely to perceive income inadequacy or lower rates of satisfaction with their economic status in many countries in the region, including Indonesia, Malaysia, Thailand, and Vietnam (Friedman et al. 2003; Knodel and Chayovan 2008; Masud, Haron, and Gikonyo 2008; Ofstedal, Reidy, and Knodel 2004).

Elderly women have access to fewer personal income sources and less-diversified income portfolios than elderly men. The lack of association between gender and poverty in the elderly population is partly because women, and in particular unmarried women, are more likely to receive money from children and other relatives than men and less likely to have their own sources of income (Masud, Haron, and Gikonyo 2008). Furthermore, men's income is more dispersed across a number of sources than women's income (Ofstedal, Reidy, and Knodel 2004). Women's greater reliance on transfers from informal support networks is likely to place them in a more tenuous position, both within their families and in society.

As migration and urbanization continue and fertility declines, there is concern that informal safety nets may decline and that transfers and care from children will erode (UNESCAP 2004; World NGO Forum on Ageing 2002).[17] A decline in transfers is likely to have a gender-differentiated impact, since elderly women are more likely than elderly men to receive support from children and other family members. Evidence suggests that attitudes of respect and responsibility toward the elderly population may also be changing in parts of the region. Studies have reported that children from single-child families in China are less committed to elder care in more recent generations than in previous ones (Wang 2010; Zhan 2004). People in the baby boom generation have been found to have more positive attitudes toward the elderly than those in more recent generations, regardless of gender (Xie, Xia, and Liu 2006).[18]

Older women are worse off than their male peers along a number of nonconsumption dimensions, many of which are associated with their age. They are less educated, have fewer assets, and are more likely to be illiterate (Knodel 2009; Long and Pfau 2008); they also have lower access to care (Knodel 2009; Long and Pfau 2008; Magnani and Rammohan 2009) and more health problems (Chen and Standing 2007). Some of these differences are likely to persist without targeted reform, whereas others are likely to narrow over time. For example, since education gender gaps have declined over time across birth cohorts, these differences are likely to be smaller among the future elderly population.

Because women live longer than men, the evidence suggests that they are also more likely to suffer from disabilities. The barriers to participating in social life for the elderly directly affect their quality of life, but such barriers also impose additional costs on

their families. In Vietnam, one study showed that having a disabled household member increased the cost of living over 11 percent (Braithwaite and Mont 2009), and adjusting for these differences would raise the poverty rate of households with disabled members significantly.[19] This is more likely to be an issue for women—the rate of disability in Vietnam is 8.5 percent for women compared to 6.6 percent for men, with the gap explained to a significant measure by the difference in longevity (Mont and Cuong 2011).

Finally, the elderly and, in particular, widows might be especially vulnerable to shocks. Men and women have different capacities to cope with economic shocks given differences in incomes and asset endowments. However, evidence from six provinces in Thailand and Vietnam suggests that rural households headed by widowed, divorced, or single women are *not* more likely to be affected by shocks than their male counterparts (Klasen, Lechtenfeld, and Povel 2011).

Caregiving and its repercussions for working-age women

Population aging may have a gender-differentiated effect among the working-age population because the increasing dependency ratios, particularly when accompanied by falling fertility rates, will raise the burden of caregiving. Gender differences in the time spent caring for the elderly imply that women are more likely to accommodate the increased demand for nonmarket time.[20]

Within households, younger women are likely to bear a greater responsibility for caring for elderly parents than younger men, which may reduce the time they spend on income-generating activities.[21] Data from the region suggests that although working age women spend more time on housework and caring than men, women residing with elderly members do not necessarily spend more time on these activities than women who do not reside with elderly members (figure 5.14). This may reflect a compensating grandparent effect in households with children: elderly household members contribute to caring for children, which thereby leaves the total time that younger women spend on housework and caring unchanged or diminished. For example, staff analysis of household data from Timor-Leste finds that living with elderly members increases the time devoted to housework and caring activities by 40 minutes a day among women without young children (NSD Timor-Leste 2007). In contrast, in households with young children the presence of elderly members reduces the time devoted by younger women to housekeeping and caring activities, who spend 30 fewer minutes on these activities than women with children who do not live with elderly members.

Implications of aging populations for gender policy

The population of the East Asia and Pacific region is aging rapidly, in many countries at a faster pace than that in the rest of the world. This demographic change will likely have significant repercussions for women since they live longer than men but have access to fewer economic assets and resources and are less vested in formal social security systems. Although women are more likely to have access to informal caregiving and support systems, traditional systems of caregiving are evolving and exposing more elderly individuals to lower levels of familial support. Because older women are less likely than men to be covered by insurance or to have accumulated assets, the burden of their care is still likely to fall on younger women.

Policy makers can address the burden of aging in several ways. Policies to close gender gaps in access to human and physical assets will affect both older and younger women, with implications for gender gaps among future generations of older women. Policies that focus on ensuring that health care and social security systems cover those most in need are likely to benefit both elderly men and women. The design of old-age security systems becomes particularly important in the context of the rapid demographic transition and cultural change occurring in many countries in the region. Old-age income security

FIGURE 5.14 Women's time devoted to housework and caregiving activities is not significantly greater in households with elderly members

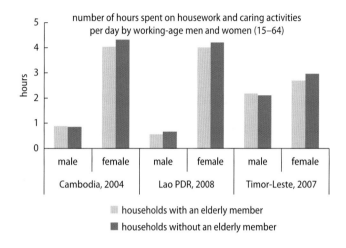

Source: World Bank staff estimates using CSES (NIS Cambodia), 2004 data; LECS (LSB Lao PDR), 2008 data; and Timor-Leste SLS (NSD Timor-Leste) 2007.

programs can protect women from destitution and reduce their reliance upon familial transfers. For example, joint annuities and survivor pensions are likely to be effective means to compensate and secure women's incomes in their old age. Finally, policy makers may consider ways to strengthen elder care to ensure that younger women do not face the disproportionate burden of care giving.

Information and communication technologies

The literature contains little evidence on how access to and use of ICTs vary by gender. Therefore, this section provides predominantly anecdotal or project-specific global evidence on the gender implications of the spread of ICTs.

The exponential growth of information and communication technologies in the developed and developing world is widely lauded as the defining economic and social force of the late 20th and early 21st centuries. ICTs encompass a plethora of technological advances, including radio, cell phones, computers, e-mail, social networking sites, and the Internet. Figures 5.15 and 5.16 show

FIGURE 5.15 **Internet use has increased substantially in the East Asia and Pacific region and around the developing world since 2000**

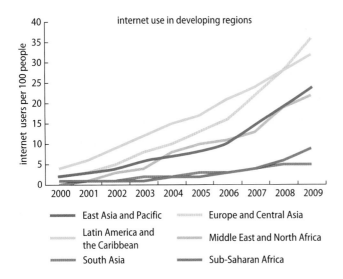

Source: WDI database.

FIGURE 5.16 **The number of cell phone subscribers per 100 people in the East Asia and Pacific region has grown substantially since 2000**

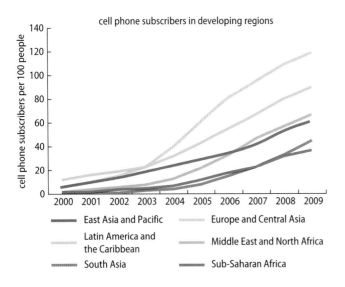

Source: WDI database.

the rapid expansion of access to the Internet and ownership of cell phones in developing countries around the world. The East Asia and Pacific region falls behind both Latin America and the Caribbean and Europe and Central Asia in both of these indicators and ties with the Middle East and North Africa.

The spread of ICTs has increased access to information by enhancing knowledge sharing and gathering and by changing business practices and production structures. ICTs have affected many aspects of men's and women's lives, from economic opportunities and health outcomes to women's empowerment.

ICTs and Gender

The reach of technology, at the national level, has been growing in parts of the East Asia and Pacific region. In 2000, the majority of countries in the region had fewer than 2 Internet users per 100 people, with the exception of Malaysia, where there were already 20 Internet users per 100 people. Internet usage increased substantially in a number of countries over the 2000s. By 2009, Malaysia had approximately 58 Internet users per 100 people, and China had overtaken Thailand as the country with the second greatest number of Internet users—28 per 100 people—followed very closely by Vietnam, with 27 users per 100 people (figure 5.17). However, rapid growth of Internet use has not occurred in all countries. In Kiribati, Lao PDR, Papua New Guinea, and the Solomon Islands, Internet use is still close to its levels in 2000. Similar but starker patterns can be seen for cell phone use. In 2000, the level was fewer than 5 cell phones per 100 people in many countries in the region, and in the rest of the countries, with the exception of Malaysia and New Caledonia, it was fewer than 20 cell phones per 100 people (figure 5.18). By 2009, the number of cell phone subscribers had risen, along with disparities between countries. In Kiribati and the Marshall Islands, the number of cell phones per 100 people was still below 10, whereas in Samoa, which started from a base similar to the other Pacific countries, the number of cell phones per 100 people had risen to 84.

Where gender-disaggregated data are available, they indicate gender differences in access to ICTs in the East Asia and Pacific region. Figure 5.19 displays the proportion of male and female subscribers to mobile phones in the population. Many regions across the world show a gender gap in cell phone subscriptions. Furthermore, gender differences in access to technology are greater among certain subgroups of the population. In other words, gender is likely to interact with socioeconomic characteristics such as income, education, location, and social and cultural constraints to determine access to technology (Huyer and Hafkin 2007).

In China, the number of Internet users has increased from approximately 20 million users in 2000 to over half a billion in 2011, a remarkable increase over time (CNNIC 2011). Although the fraction of female users has risen since 2000, data suggest that men were more likely than women to be Internet users in China in the early to mid-2000s (figure 5.20). More recent data suggest that the gap between the number of men and the number of women using the internet in China has broadened as Internet use has risen: in 2011, 287 million men used the internet in China compared to 226 million women (CNNIC 2011). In percentage terms, however, the gender gap has narrowed: in 2011, women made up 44 percent of users, up from 30 percent of users in 2000 (CNNIC 2011).

ICTs can improve the welfare of both men and women in several different ways and can also have gender-differentiated effects by reducing constraints, such as time and transportation, that influence the decisions of women (Melhem and Tandon 2009). First, ICTs can enhance women's economic opportunities by reducing the transaction costs of reaching markets. Second, the widespread dissemination of information on how men and women live their lives in other parts of the world may gradually change attitudes about gender norms and roles. Third, access to television, radio, and the Internet enables men and women to increase their knowledge and training through distance learning. Finally, rural populations can use ICTs to gain access to better health care through increased access to medical advice. These final two opportunities

FIGURE 5.17 Internet use has grown quickly in many countries in the region but has grown slowly in others

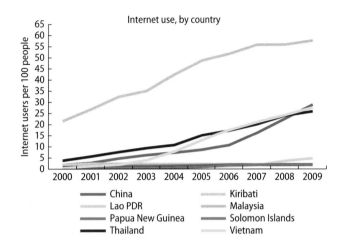

Source: WDI database.

FIGURE 5.18 The number of cell phone subscribers in the population has grown across most of East Asia but has remained limited in some Pacific countries

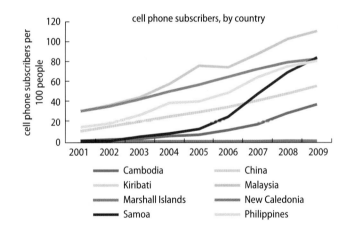

Source: WDI database.
Note: The data points for Kiribati are closely matched by those of the Marshall Islands and show as a single line.

may particularly benefit women, since evidence suggests that they face greater time and transportation constraints in accessing basic public services.

ICTs can enhance women's economic opportunities by increasing their access to markets through reduced information barriers and time and transportation constraints,

FIGURE 5.19   Women in the region are less likely to be mobile phone subscribers than men

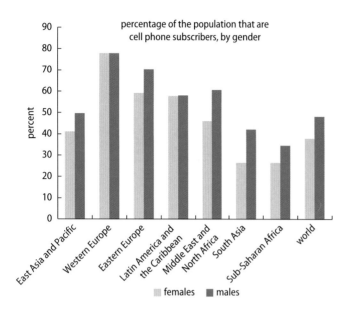

Source: World Bank estimates using GSMA Development Fund, the Cherie Blair Foundation, and Vital Wave Consulting, 2010.

FIGURE 5.20   Internet use has grown for both men and women in China, although a persistent gender gap in access remains

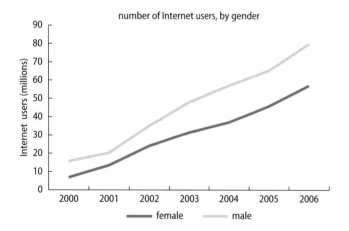

Source: CNNIC (China Internet Network Information Center) Internet Statistics.

television programs—have been used for decades in the developing world to provide advisory services to farmers in agricultural areas (Goyal 2010), and telecommunication centers and Internet kiosks allow farmers to access e-learning programs and search for agriculture-related information (Aker 2011).[23] Although reducing information barriers and transaction costs is likely to affect the economic opportunities of both men and women, it is likely to have a greater effect on women, whose economic decisions are more likely to be constrained by transaction costs because of their multiple roles in households. For example, the e-Homemakers initiative in Malaysia reduces barriers for women entering the labor market; the initiative promotes home-based entrepreneurship and reduces time constraints by allowing women to work from home and to telework using ICTs.

Second, female empowerment and autonomy may be raised through the representation of lifestyles, autonomy, and gender roles in other parts of the world.[24] One of the earliest examples of the power of entertainment-education to change behavior occurred in Mexico in 1977. The soap opera *Acompaname* promoted family planning, a socially sensitive topic at the time. The program is credited to have increased both awareness and use of birth control among Mexicans (Singhal and Rogers 1999). In another example, approximately 150 million individuals gained access to cable television service in India between 2001 and 2006 (National Readership Studies Council 2006, cited in Jensen and Oster 2009). This development had a large and swift effect on attitudes toward domestic violence and women's participation in household decision making (Jensen and Oster 2009). Studies have also found that exposure to cable television increased school enrollment for girls (but not for boys) and decreased fertility, an outcome that was associated with the increase in female autonomy (see chapter 4).

In societies in which social customs limit interactions between males and females, television may reduce stereotypes by increasing interaction or knowledge of others. For

in particular those in rural areas. For example, cell phones allow farmers to get better prices, by benefiting from market arbitrage opportunities (Aker 2010; Jensen 2007).[22] Traditional technologies—such as radio and example, in Rwanda, a radio program aimed at discouraging blind obedience and promoting independent thought and collective action in problem solving was found to have increased listeners' willingness to express dissent and changed the ways in which communal problems were resolved (Paluck and Green 2009).

Technology can also be used to give women more control over their money and their actions. For example, in Bangladesh before mobile phones were common, a migrant husband's relatives had greater control over his remittances than his wife did; the advent of cell phones has allowed closer communication between husband and wife on how best to use the remittances (Schuler, Islam, and Rottach 2010).

Third, education programs conducted through ICTs can help to overcome the constraints faced by women in higher education by allowing them to pursue their education on a more flexible basis than normal programs allow. It also has the potential to increase access to higher education for women living in restrictive social situations, who may not be able to travel or live away from home to engage in higher education.[25] Cell phones can also be used to increase adult literacy. Because sending text messages is cheaper than making voice calls, cell phone users have a financial incentive to send text messages and, thus, practice their reading and writing skills.[26]

Finally, ICTs can also be used to improve the provision and quality of health care services. In remote communities where access to high-quality medical care is limited, women often have less access to health care than men (see chapter 2). Therefore, increasing access to health care through ICTs may diminish existing health care disparities, but only if women are able to access ICTs to the same degree as men. For example, Sehat First provides health care and pharmaceutical services across Pakistan through telehealth centers where local clinic staff can seek advice from qualified physicians and specialists to whom they would not normally have had access (Sehat case study, cited in Melhem and Tandon 2009).

Implications for policy related to ICTs

ICTs can enhance the lives and economic opportunities of women in many ways. The widespread dissemination of information in the media on how men and women live their lives in other parts of the world may gradually change gender norms and roles, thus giving women in many countries the chance to live less restricted, more empowered lives. Technology can enhance women's economic opportunities as well as give them more control over their money. However, for this to happen, women must have access to these technologies, and the little evidence that is available shows that women are lagging behind men in their access to and use of ICTs. This gap implies that women may be less able to reap the direct benefits of having access to, for example, distance-learning opportunities and online agricultural extension services. More research is needed to explore whether and why women in the East Asia and Pacific region may be failing to benefit from these technologies to the same extent as men.

Notes

1. Exports as a proportion of GDP fell to 68 percent in Vietnam in 2009 due to the global financial crisis, and in Indonesia they rose to over 50 percent during the Asian crisis in the late 1990s.

2. According to neoclassical trade theory, trade liberalization should encourage countries to specialize in the production of goods in which they have a comparative advantage. Trade liberalization can lead to the expansion of employment in labor-intensive industries in developing economies. Since women are overrepresented in labor-intensive sectors, trade liberalization may benefit women by increasing demand for their skills more than those of men, thereby reducing pay differentials between men and women. International trade can also affect women's relative pay by reducing the power of producers to discriminate against female employees, which producers do by paying women less for the same work as men (Becker 1957; Berik, van der Meulen Rodgers, and Zveglich 2004; Schultz 2003).

3. One cannot draw strong conclusions from the literature relating trade liberalization and wage gaps because gender-differentiated employment impacts after trade liberalization are also likely to directly affect productivity in the wage sector (Schultz 2003).

4. The results of the analysis by Bhalotra and Umana-Aponte (2010) indicate that, in Africa, women lose employment during recessions but, in South Asia, East Asia and the Pacific, and Latin America and the Caribbean, women's employment increases during recessions. This suggests that households use female employment as a financial coping strategy.

5. Evidence shows multiple examples of women working harder during crises in East Asian and Pacific countries. In Indonesia, the 1997–98 economic crisis increased employment among women, whereas male employment rates fell. These additional female workers were predominantly employed in the informal sector (Frankenberg, Smith, and Thomas 2003). During the 1997–98 East Asian crisis, the initial difference between the total number of hours worked by men and by women widened, meaning that women were increasingly overworked and men underworked in the Philippines (Lim 2000). Turk and Mason (2009) reported that, during the 2009 financial crisis, many women worked longer hours to maintain their household income. In Thailand and Vietnam, women searched for additional work to supplement their primary job, with child care sometimes being taken over by elderly household members or older children.

6. Mongolia is an exception. Women reported an improvement in their domestic relationships because men were working such long hours that they had little time left for fighting (Turk and Mason 2009).

7. For example, in Vietnam's apparel sector, better jobs with higher skill levels are largely held by male workers while sewing jobs are largely done by female workers (Kabeer and Tran 2003).

8. In internal rural to urban migration in China, differences in socioeconomic factors by gender do seem to exist. Migrant women tend to be less educated than rural migrant men, owing to their lower average age and to gender inequalities in educational attainment across rural China (Jacka 2009, cited in IOM 2009).

9. Within the growing export–based, large-scale manufacturing and assembly activities in Asian cities, particularly in the garment sector, studies have suggested that employers prefer women because they accept lower wages, are considered more easily controlled, and are considered better able to undertake tasks that involve delicate and intricate finger work as well as repetitive tasks (Hugo 2003).

10. Female migration is complex and context-specific, and studies show that the effect of migration on women's position is neither wholly positive nor wholly negative. Parrado, Flippen, and McQuiston (2005) argue that migration entails a large change in "structural context." Depending on the "context of reception, the degree of labor market segmentation and the extent to which migrants are isolated in the receiving society, migration may mitigate or reinforce patriarchal gender inequality" (Parrado et al. 2005, page 8).

11. This increase is in part due to the greater number of female migrants than male migrants from Indonesia. The difference in per capita consumption between households with male migrants and female migrants at the outset is not statistically significant, indicating that the difference is not because female migrants come from poorer households. No hard evidence suggests that female migrants earn more than male migrants; instead, female migrants from Indonesia likely earn more now than in the past because of high demand for their labor in the Gulf countries.

12. The numbers may actually be higher because respondents are often reluctant to report this form of abuse.

13. Human trafficking estimates may underrepresent actual numbers because of the elusiveness of the industry.

14. Work opportunities may, however, be low-skilled and precarious and may not be covered by formal labor market regulations. Many women in urban areas are self-employed in the informal sector and in activities that yield low incomes (UNFPA 2007).

15. In Indonesia, men and women ages 15–19 were commuting in equal numbers; beyond the age of 30, however, the fraction of female commuters decreased (Rachmad, Adji, and Handiyatmo 2010). The change in commuting pattern implies that, during childbearing years, women's labor force participation concentrates in or around the home.

16. Age dependency ratio is the ratio of elderly dependents (those older than 65) to the working-age population (ages 15–64).

17. However, some question this conclusion (Aboderin 2004; Hermalin 2003). Effects may differ across countries, depending on the interplay between culture, the economy, and the geography of development. A recent study in Thailand, for example, shows that as children migrate to the city, they do not abandon their rural elders (Knodel 2009). As noted earlier, some evidence from China suggests a generational shift in support may be occurring.

18. However, this divergence in views occured only in higher income groups. No generational differences existed among the lower income groups.

19. Qualitative evidence from Cambodia also illustrates how disability status increases vulnerability. In a 2008 World Bank participatory poverty and gender assessment in Siem Reap province, of the 107 households ranked in a participatory wealth ranking exercise by male and female villagers in Dour Dantrei village, 5 were identified as headed by disabled men, with 4 of the 5 ranked as "poor" or "very poor"—a poverty rate of 80 percent of disabled households, compared to a poverty rate of 50 percent for all households (Kuriakose and Kono, 2008).

20. Studies in the United States have consistently found that daughters play a central role in the more time-intensive day-to-day care of elderly parents, whereas sons take on a more managerial role (Dwyer and Coward 1992; Ofstedal, Knodel, and Chayovan 1999).

21. In China evidence shows that the effect of caring for elderly parents on women's labor market experience depended on whether the parents being cared for were the woman's parents (where the effect was insignificant) or her in-laws (where the effect was substantial), suggesting the important role of culture on family relations and obligations (Liu et al. 2008). Cultural differences also affect cross-generational support across different regions in Indonesia and Vietnam (Friedman et al. 2003; Kreager and Schroder-Butterfill 2009).

22. For example, in Uganda, the Market Information Service Project collected data on the prices of the main agricultural commodities in major market centers and distributed this information to local radio stations to broadcast. Svensson and Yanagizawa (2009) found that better-informed farmers were able to bargain for higher farm-gate prices on their surplus production.

23. Within the region, three self-employment initiatives demonstrate how ICT access can enhance economic opportunities, particularly of women (Melhem and Tandon 2009). In the Philippines, the Sharing Computer Access Locally and Abroad project has set up computer livelihood centers to help give underprivileged young people access to employment opportunities. The project is reported to have resulted in increased Internet access and greater self-esteem and self-confidence, and 60 percent of the beneficiaries were women. In Malaysia, the e-Homemakers initiative aimed to boost home entrepreneurship by increasing the self-esteem of mothers. In Fiji, the Foundation for Rural Integrated Enterprises and Development aims to increase economic empowerment in marginalized communities through three programs that focus on income generation, saving, and governance. The program uses the Internet to market products developed by the participants of the program, the majority of whom are women.

24. Exposure to television can also have less positive outcomes. For example, it can increase discontent within marriages, resulting in divorce. In Brazil, exposure to soap operas that feature modern values such as female empowerment and emancipation have increased rates of separation and divorce (Chong and La Ferrara 2009). In Indonesia, greater exposure to television and radio was found to reduce social capital, as measured by lower participation in social organizations and lower self-reported community-level trust (Olken 2009).

25. For example, in the United Kingdom, the Open University, which broadcasts lectures over the television and Internet, enables men and women to study on an individually determined schedule. In the United States, 60 percent of the students over age 25 who are engaged in distance learning are women (Kramarae 2001).

26. A cell phone–based literacy and numeracy program implemented in Niger was found to increase adults' math test scores, with the greatest increases being achieved by younger participants, who can acquire new skills using ICT more easily than older people (Aker, Ksoll, and Lybbert. 2010).

References

Aboderin, Isabella. 2004. "Intergenerational Family Support and Old Age Economic Security in Sub-Saharan Africa: The Importance of Understanding Shifts, Processes and Expectations." In *Living Longer: Ageing, Development and Social Protection*, edited by Peter Lloyd-Sherlock. United Nations Research Institute for Social Development (UNRISD) and Zed Books.

ADB (Asian Development Bank). 2004. *A Fair Share for Women: Cambodia Gender Assessment*. Phnom Penh, Cambodia: United Nations Development Fund for Women (UNIFEM), World Bank, ADB, United Nations Development Programme (UNDP), and the U.K. Department for International Development (DFID).

Aker, Jenny. 2010. "Information from Markets Near and Far: Mobile Phones and Agricultural Markets in Niger." *American Economic Journal: Applied Economics* 2 (3): 46–59.

———. 2011. "Mobile Phones and Economic Development in Africa." *Journal of Economic Development* 24 (3): 207–32.

Aker, Jenny, Christopher Ksoll, and Travis Lybbert. 2010. "ABC, 123: The Impact of a Mobile Phone Literacy Program on Educational Outcomes." Working Paper 223, Center for Global Development, Washington, DC.

Atkin, David. 2011. "Endogenous Skill Acquisition and Manufacturing in Mexico." Paper presented at the "3rd IZA Workshop: Child Labor in Developing Countries," Mexico City, June 1.

Becker, Gary S. 1957. *Economics of Discrimination*. Chicago: University of Chicago Press.

Berik, Gunseli. 2000. "Mature Export-Led Growth and Gender Wage Inequality in Taiwan." *Feminist Economics* 6 (3): 1–26.

———. Forthcoming. "Gender Aspects of Trade." In *Trade and Employment: From Myths to Facts*, edited by M. Jansen, R. Peters, and J. M. Salazar-Xirinachs. Geneva: International Labour Organization–European Commission.

Berik, Gunseli, Yana van der Meulen Rodgers, and Joseph E. Zveglich. 2004. "International Trade and Gender Wage Discrimination: Evidence from East Asia." *Review of Development Economics* 8 (2): 237–54.

Bernanke, Ben S. 1983. "Nonmonetary Effects of the Financial Crisis in Propagation of the Great Depression." *American Economic Review* 73 (3): 257–76.

Bhalotra, Sonia R., and Marcela Umana-Aponte. 2010. "The Dynamics of Women's Labour Supply in Developing Countries." IZA Discussion Paper 4879, Institute for the Study of Labor (IZA), Bonn, Germany.

Black, Sandra, and Elizabeth Brainerd. 2002. "Importing Equality? The Impact of Globalization on Gender Discrimination." *Industrial and Labor Relations Review* 57 (4): 540–59.

Blank, Lorraine. 2008. "Rapid Youth Assessment in Port Moresby, Papua New Guinea." September 4. http://siteresources.worldbank .org/INTEAPREGTOPSOCDEV/Resources/ 080904PNGUYEPPNGRapidYouthAssessm ent.pdf.

BPS (Badan Pusat Statistik). Indonesia National Labor Force Survey (SAKERNAS). Jakarta, Indonesia.

———. Indonesia National Socioeconomic Survey (SUSENAS). Jakarta, Indonesia. http://dds.bps.go.id/eng/aboutus. php?id_subyek=29&tabel=1&fl=3.

Braithwaite, Jeanine, and Daniel Mont. 2009. "Disability and Poverty: A Survey of World Bank Poverty Assessments and Implications." ALTER, *European Journal of Disability Research* 3: 219–32.

Bruni, Lucilla, Andrew Mason, Laura Pabon, and Carrie Turk. 2012. "Gender Impacts of the Global Financial Crisis in Cambodia." Photocopy, World Bank.

Cabegin, E., and M. Alba. 2011. "More or Less Consumption? The Effect of Remittances on Filipino Household Spending Behavior." Unpublished manuscript, University of the Philippines, Manila.

CNNIC (China Internet Network Information Center) Internet Statistics. Beijing, China. http://www1.cnnic.cn/en/index/0O/index .htm.

———. 2011. "Statistical Survey Report on the Internet Development in China. 29th Survey Report." Beijing, China. http://www1.cnnic .cn/en/index/0O/index.htm.

Chen, Lanyan, and Hilary Standing. 2007. "Gender Equity in Transitional China's Healthcare Policy Reforms." *Feminist Economics* 13 (3–4): 189–212.

Chong, Alberto, and Eliana La Ferrara. 2009. "Television and Divorce: Evidence from Brazilian Novelas." *Journal of the European Economic Association* 7 (2–3): 458–68.

Connelly, Rachel, Kenneth Robert, Zhenzhen Zheng, and Zhenming Xie. 2010. "The Impact of Migration on the Position of Women in Rural China." *Feminist Economics* 16 (1): 3–41.

Constable, Nicole. 2009. "Migrant Workers and the Many States of Protest in Hong Kong." *Critical Asian Studies* 41 (1): 143–64.

Davin, Delia. 2004. "The Impact of Export-Oriented Manufacturing on the Welfare Entitlements of Chinese Women Workers." In *Globalisation, Export-Oriented Employment and Social Policy: Gendered Connections*, edited by Shahra Razavi, Ruth Pearson, and Caroline Danloy. London: Palgrave.

de Brauw, Alan, Jikun Huang, Scott Rozelle, Linxiu Zhang, and Yigang Zhang. 2002. "The Evolution of China's Rural Labor Markets During the Reforms." *Journal of Comparative Economics* 30 (2): 329–53.

Dercon, Stefan, and Pramila Krishnan. 2000. "In Sickness and in Health: Risk Sharing within Households in Rural Ethiopia." *Journal of Political Economy* 108 (4): 688–727.

Dumont, Jean-Christophe, John P. Martin, and Gilles Spielvogel. 2007. "Women on the Move: The Neglected Gender Dimension of the Brain Drain." IZA discussion paper No. 2920. Institute for the Study of Labor, Bonn, Germany.

Dwyer, Jeffrey, and Raymond Coward. 1992. "Gender and Family Care of the Elderly: Research Gaps and Opportunities." In *Gender, Families, and Elder Care*, edited by Jeffrey W. Dwyer and Raymond T. Coward, 151–62. London: Sage.

Enterprise Surveys (database). World Bank, Washington, DC. http://data.worldbank.org/data-catalog/enterprise-surveys.

Frankenberg, Elizabeth, Victoria Beard, and Mudu Saputra. 1999. "The Kindred Spirit: Ties Between Indonesian Children and Their Parents." *Southeast Asian Journal of Social Sciences* 27 (2): 65–86.

Frankenburg, Elizabeth, Jed Friedman, Thomas Gillespie, Nicholas Ingwersen, Robert Pynoos, Iip Umar Rifai, Bondan Sikoki, Alan Steinberg, Cecep Sumantri, Wayan Suriastini, and Duncan Thomas. 2008. "Mental Health in Sumatra after the Tsunami." *American Journal of Public Health* 98 (9): 1671–77.

Frankenberg, Elizabeth, James P. Smith, and Duncan Thomas. 2003. "Economic Shocks, Wealth, and Welfare." *Journal of Human Resources* 38 (2): 280–321.

Frederick, Stacey, and Cornelia Staritz. 2011. "Background Global Value Chain Country Papers: Morocco." Background paper for *Sewing Success? Employment and Wage Implications of the End of the Multifibre Arrangement*, World Bank, Washington, DC.

Friedman, Jed, John Knodel, Bui The Cuong, and Truong Si Anh. 2003. "Gender Dimensions of Support for Elderly in Vietnam." *Research on Aging* 25 (6): 587–630.

Friedman, Jed, and Duncan Thomas. 2007. "Psychological Health before, during and after a Crisis: Results from Indonesia, 1993–2000." Policy Research Working Paper 4386, World Bank, Washington, DC.

Gaetano, Arianne, and Tamara Jacka, eds. 2004. *On the Move: Women and Rural-to Urban Migration in Contemporary China*. New York: Columbia University Press.

Gamberoni, Elisa, and José Guilherme Reis. 2011. "Gender-Informing Aid for Trade: Entry Points and Initial Lessons Learned from the World Bank." Economic Premise Note 62, World Bank, Washington, DC.

Goyal, Aparajita. 2010. "Information, Direct Access to Farmers, and Rural Market Performance in Central India." *American Economic Journal: Applied Economics* 2 (3). doi:10.1257/app.2.3.22.

GSMA Development Fund, the Cherie Blair Foundation, and Vital Wave Consulting. 2010. "Women & Mobile: A Global Opportunity: A Study on the Mobile Phone Gender Gap in Low- and Middle-Income Countries." London, GSMA.

GSO Vietnam (General Statistics Office of Vietnam). Vietnam Household Living Standards Surveys. General Statistics Office, Hanoi, Vietnam. http://www.gso.gov.vn.

Guzmán, Juan Carlos. 2006. "Trends in International Migration: Is There a Feminization of Migration Flows?" Unpublished manuscript, Gender and Development Group, World Bank, Washington, DC.

Hermalin, Albert. 2003. "Theoretical Perspectives, Measurement Issues, and Related Research." In *The Well-Being of the Elderly in Asia: A Four-Country Comparative Study*, edited by Albert Hermalin, 101–41. Ann Arbor, MI: University of Michigan Press

HNPStats (Health Nutrition and Population Statistics) database, World Bank, Washington, DC. http://databank.worldbank.org.

Hooyman, Nancy. 1999. "Research on Older Women: Where Is Feminism?" *Gerontologist* 39 (1): 115–18.

Hugo, Graeme. 2003. "Urbanisation in Asia: An Overview." Paper prepared for the "Conference on African Migration in Comparative Perspective," Johannesburg, South Africa, June 4–7.

——. 2005. "Migration in the Asia-Pacific Region," paper prepared for the Policy Analysis and Research Programme of the Global Commission on International Migration, Geneva.

Human Rights Watch. 2004. "Help Wanted: Abuses against Female Migrant Domestic Workers in Indonesia and Malaysia." Human Rights Watch, New York.

Huyer, Sophia, and Nancy Hafkin. 2007. *Engendering the Knowledge Society: Measuring Women's Participation*. Montreal: Orbicom.

IFC (International Finance Corporation). 2008. "Special Economic Zones: Performance, Lessons Learned and Implications for Zone Development." IFC Financial and Private Sector Development, Washington, DC.

ILO (International Labour Organization). 2008. "Women's Entrepreneurship Development in Aceh: Impact Assessment." ILO, Jakarta, Indonesia.

INSTRAW (International Research and Training Institute for the Advancement of Women). 1999. *Ageing in a Gendered World: Women's Issues and Identities*. Santo Domingo, Dominican Republic: INSTRAW.

IOM (International Organization for Migration). 2009. *Gender and Labour Migration in Asia*. Geneva: IOM.

Jacka, Tamara. 2009. "The Impact of Gender on Rural-to-Urban Migration in China." In *Gender and Labour Migration in Asia*, edited by IOM, 263–292. Geneva: IOM.

Jampaklay, Aree, John Bryant and Rita Litwiller 2009. "Gender and migration from Cambodia, Laos and Myanmar to Thailand," in *Gender and Labour Migration in Asia*, edited by IOM, 193–216. Geneva: IOM.

Jensen, Robert. 2007. "The Digital Provide: Information (Technology), Market Performance and Welfare in the South Indian Fisheries Sector." *Quarterly Journal of Economics* 122 (3): 879–924.

Jensen, Robert, and Emily Oster. 2009. "The Power of TV: Cable Television and Women's Status in India." *Quarterly Journal of Economics* 124 (3): 1057–94.

Jones, Nicola, Rebecca Holmes, Hannah Marsden, Shreya Mitra, and David Walker. 2009. "Gender and Social Protection in Asia: What Does the Crisis Change?" Paper presented at the Conference on the Impact of the Global Economic Slowdown on Poverty and Sustainable Development in Asia and the Pacific, Hanoi, Vietnam, September 28–29.

Kabeer, Naila, and Tran, T.V.A. 2003. "Global Production, Local markets: Gender, Poverty and Export Manufacture in Vietnam," Institute of Development Studies, Brighton.

Kang, E. 1999. "Economic Crisis and Trend of Female Labour Market Change." *Trend and Prospect* 40: 89–111 (in Korean).

Kim, Andrew, and Innwon Park. 2006. "Changing Trends of Work in South Korea: The Rapid Growth of Underemployment and Job Insecurity." *Asian Survey* 46 (3): 437–56.

Kim, Haejin, and Paula Voos. 2007. "The Korean Economic Crisis and Working Women." *Journal of Contemporary Asia* 37 (2): 190–208.

Klasen, Stephan, Tobias Lechtenfeld, and Felix Povel. 2011. "What about the Women? Female Headship, Poverty and Vulnerability in Thailand and Vietnam." Discussion Paper 76, Courant Research Centre Poverty, Equity and Growth, Göttingen, Germany.

Knodel, John. 2009. "Is Intergenerational Solidarity Really on the Decline? Cautionary Evidence from Thailand." Seminar on "Family Support Networks and Population Ageing," Doha, Qatar, June 3–4.

Knodel, John, and Napaporn Chayovan. 2008. "Population Ageing and the Well-Being of Older Persons in Thailand." United Nations Population Fund (UNFPA), New York.

Knodel, John, and Zachary Zimmer. 2009. "Gender and Well-Being of Older Persons in Cambodia." Research Report 09-665, University of Michigan, Population Studies Center, Ann Arbor, MI.

Kong, Mee-Hae. 2007. "Rethinking Women's Status and Liberation in Korea." http://unpan1. un.org/intradoc/groups/public/documents/ APCITY/UNPAN007144.pdf , Retrieved September 9, 2011.

Kong, Mee-Hae. 2007. "Rethinking Women's Status and Liberation in Korea." Unpublished paper. http://unpan1.un.org/intradoc/groups/ public/documents/APCITY/UNPAN007144. pdf.

Koolwal, Gayatri, and Dominique van de Walle. Forthcoming. "Access to Water, Women's Work, and Child Outcomes." *Economic Development and Cultural Change*. (Published previously as Policy

Research Working Paper 5302, World Bank, Washington, DC, 2010.)

Kramarae, Cheris. 2001. "The Third Shift: Women Learning Online." American Association of University Women Educational Foundation, Washington, DC.

Kreager, Philip, and Elisabeth Schroder-Butterfill. 2009. "Ageing and Gender Preferences in Rural Indonesia." CRA Discussion Paper 0905, University of Southampton Centre for Research on Ageing, England.

Kuriakose, Anne T., and Satoko Kono. 2008. "Empowerment of the Poor in Siem Reap EPSR), Cambodia: Preliminary Participatory Poverty and Gender Assessment (PPGA)." Draft report. World Bank, Washington, DC.

Kusago, Takayoshi, and Zafiris Tzannatos. 1998. "Export Processing Zones: A Review in Need of Update." Social Protection Discussion Paper 9802, World Bank, Washington, DC.

Lee, June. 2005. "Human Trafficking in East Asia: Current Trends, Data Collection, and Knowledge Gaps." International Migration 43: 165–201.

Lim, Joesph Y. 2000. "The Effects of the East Asia Crisis on the Employment of Women and Men: The Philippines Case." World Development 28 (7): 1285–1306.

Liu, Lan, Xiao-yuan Dong, and Xiaoying Zheng. 2010. "Parental Care and Married Women's Labor Supply in Urban China." Feminist Economics 16 (3): 169–192.

Long, Giang Thanh, and Wade D. Pfau. 2008. "The Vulnerability of the Elderly Households to Poverty: Determinants and Policy Implications for Vietnam." VDF Working Paper Series 87, Vietnam Development Forum, Hanoi.

LSB (Lao Statistics Bureau). Lao Expenditure and Consumption Survey. Vientane, Lao PDR. http://www.nsc.gov.la/index.php?option=com_content&view=article&id=50&Itemid=73&lang=.

Magnani, Elisabetta, and Anu Rammohan. 2009. "Ageing and the Family in Indonesia: An Exploration of the Effect of Elderly Care-Giving on Female Labor Supply." Journal of Income Distribution 18 (3–4): 110–30.

Mason, Karen, and Herbert Smith. 2003. "Women's Empowerment and Social Context: Results from Five Asian Countries." Unpublished paper, World Bank, Washington, DC.

Masud, Jariah, Sharifah Azizah Haron, and Lucy Wamuyu Gikonyo. 2008. "Gender Differences in Income Sources of the Elderly in Peninsular Malaysia." Journal of Family Economic Issues 29: 623–33.

McKay, Steven. 2006. "The Squeaky Wheel's Dilemma: New Forms of Labor Organizing in the Philippines" Labor Studies Journal 30 (4).

Melhem, Samia and Nidhi Tandon, with Claudia Morrell. 2009. Information and Communication Technologies for Women's Socio-Economic Empowerment. Washington, DC: World Bank.

Menon, Nidhiya, and Yana van der Meulen Rodgers. Forthcoming. "Impact of the 2008–2009 Twin Economic Crises on the Philippines Labor Market." World Development.

Mont, Daniel, and Nguyen Viet Cuong. 2011. "Disability and Poverty in Vietnam." World Bank Economic Review 25 (2): 323–59. doi: 10.1093/wber/lhr019.

Moser, C. 1993. Gender Planning and Development: Theory, Practice and Training. New York: Routledge.

Mu, Ren, and Dominique van de Walle. 2009. "Left behind to Farm? Women's Labor Re-allocation in Rural China." Policy Research Working Paper 5107, World Bank, Washington, DC.

Munshi, Kaivan and Mark Rosenzweig. 2004. "Traditional Institutions Meet the Modern World: Caste, Gender, and Schooling Choice in a Globalizing Economy," American Economic Review 96 (4): 1225–52.

Murayama, Mayumi, and Nobuko Yokata. 2008. "Revisiting Labour and Gender Issues in Export Processing Zones: The Cases of South Korea, Bangladesh and India." IDE Discussion Paper 174, Institute of Developing Economies, Chiba, Japan.

Ngai, Pau. 2004. "Women Workers and Precarious Employment in Shenzhen Special Economic Zone, China." Gender and Development 12 (2): 29–36.

Nguyen, Trang, and Ririn. Purnamasari. 2011. "Impacts of International Migration and Remittances on Child Outcomes and Labor Supply in Indonesia: How Does Gender Matter?" Policy Research Working Paper 5591, World Bank, Washington, DC.

Noel, Stacey, John Soussan, and Jennie Barron. 2007. "Water and Poverty Linkages in Africa: Tanzania Case Study." Stockholm Environment Institute, Stockholm.

NSCB (National Statistical Coordination Board) Philippines. 2006. Family Income and Expenditures Survey (FIES) 2006. National

Statistical Coordination Board, Makati City, Philippines. http://www.nscb.gov.ph/fies/default.asp.

NSD (National Statistics Directorate) Timor-Leste. Timor-Leste Survey of Living Standards. Ministry of Planning and Finance, National Statistics Directorate, Dili, Timor-Leste. http://dne.mof.gov.tl/TLSLS/AboutTLSLS/index.htm.

NSO Mongolia (National Statistical Office of Mongolia). Living Standards Measurement Survey (LSMS). National Statistical Office, Ulaanbaatar, Mongolia. http://www.nso.mn.

NSO Thailand (National Statistical Office Thailand). Thailand Socio-Economic Survey. Bangkok, Thailand: National Statistical Office. http://web.nso.go.th/tnso.htm.

Ofstedal, Mary Beth, John E. Knodel, and Napaporn Chayovan. 1999. "Intergenerational Support and Gender: A Comparison of Four Asian Countries." *Southeast Asian Journal of Social Sciences* 27 (2): 21–41.

Ofstedal, Mary Beth, Erin Reidy, and John E. Knodel. 2004. "Gender Differences in Economic Support and Well-Being of Older Asians." *Journal of Cross-Cultural Gerontology* 19: 165–201.

Olken, Benjamin. 2009. "Do TV and Radio Destroy Social Capital? Evidence from Indonesian Villages." *American Economic Journal: Applied Economics* 1 (4): 1–33.

Oostendorp, Remco. 2009. "Globalization and the Gender Wage Gap." *World Bank Economic Review* 23 (1): 141–61.

Oster, Emily, and Bryce Millet. 2010. "Do Call Centers Promote School Enrollment? Evidence from India." Working Paper Series 15922, National Bureau of Economics Research, Cambridge, MA.

Paluck, Betsy L., and Donald P. Green. 2009. "Deference, Dissent, and Dispute Resolution: An Experimental Intervention Using Mass Media to Change Norms and Behavior in Rwanda." *American Political Science Review* 103: 622–44.

Parrado, Emilio A., Chenoa Flippen, and Chris McQuiston. 2005. "Migration and Relationship Power among Mexican Women." *Demography* 42 (2): 347–72.

Pierre, Gaelle. 2011. "Recent Labor Market Performance in Vietnam through a Gender Lens." Policy Research Working Paper 6056, World Bank, Washington, DC.

Piper, Nicola 2009. "Overview." In *Gender and Labour Migration in Asia*, 21–42. Geneva: IOM.

Punpuing, Sureeporn. 2006. "Female Migration in Thailand: A Study of Migrant Domestic Workers." In *Perspectives on Gender and Migration.* Bangkok: United Nations Economic and Social Commission for Asia and the Pacific.

Punpuing, Sureeporn, Therese Caouette, Awatsaua Panam, and Khaing Mar Kyaw Zaw 2005. "Migrant Domestic Workers: From Burma to Thailand." Publication 286, Institute for Population and Social Research, Mahidol University, Thailand.

Rachmad, Sri H., Ardi Adji, and Dendi Handiyatmo. 2010. "Gendered Patterns of Urban Commuting with Better Connectivity in Jakarta Megapolitan Area." In *Gender, Roads and Mobility in Asia*, edited by Kyoko Kusakabe. Rugby, U.K.: Practical Action.

Rammohan, Anu, and Meliyanni Johar. 2009. "The Determinants of Married Women's Autonomy in Indonesia." *Feminist Economics* 15 (4): 31–55.

Rodriguez, E. R., and E. R. Tiongson. 2001. "Temporary Migration Overseas and Household Labor Supply: Evidence from Urban Philippines." *International Migration Review* 35 (3): 709–25.

Sabarwal, Shwetlena, Nistha Sinha, and Mayra Buvinic. 2010. "How Do Women Weather Economic Shocks? A Review of the Evidence." Policy Research Working Paper 5496, World Bank, Washington, DC.

Schuler, Sidney Ruth, Farzana Islam, and Elisabeth Rottach. 2010. "Women's Empowerment Revisited: A Case Study from Bangladesh." *Development in Practice* 20 (7): 840–54.

Schultz, T. Paul. 2003. "Does Globalization Advance Gender Equality?" Paper prepared for the conference "The Future of Globalization: Explorations in Light of Recent Turbulence," New Haven, CT, October 10–11.

Seguino, Stephanie. 1997. "Gender Wage Inequality and Export-Led Growth in South Korea." *Journal of Development Studies* 34 (2): 102–32.

———. 2000. "The Effects of Structural Change and Economic Liberalisation on Gender Wage Differentials in South Korea and Taiwan." *Cambridge Journal of Economics* 24 (4): 437–59.

Seguino, Stephanie, and Caren Grown. 2006. "Gender Equity and Globalization: Macroeconomic Policy for Developing Countries." *Journal of International Development* 18 (8): 1081–104.

Shastry, Gauri Kartini. 2010. "Human Capital Response to Globalization: Education and Information Technology in India." Working paper, Wellesley College. http://www.wellesley.edu/Economics/gshastry/shastry-ITandEducation.pdf.

Singhal, Arvind, and Everett M. Rogers. 1999. *Entertainment-Education: A Communication Strategy for Social Change.* Mahwah, New Jersey: Lawrence Erlbaum.

Sørensen, Ninna Nyberg. 2005. "Migrant Remittances, Development and Gender." DIIS Brief, Dansk Institut for Internationale Studier, Copenhagen.

Standing, Guy. 1999. "Global Feminization Through Flexible Labor." *World Development* 27 (3): 583–602.

Strauss, John, and Duncan Thomas. 2008. "Health over the Life Course." In *Handbook of Development Economics*, edited by T. Paul Schultz and John A. Strauss, Vol. 4. Amsterdam: North Holland.

Summerfield, Gale. 1995. "The Shadow Price of Labour in the Export-Processing Zones. A Discussion of the Social Value of Employing Women in Export Processing in Mexico and China." *Review of Political Economy* 7 (1): 28–42.

Svensson, Jakob, and David Yanagizawa. 2009. "Getting Prices Right: The Impact of the Market Information Service in Uganda." *Journal of the European Economic Association* 7 (2–3): 435–45.

Tan, Peck-Leong, and John Gibson. 2010. "The Impacts of Temporary Emigration of Lower-Skilled Females on Sending Households in Indonesia." Paper presented at the Sixth Australasian Development Economics Conference.

Tejani, Shirin, and William Milberg. 2010. "Global Defeminization? Industrial Upgrading, Occupational Segmentation and Manufacturing Employment in Middle-Income Countries." Working paper, Schwartz Center for Economic Policy Analysis, New York.

Thailand Household Socioeconomic Survey 2009, Thailand Household Socioeconomic Surveys, National Statistical Office of Thailand, http://web.nso.go.th/tnso.htm

Thomas, Duncan, Kathleen Beegle, Elizabeth Frankenberg, Bondan Sikoki, John Strauss, and Graciela Teruel. 2004. "Education in a Crisis" *Journal of Development economics* 74: 53–85.

Turk, Carolyn, and Andrew Mason. 2009. "Impacts of the Economic Crisis in East Asia: Findings from Qualitative Monitoring in Five Countries." In *Poverty and Sustainable Development in Asia: Impacts and Responses to the Global Crisis*, edited by A. Bauer and M. Thant. Manila: Asian Development Bank.

UN (United Nations). 2007. "Population Estimates and Projections Section." Fact Sheet Series A. 7 March 2007. UN Department of Economic and Social Affairs (DESA) Population Division, New York.

UN DESA (UN Department of Economic and Social Affairs), Population Division. 2010. *World Urbanization Prospects: The 2009 Revision.* (CD-ROM). New York: UN.

UNESCAP (Economic and Social Commission for Asia and the Pacific). 2004. "Population Ageing, Focus of Regional Seminar in Macau." *Population Headliners* No. 302, UNESCAP, New York.

———. 2007. "Perspectives on Gender and Migration, Economic and Social Commission for Asia and the Pacific." UNESCAP, New York.

UNFPA (United Nations Population Fund) 2006. *State of the World Population 2006: A Passage to Hope—Women and International Migration.* New York: UNFPA.

———. 2007. *State of the World Population 2007: Unleashing the Potential for Urban Growth.* New York: UNFPA. http://www.unfpa.org/swp/2007/.

UNIFEM (United Nations Development Fund for Women). 2005. "Good Practices to Protect Women Migrant Workers." High-Level Government Meeting of Countries of Employment, Co-Hosted by the Ministry of Labor, Royal Thai Government, and UNIFEM, East and South-East Asia, Bangkok.

UNPD (United Nations Population Division). 2008. Trends in Total Migration Stock, 2008 Revision (CD-ROM). http://www.un.org/esa/population/publications/migration/migration2008.htm.

Wang, Danning. 2010. "Intergenerational Transmission of Family Property and Family Management in Urban China." *China Quarterly* 204: 960–79.

Wood, Adrian. 1991. "How Much Does Trade with the South Affect Workers in the North?" *World Bank Research Observer* 6 (1): 19–36.

World Bank. 2006a. "Cambodia: Women and Work in the Garment Industry." World Bank, Washington, DC.

———. 2006b. "Expanding Job Opportunities for Pacific Islanders through Labour Mobility: At Home and Away." World Bank, Washington, DC.

———. 2006c. "Labor Migration in the Greater Mekong Sub-Region: Synthesis Report Phase I," World Bank, Washington, DC.

———. 2008. *World Development Report 2009: Reshaping Economic Geography.* Washington, DC: World Bank.

———. 2010b. "Mainstreaming Gender in Road Transport: Operational Guidance for World Bank Staff." Transport Paper 28, World Bank, Washington, DC.

———. 2011a. "Gender Disparities in Mongolian Labor Markets and Policy Suggestions." World Bank, Washington, DC.

———. 2011b. *World Development Report 2012: Gender Equality and Development.* Washington, DC: World Bank.

———. Forthcoming. *International Migration and Development in the East Asia and Pacific Region.* Washington, DC: World Bank.

World Development Indicators (WDI) database. World DataBank, World Bank, Washington, DC. http://databank.worldbank.org.

World NGO Forum on Ageing. 2002. "Final Declaration and Recommendations of the World and GOP Forum on Aging." Global Action on Aging. http://www.globalaging.org.

Xie, Xiaolin, Yan Xia, and Xiaofan Liu. 2006. "Family Income and Attitudes Toward Older People in China: Comparison of Two Age Cohorts." *Journal of Family Economic Issues* 28: 171–82.

Yamanaka, Keiko, and Nicola Piper, eds. 2003. "Gender, Migration and Governance in Asia." Special issue, *Asian and Pacific Migration Journal* 12 (1–2).

Yamanaka, Keiko, and Nicola Piper. 2005. "Feminised Cross-Border Migration, Entitlements and Civil Action in East and Southeast Asia." Background paper for *Gender Equality: Striving for Justice in an Equal World,* Geneva: United Nations Research Institute for Social Development (UNRISD).

Yang, Xiushi, and Guomei Xia. 2005. "Risky Sexual Behavior among Female Entertainment Workers in China: Implications for HIV/STD Prevention Intervention." *AIDS Education and Prevention* 17 (2): 143–56.

Zhan, Heying Jenny. 2004. "Socialization or Social Structure: Investigating Predictors of Attitudes toward Filial Responsibility among Chinese Urban Youth from One and Multiple-Child Families." *International Journal of Aging and Human Development* 59 (2): 105–24.

Zhang, Heather. 1999. "Female Migration and Urban Labour Markets in Tianjin." *Development and Change* 30 (1): 21–41.

Promoting Gender Equality in East Asia and the Pacific: Directions for Policy

6

The evidence presented in the preceding chapters has shown that growth and development cannot by themselves end gender disparities in all their dimensions. The evidence has also shown that persistent gender inequalities can exact costs on countries' productivity, income growth, and quality of development. Thus, public policy needs to play a role to reduce gender gaps. East Asian and Pacific countries will benefit from adopting appropriate policies to promote gender equality in endowments, in access to economic opportunities, and in agency. This chapter examines policy approaches to addressing the most serious, persistent, and costly gender disparities. In doing so, it attempts to answer the question: How can public policies and investments promote both gender equality and more effective development?

To show the types of policies and investments that can be most productive, this report has provided evidence on which dimensions of gender disparity tend to close with growth and development in East Asian and Pacific countries and which do not. Rapid growth and development in East Asian and Pacific countries have been accompanied by reduced gender inequalities in the areas of education, health, and female labor force participation, particularly among younger women. Progress toward gender equality in many domains is not automatic. Progress in education, health, and labor force participation has often been uneven, both within and across countries. Several East Asian and Pacific countries still experience the phenomenon of "missing" girls at birth. Women continue to work in less stable employment with lower rates of remuneration. They still possess weaker voice and influence than men. Women remain underrepresented in leadership positions in business and government. In a number of countries, gender-based violence remains prevalent.

The collection of evidence in this report suggests the following broad outlines of a regional strategy to promote gender equality and more effective development:

- Promoting human development remains a priority where gender gaps in education remain large or where health outcomes are poor. Investments in these areas are likely to yield high returns.
- Taking active measures to promote gender equality in economic opportunity can address the disparities that often persist with development. Such measures can

often yield positive returns to economic productivity, though specific policy priorities depend on the structure of the economy and on which specific constraints are most binding.

- Initiating active measures to close gender gaps in agency across the region—and to protect women from discrimination and violence—will contribute to better developmental decision making and thus to development more broadly.
- Increasing economic integration, increased access to information and communication technologies (ICTs), migration, rapid urbanization, and population aging all bring with them new opportunities and new risks with respect to gender equality. An important role for public policy will be to foster the opportunities while managing the risks.

Interpretation of this proposed strategy should consider two points: that policies can simultaneously influence several domains—endowments, economic opportunity, and agency—and that policies should be context-specific. First, policy approaches to promote better outcomes with respect to endowments, access to economic opportunities, and agency are likely to be mutually reinforcing. Individual outcomes in these areas are often interdependent. Basic human capital, for example, represents a critical factor in women's (and men's) access to economic opportunity and agency. Therefore, policies and investments that promote equality in education will also contribute to enhanced economic opportunity and agency. Similarly, measures to close gender gaps in voice and influence will enable women (and men) to be more effective agents on their own behalf, whether in pursuit of education, access to resources, or productive employment.

Second, the choice and design of policies should consider national and local realities. The great diversity of the East Asia and Pacific region with respect to gender inequalities will influence policy priorities. For example, even in countries where gender disparities in access to basic education and

health services are no longer a dominant concern, special policies and programs may be warranted to address large and persistent gender gaps among specific subgroups of the population (for example, for poor, indigenous, or remote communities). Similarly, current policy priorities for promoting gender equality in economic opportunity will differ across countries, depending on the specific obstacles faced by women and on the structure of the economy (for example, the fraction of the population in agriculture or in the informal sector). Some aspects of women's agency improve with development in East Asian countries, but women in the Pacific Islands tend to experience a severe and persistent lack of agency. As a result, closing gender gaps in agency is a policy priority across the region, and effective measures will differ as a function of a number of factors, including local social norms regarding women's voice and influence, differences in countries' legal systems, societies' tolerance for violence, and so on.

This chapter examines each part of the proposed strategy in turn. To be most effective in reducing gender disparities, policy makers may need to address both systemic problems and gender-specific problems. To the extent possible, the discussion in this chapter will account for the potential costs as well as the benefits of policies. Such trade-offs can be particularly important in the context of labor market interventions. For example, although raising the minimum wage may close gender wage gaps among those employed in the formal sector, the numbers of women the sector can employ may decline. Similarly, affirmative action policies may help to promote female employment, but they can also raise questions about worker productivity and may be negatively perceived in the workplace. Although the evidence reviewed and generated for this report helps to answer many questions about gender and development in the East Asia and Pacific region, many knowledge gaps remain. To that end, the chapter concludes by highlighting a number of areas that deserve additional analytical work.

Promoting gender equality in human development

East Asian and Pacific countries vary in the types of challenges they face with respect to gender equality in endowments, as analyzed in chapter 2. Where policies may be used and can have an impact also varies. Using that body of evidence as a basis, this section is organized around four themes: (a) closing persistent gender gaps in human development, (b) reducing gender streaming in education, (c) promoting balanced sex ratios at birth, and (d) addressing male-specific gender issues. Because basic human capital is an important factor for access to economic opportunity and agency, the discussion below will highlight examples where policies to promote equality in education also contribute to enhanced economic opportunity and agency.

Closing persistent gender gaps in human development

For the few East Asian and Pacific countries with overall low and unequal gender outcomes in education and health (for example, Cambodia, the Lao People's Democratic Republic, and Papua New Guinea), promoting human development remains a policy priority. Education and health outcomes are likely to improve as these countries become wealthier, with higher household income, better service delivery, and better expected employment opportunities. However, countries must act now, because human capital itself is an important engine for growth and can have long-term effects on well-being. For countries with localized gender disparities among certain ethnic groups or low-income regions, growth in aggregate is not enough to improve outcomes for all subpopulations. Policy makers need to target interventions to these groups. For example, the efforts that have successfully improved human capital for both genders elsewhere in these countries can be replicated for these subpopulations, in combination with complementary interventions that address their specific constraints.

What can be done in this group of countries to provide equal access to education and reduce maternal mortality? Chapter 2 shows that demand-side and supply-side factors are responsible for these countries' poor human capital outcomes. Therefore, the set of education and health policies can range from improving service delivery (for example, through infrastructure, staffing, incentives, and use of information and communication technologies) to demand-side interventions (for example, cash transfers, information campaigns, and accountability). The exact constraints in each country context will influence the choice and sequencing of policies.

Providing equal access to education
A few countries in the East Asia and Pacific region still face overall low and gender-unequal access to education, particularly starting at the secondary level. This problem requires solutions to increase the demand for schooling as well as the capacity and incentives of the education system. Policies to improve education outcomes in general— that is, addressing systemic constraints—are expected to also improve gender equality as the enrollment gap between genders narrows.

Supply-side solutions can play an important role. Although primary schools are widely available in East Asian and Pacific countries, secondary school availability is a problem in a few countries, including Cambodia and Lao PDR. Those countries should consider building more schools and improve infrastructure in poor and remote areas where distance to school has been identified as a key constraint. These types of interventions, though usually with high up-front costs, can have significant impacts on increasing education and future earnings, as shown in an evaluation of Indonesia's school construction program in the 1970s (Duflo 2000). Supply-side solutions need to be customized to address the context-specific constraints. Making special provisions for girls, such as private female latrines, may be necessary to ease parents' concern for girls' privacy and safety in school. Particular excluded groups such as ethnic minorities may need additional support. Teachers who speak the local language or who come from the relevant ethnic group can serve as role models and relate to the local social norms about educating children.

Innovative use of information and communication technologies (ICTs) in delivering services can enhance learning opportunities for girls and boys and improve the provision of education. The availability of distance learning, for example, can circumvent the constraints of travel for schooling and deliver education to youths who have no access to formal schooling or who have left the formal system. The Flexible Open and Distance Education (FODE) program in Papua New Guinea presents an interesting approach. It helps public distance learning centers provide upper primary- and secondary-level courses in English, mathematics, science, and social sciences. One key component of the project aims to provide FODE centers with appropriate information technology and to train staff members to effectively use that technology in administration, curriculum development, materials production, and teaching and learning activities.

Several demand-side solutions have been shown to work effectively. As discussed in chapter 2, parents in low-income families in Cambodia have difficulty covering the direct and indirect costs of schooling. One way to expand access is to provide financial incentives to attend school. In many developing countries worldwide, evaluation of the impacts of conditional and unconditional cash transfers indicates a significant gain in school enrollment (Fiszbein and Schady 2009). In Cambodia, a scholarship program targeted to girls and a related program targeted to boys and girls from low-income households led to an increase in school enrollment of at least 20 percentage points for both genders (Filmer and Schady 2008, 2009). Cash transfers and scholarship programs for school-age girls can also positively influence their long-term agency. For example, in Malawi and Pakistan, conditional cash transfers and school scholarship programs have

been found to delay girls' age of marriage and increase their earning opportunities, which in turn strengthens women's bargaining power within the household (Baird et al. 2009; IEG 2011).

Aside from affordability, school enrollment may be discouraged by the weak incentives of service providers and lack of accountability of the school systems. The 2004 *World Development Report* called for improving accountability in service delivery, and follow-up work has demonstrated the positive impacts of various measures to encourage and enable parents to hold the school and teachers accountable for their childrens' schooling outcomes (World Bank 2003). Evidence on this aspect has been abundant in other developing countries but is much needed in the East Asia and Pacific region.[1]

Reducing maternal mortality

The slow improvements in maternal health outcomes in several East Asian and Pacific countries, including Cambodia, Lao PDR, parts of Indonesia, and Timor-Leste, suggest the importance of improving service delivery. Access to pre- and postnatal health facilities—including sufficient infrastructure, especially in rural areas—and improved access to sanitation and potable water are fundamental to improving maternal mortality rates. Therefore, this report recommends continuing efforts to ensure the basic infrastructure and to provide access to safe water. The low rate of births attended by skilled professionals, especially in poor and rural areas in these countries, calls for interventions related to staffing. Impact evaluation evidence shows that performance-based contracting increases staff attendance and, consequently, the use of antenatal care, as shown in the case of Cambodia (see box 6.1). In addition, training and increased allocation of midwives to villages in Indonesia have increased women's body mass index (BMI) and children's birthweight and neonatal health (Frankenberg and Thomas 2001). However, just having more midwives is not enough to ensure and sustain service quality. Malaysia's experience in successfully reducing maternal mortality

---

BOX 6.1 Reducing maternal mortality rates through improved staffing and quality of service delivery

Cambodia Basic Health Service Project. In the late 1990s, Cambodia initiated a performance-based contracting scheme with three service delivery forms: contracting-out, in which the whole service provision was allocated to the contractor; contracting-in, in which an external contractor was hired to improve the existing structure; and a comparison group in which no contracting took place (OECD 2009). Using the initial randomization of the program over 12 districts, Bloom et al. (2006) found that contracting-in significantly increased the likelihood of 24-hour service provision. Both contract forms increased the probability of staff attendance and the existence of supplies and equipment, management quality, and use of public providers. Among other positive health effects, antenatal health care increased by 36 percentage points for pregnant women in contracting-in districts, and both contract forms increased the use of health facilities for delivery (Bloom et al. 2006).

Indonesia Village Midwife Program. Initiated in 1989, the program aimed at allocating more midwives to villages and exempted delivery fees for poor families (Makowiecka et al. 2008). Using panel data from 1993 and 1997, Frankenberg and Thomas (2001) found that the introduction of a midwife increased the BMI of women in reproductive age relative to the comparison groups. Also, the program has been found to increase birthweight (Frankenberg and Thomas 2001) and reduce neonatal mortality (Shrestha 2007). Although midwife coverage increased in Indonesia, along with the number of safe deliveries, progress in reducing maternal mortality rates has been argued to be below expectations (Dawson 2010). Using data on midwives in Serang and Pandeglang districts, Makowiecka et al. (2008) found that village-based midwives were more likely to have temporary contracts, were less experienced,

and had less training than health center midwives. Also, midwives were found to be attracted to urban areas. In addition, Shankar et al. (2008) argued that the lack of practical experience and of training, supervision, and guidance have limited midwife retention and, more important, service quality.

Malaysia. With a reduction of maternal mortality from 1,500 per 100,000 live births in 1930 to less than 58 in 2005 (Dawson 2010), Malaysia constitutes one of the success stories of improving maternal health. Among other factors, the quality and the strict supervision of the midwives by certified nurses, support and guidelines, and training for emergency situations, such as improving the communication between the rural service and hospital staff (Dawson 2010), may explain this success. In addition, traditional birth attendants were also trained, registered, and given tools, and midwives were briefed on the integration with traditional practitioners (Dawson 2010).

Mongolia Maternal Mortality Reduction Strategy. Implemented between 2001 and 2010, this approach is a package of incentives, staffing, and demand-side information campaigns. The package included awareness-raising campaigns, such as radio and television programs about maternal health needs, that were broadcast multiple times a week (Yadamsuren et al. 2010). Initiatives such as the "mother-friendly governor" aimed to strengthen the commitment and rewards of improving maternal mortality–related indicators at the local government level. Several additional measures, such as the provision of attractive contracts to doctors for relocating into remote areas, maternal waiting homes, partial service and transportation cost coverage, and training of doctors and health practitioners, contributed to improving maternal health provision (Yadamsuren et al. 2010).

---

suggests that rigorous training, supervision, and communication support for midwives are also essential factors for quality (Dawson 2010). Box 6.1 discusses these examples in further detail.

Service provision should also respond to noneconomic factors: maternal mortality may be reduced by making service delivery culturally sensitive. As described in chapter 2, norms about birthing practices play a role in birthing decisions and outcomes. Programs such as those in Malaysia and Tonga, which provide guidance to traditional birth attendants on hygiene practices, diagnosis of complicated cases, and the importance of prenatal care, are likely to be beneficial. In Malaysia, hospital services have adapted to allow naming rituals for Muslims (Hillier 2003). Studies in Cambodia and Lao PDR recommend building birth huts in the health center to allow privacy and to accommodate the presence of family and relatives (UNFPA 2005).

Financial assistance, such as subsidies or cash transfers, has been shown to improve the use of health services and reduce maternal mortality, given the financial barriers faced by poor women. The Cambodia Health Equity Fund (HEF) combined financial assistance to the poor and an incentive scheme for service providers. In 2007, the fund launched a voucher program to provide poor pregnant women with access to free prenatal and postnatal health care visits and delivery, as well as to cover the transportation costs. At the same time, the government implemented a nationwide incentive system that paid midwives and other health care staff an additional US$12.50 and US$15.00 per live birth in hospitals and health care centers, respectively (Ir et al. 2010). Using data from three health districts between 2006 and 2008, Ir et al. (2010) found that deliveries in health facilities increased by 196 percent for HEF voucher beneficiaries.

Policy interventions to improve accountability in service delivery can lead to increased use of maternal care services, as indicated by evidence from other parts of the world. In Uganda, community-based monitoring improved the quality and quantity of primary health care services (Björkman and Svensson 2009). In Peru, incentives to service providers, coupled with a mechanism giving citizens a voice to reach policy makers, proved useful: professional attendance for deliveries increased from 58 percent of births in 2000 to 71 percent in 2004 (Cotlear 2006).

Reducing gender streaming in education

For many East Asian and Pacific countries, most gender issues regarding basic access in education and health have abated with growth and development. If gender disparities in human capital persist in subgroups such as ethnic minorities or low-income regions, interventions such as those discussed earlier need to address the specific constraints faced by these subpopulations. However, visible gender streaming in education remains a direct barrier to gender equality in productive activities, voice, and participation.

Countries can make concerted efforts in education and labor market policies to tackle the current equilibrium of females sorting into lower-paying occupations and lower-productivity sectors, which is partly due to education streaming. The evidence presented in the preceding chapters shows that females shy away from certain fields of study, despite the high economic return, for reasons such as gender norms about school and work, stereotyping in school curricula, lack of role models, or lack of information. Although changing norms is very difficult, education systems can at least limit the perpetuating influence of gender-biased norms. What follows is a discussion about promising approaches, but future research needs to provide more rigorous evaluations of these policies.

One possible approach on this agenda is to reduce gender stereotyping in school curricula. Several countries in the region have begun to review and revise teaching materials from a gender perspective, though additional efforts are called for to ensure that the revisions introduced to promote gender-sensitive curricula are systematically applied (UNICEF 2009). For example, in Lao PDR, supplementary learning materials, new gender-sensitive primary education curricula, and textbooks were developed following a gender review and training of curriculum writers. In Vietnam, a review of the current curricula indicated that gender stereotyping was frequent, and isolated examples of good practice (including collaborative behavior between boys and girls) have been highlighted in teaching materials for replication. Piloting and evaluation are important since empirical evidence on the impacts of gender-sensitive curriculum reforms is very limited worldwide, let alone in the East Asia and Pacific region.[2]

Revising school and professional training curricula to include better female role models can enable greater agency in the household over time. The formal education system can educate young children on the basic principles of gender equality by integrating these principles into the curricula, introducing the value system early in children's lives, and challenging existing gender-related social norms that are learned outside the classroom. In Indonesia, a recent textbook evaluation shows strong gender bias starting in the sixth grade, with stereotypical gender roles that portray men as income earners and women as homemakers. The evaluation resulted in a recommendation that the government should promote the incorporation of gender equality into the curricula for primary and secondary school to address gender-based violence as well as other social norms that perpetuate gender inequality (Utomo et al. 2009).

Alternatively, policy interventions can actively promote entrance into nontraditional fields of study. Offering scholarships to girls and women to study fields such as engineering or law is one option. Not enough is known about the impact of such scholarships on breaking traditional patterns that are deeply rooted in norms, and small financial incentives may lead to changes only at the margin. Aside from financial incentives, other interesting approaches use female role models and encourage female pupils in nontraditional careers, though evidence of their impacts is usually not available either.[3] In the United States, the Science Connections (SciCon) program offered monthly science workshops for girls plus a summer science weekend for the family to foster girls' interest in science. After the program, the retention rate from the first to the second workshop rose from only 10 percent to 25–50 percent for different cohorts. The program reportedly also increased girls' knowledge about nontraditional careers in science, self-confidence, interest in science, and motivation to increase their efforts in science-related courses.[4] The Technical/Engineering Education Quality Improvement Project in India takes another approach to create a gender-friendly environment in science and

engineering schools and department. The project expands access to training for female engineering faculty and includes refurbishments to enhance the ability of campuses to serve women students.

Activities to reduce traditional patterns of gender streaming in education are likely to become increasingly important as economies move up the value chain away from labor-intensive production. In middle- and higher-income countries in the East Asia and Pacific region, movements up the value chain have typically been accompanied by reductions in the fraction of females working in Special Economic Zones (SEZs). For example, in Malaysia only 40 percent of workers in SEZs are now female, compared to 60 percent two decades ago (Simavi, Manuel, and Blackden 2010). Addressing the skill constraints of workers as countries make the transition up the value chain will require policies ranging from appropriate technical and vocational training among school-age girls and boys to training programs to help retrain workers whose skills are in lower demand.

Promoting balanced sex ratios at birth

A few countries—China and Vietnam, in particular—should continue their efforts to address the phenomenon of "missing girls" at birth. This problem does not have easy policy solutions since preference for sons is the underlying cause, and certain population policies or access to technology that comes with economic growth can exacerbate the problem. Countries can enact and enforce legal clauses against sex-selective abortion; however, those measures are extremely difficult to implement when societies have strong incentives to select the preferred gender and bypass the law. More promising policy approaches could alter the incentives themselves by enhancing the relative value of daughters as perceived by families.

Relying on economic growth to raise female education and participation in the labor market alone may not suffice, as the Republic of Korea's recent experience suggests. Chung and Das Gupta (2007) argued that the impacts of development in Korea since the 1950s influenced son preference through many factors: (a) higher earning prospects increased individuals' independence of family lineage, (b) retirement savings reduced financial dependence on children in old age, (c) urban life setting reduced the focus on traditional filial duty and promoted female-inclusive social networks, (d) women's greater economic and physical mobility enhanced the value of daughters, and (e) urban life, with assets associated with nonfarm activities and less pressure from customary laws, facilitated gender equity in inheritance. The authors attributed the observed reduction in son preference in larger part to changing social norms (changes in son preference within all education and urban and rural population groups) and in smaller part to increased urbanization and education (changes due to movements between education groups and between urban and rural populations). Their findings suggest that interventions to influence social norms and facilitate the spread of new values work better than relying on raising female education and labor force participation alone. Therefore, policy makers should consider information campaigns, financial incentives, and improved social security for the elderly to reduce the imbalance in the sex ratio at birth.

China has been taking public actions to address the unbalanced sex ratio at birth with a package of approaches. Aside from regulations outlawing sex-selective abortion, the nationwide Care for Girls program in China is a well-known media campaign. It started with a pilot in Chaohu from 2000 to 2003 (Li 2007), which combined advocacy and training, along with punishment for discrimination in the form of sex-selective abortions and infanticide. The program achieved its main objectives: the sex ratios at birth in Chaohu fell from 125 boys per 100 girls in 1999 to 114 per 100 in 2002. Some of the program activities asked men to participate in the discussion as well, and encouraged them to help improve women's status in the home and in society (Li 2007). The National Population and Family Planning Commission

then scaled up the Chaohu pilot through the national Care for Girls campaign in 24 counties of 24 provinces with severe gender imbalances. This campaign went beyond advocacy and media publicity alone. Direct financial incentives for parents to raise daughters have also been introduced as part of the Care for Girls campaign: families with elderly parents without sons receive an annual allowance of 600 yuan; daughter-only families receive preferential loans; daughters from these families also receive bonus points in college entrance exams. Although the exact attribution to the program's effects has not been evaluated, the sex ratios at birth in these 24 counties fell from 133.8 in 2000 to 119.6 in 2005 (Li 2007).

Improving social security for the elderly can also affect the sex ratio, since families might prefer to have sons in order to receive financial support in old age. Ebenstein and Leung (2010) assessed the impact of the introduction of a rural old-age pension program on sex ratios at birth in China. Using cross-sectional data, they found that the number of sons decreases the likelihood of participating in the pension program. Using a measure of availability of the program at the village level, the authors compared changes in the sex ratios at birth for cohorts born before and after the introduction of the program in 1991. They found that counties in which the pension program is available exhibit smaller increases in the sex ratio at birth.

Addressing male-specific gender issues

Male gender issues may also hamper countries' growth and development. First, in many parts of the East Asia and Pacific region (for example, China, Fiji, Malaysia, Mongolia, the Philippines, Samoa, and Thailand), countries need to monitor the initial signs of the reverse gender gap in education. Second, in many countries, being born male is the single greatest determinant for tobacco use and a main determinant for harmful alcohol use (WHO 2005, 2007). Such use is harmful for households and societies and warrants the attention of sound public policy, since the social costs are usually higher than private costs, which are externalized to other members of the society. Costs to economic growth can be even more severe when combined with the effects of an aging population.

Although this report does not exhaustively cover harmful tobacco and alcohol use, table 6.1 highlights several policy approaches to tackle the challenge inside and outside the region. These measures include policies to reduce the prevalence of alcohol and tobacco use and other policies to reduce the harmful impacts of such use. Taxation and restrictions on sales are usually not very effective in countries where alcohol and tobacco products are mostly sold through informal markets (WHO 2005, 2008).

Policies against tobacco use and excessive alcohol use should consider gender dimensions, gender norms, and cultural values. Gender-specific education and communication approaches may be used to more effectively target men (WHO 2007). Programs expected to decrease the prevalence of smoking include warnings about male-specific health risks of smoking (including reduced potency and fertility), messages about the harmful effects of passive smoking, and campaigns targeting men who are role models to boys, such as fathers, teachers, and peers (WHO 2007). Research from Cambodia, Malaysia, and Vietnam suggest the usefulness of messages that resonate with core social values of responsibility for family welfare, such as reminding the smoker that smoking harms children's health and stressing the role of setting a good example for children (Efroymson and Velasco 2007).

Taking active measures to promote gender equality in economic opportunity

Gender inequalities in economic opportunities, such as gender wage gaps and the concentration of women and female-led enterprises

TABLE 6.1 Selected policy approaches to tackle excessive tobacco and alcohol use

Policy approach	Examples and impacts
Tobacco and alcohol taxation	Analysis shows that if China transferred its 2009 tobacco tax adjustment from the producer to the smoker, the retail price would have increased 3.4 percent and would have resulted in 640,000 to 2 million smokers quitting and between 210,000 and 700,000 quitters avoiding smoking-related premature death (Hu, Mao, and Shi 2010).
Information campaigns on the related health risks	The Thai Health Promotion Foundation (ThaiHealth) uses alcohol excise tax revenues to promote health. ThaiHealth's activities include supporting the establishment of a surveillance center to enforce alcohol control regulations; a research center on alcohol consumption; and advertising campaigns to reduce alcohol-related traffic accidents, promote abstinence, and increase knowledge about the links between alcohol use and domestic violence. The agency also played a vital role in the successful passage of a national policy to control alcohol advertising and in launching a National Committee for Alcohol Consumption Control (WHO 2011).
Sales restrictions and advertising bans	Restrictions on sales may include regulating the types of retail establishments that can sell these products, requiring and enforcing licensing, and limiting the hours and days of sale. Advertising against tobacco and alcohol use may encourage changes in the perception of masculinity to break the link between masculinity and substance use (WHO 2005, 2007). In a study in Norway, youths ages 13–15 years in 1990 and 1995 that were exposed to tobacco marketing were significantly more likely to be current smokers (controlling for important social influence predictors) than those adolescents who were not exposed to tobacco marketing. Such evidence suggests that bans on advertising would be useful to limit the promotion of adolescent smoking (Braverman and Aarø 2004).
Bans on smoking in public places	China's ban on smoking in public places became effective in May 2011, after successful trials during the 2008 Beijing Olympic Games and the 2010 Guangzhou Asian Games (World Bank 2011e).
Appropriate drunk-driving policies	Establishing maximum blood alcohol concentrations of drivers and enforcing these with sobriety checks and random breath testing can decrease alcohol-related motor vehicle crashes by roughly 20 percent and are highly cost-effective (Elder et al. 2002 and Peek-Asa 1999, cited in WHO 2011). Enforcement of the laws and severity of punishment are imperative for the success of such initiatives (WHO 2008, 2011). In Brazil, infringement of the law culminates in penalties of up to three years in prison, a hefty fine, and suspension of the drunk driver's license for one year (WHO 2011).

in less-remunerative jobs and sectors, exist in all countries in the region, although the gravity of inequalities and hence policy priorities vary by country. The focus of policies in all countries, however, has a common theme— to provide an enabling environment in which women can balance their multiple roles and flourish in their income-generating activities. As discussed in chapter 3, the constraints that infringe on women's ability to flourish as economic actors occur in households, markets, and informal and formal institutions. This section highlights what policy makers can do in these three domains to improve gender equality in economic opportunity. The following sections focus on three primary areas: (a) reducing the time constraints associated with women's household roles, (b) increasing women's access to resources, and (c) establishing a level playing field by reforming institutions. Because the most serious constraints faced by women vary across economies according to economic structure (see chapter 3), this section ends with a brief discussion of countries' appropriate policy priorities.

Reducing trade-offs between women's household and market roles

This report finds that, in all countries, women's household roles affect their decisions as economic actors—for example, they affect the amount of time women are able to devote to market-orientated activities, the types of occupations that they enter into, and the types of enterprises they run. A key policy priority across countries is to reduce the impact of household roles on women's economic lives, a priority that will become increasingly important as aging continues and the demands of caring for elderly family members rise. Addressing women's dual roles is likely to be the most important in contexts where female labor force participation is severely affected by the competing demands for women's time (for example, in Malaysia, the Philippines, and Fiji, where female labor force participation rates drop particularly sharply when they have children).

Although governments cannot directly change the norms that influence gender divisions of domestic responsibilities, policies can work around existing customs and thereby reduce the impact that customs have on women's economic activities. Policies may also indirectly influence divisions of labor within households, for example, by instituting interventions that raise the value of women's market time.[5]

Although household responsibilities infringe upon women's economic activities across all countries, the policy recommendations in this area vary with a country's economic structure. In countries with predominantly rural populations and infrastructural constraints that limit women's access to markets and energy and water sources, policies targeted at improving existing infrastructure will likely have the greatest impact on women's time balance. In countries with a larger urban population, governments should consider policies that increase access to affordable child care, particularly where informal mechanisms for child care are more limited. In countries where the formal sector is increasingly important, addressing parental leave policies will help to level the playing field for men and women and may ameliorate gender divisions of labor within households.

Investing in basic infrastructure and transportation

In rural or urban areas with limited infrastructure, governments can reduce the time spent on domestic activities by investing in improved water, sanitation, and energy services. These services may release time that can be spent on market work or leisure and increase the income potential or well-being of women. Evidence from around the world suggests that water and electrification projects can reduce the time that women spend collecting water and fuel, and can increase productivity by increasing the length of the working day.[6] In Lao PDR, evidence suggests that electricity extends the hours available for both productive and leisure activities, particularly for women and girls (World Bank

2011d). In the Gansu and Inner Mongolia Poverty Reduction Project, the construction of drinking water facilities was found to reduce the workload for women (World Bank 2007b).

Government investment in transport infrastructure can improve access to markets by reducing travel time constraints. Although both women and men benefit from better access to markets, it likely affects the well-being of women more because they have less time to devote to market-oriented activities. In Papua New Guinea, evidence suggests that a road maintenance and rehabilitation program improved the quality of life of remote disadvantaged communities and improved access to markets,

particularly for women, who were found to have increased the frequency of market visits and their own incomes generated from market activities (Jusi, Asigau and Laatunen 2007). Recent experience on infrastructure and transportation projects also suggests that the design of infrastructure projects can have important gender implications (box 6.2).

Improving access to affordable child care

Access to affordable and high-quality child-care options may reduce the amount of time that women spend on care-related activities, particularly among groups of women such as migrants, who have limited access to alternative child-care mechanisms.[7] The evidence

BOX 6.2 The design of infrastructure projects has important gender dimensions

Consultations with both men and women can ensure that infrastructure projects meet the needs of both men and women, since use and access to infrastructure varies by gender. For example, many rural women lack access to motorized transport, tend to travel on feeder roads and tracks on foot, or use intermediate means of transport such as donkey carts and bicycles. Conventional rural transport planning has tended to focus on road networks and long-distance transport, which has led to the neglect of the needs of women in rural areas. In Peru, for example, women's participation in economic activities increased in response to a program that focused on upgrading nonmotorized tracks that are predominantly used by women and are not generally covered by traditional road-upgrading programs (Valdivia 2010).

In urban areas, integrating gender-specific needs into the design of urban transport systems can increase access for women. Women are more likely to travel during nonpeak hours and to conduct multiple short trips, which makes using public transport expensive. To make public transportation more accessible to women, fee structures may compensate gender differences in use by providing tickets that allow for multiple stops on a single trip, thereby reducing costs. The transportation system can also be improved to offset social and cultural factors that hinder women's access. For example, to accom-

modate the needs of the Muslim majority, Malaysia introduced women-only trains in 2010. In a similar spirit, women-only buses during peak times were introduced to counteract sexual harassment and discomfort among female customers. Jakarta, Indonesia, introduced women-only train compartments on a busy commuter route to address sexual harassment complaints.

Because female-headed households may be subject to greater financial constraints than male-headed households, gender-blind program design may have gender-differentiated effects. In Lao PDR, the rural electrification project did not reach poor female-headed households at the initial stage. An early assessment of the project suggested that approximately 40 percent of rural households could not afford the connection fees of approximately US$100. These households were predominantly poor and were disproportionately headed by women. As a response, a pilot program targeted the poorest households, with a focus on female-headed households. A revolving loan fund provided these households with interest-free loans to cover 80 percent of their costs, and all female-headed households were eligible for support under this fund. Within a few years, electrification rates in pilot areas increased from 63 percent to 90 percent overall, and from 79 percent to 96 percent for female-headed households (Boatman and Chanthalinh 2009).

presented in chapter 3 suggests that having children affects women's economic opportunities in all countries in the region, although the effects are greatest in urban areas where informal support networks are the most limited.[8] A lack of child care can affect multiple dimensions of economic opportunities, including labor force participation, the type of job conducted, and gender wage gaps. For example, in Vietnam, a lack of child care has been found to be an important factor that pushes mothers from formal to informal employment (Heymann 2004). In China and Mongolia, reductions in the provision of subsidized child care led to a reduction in female labor force participation in urban areas (Chi and Li 2007; World Bank 2011a).

Policies to promote better access to affordable child care should target those women whose needs are the greatest and who have limited alternative access to child-care facilities. Community child-care centers, particularly those targeted at low-income neighborhoods, have been found to increase maternal employment in a number of countries across the world. Although evidence from the East Asia and Pacific region is limited, evidence from the Organisation for Economic Co-operation and Development (OECD) and the Latin America and Caribbean region suggests that the impact of increasing access to child care on female labor force participation and hours worked varies with, among other things, the availability of alternative caring arrangements and characteristics of families (Attanasio et al. 2004; Blau and Currie 2006; Lefebvre, Merrigan, and Verstraete 2009; Paes de Barros et al. 2010).

Targeting child care toward the needs of mothers is likely to increase its uptake. Programs that have identified and targeted groups with particular needs can increase women's labor force participation while addressing persistent inequalities. In Vietnam, the Community-Based Early Childhood Care and Development project aims to improve child outcomes, with a focus on ethnic minorities and migrant workers (ADB 2010). Migrant workers often have lower access to familial support

and informal child-care arrangements, and hence have a substantial need for affordable child-care assistance. In Brazil, publicly provided child care lacked flexibility and had limited facility hours, which limited the impact of child care on women's earnings (Deutsch 1998).

Public policy may also need to tackle negative perceptions about the use of child care. Location-specific social norms may initially limit women's use of child-care services. Comparative analysis of women's labor force participation in Japan and Taiwan, China, noted that disapproval of mothers that rely on alternative child-care arrangements is frequent in Japan, with women being less reluctant to leave young children in the care of others (particularly relatives) in Taiwan, China. In Korea, a study suggested that three-quarters of men and women believe that a preschool-age child will suffer if his or her mother works (Yun-Suk and Ki-Soo 2005).

Instituting parental leave policies

Parental leave policies may improve gender parity in economic opportunities by enabling and promoting a more equitable division of child-rearing responsibilities between men and women and by allowing women to have the same opportunities as men for advancing their careers. The success of parental leave policies in achieving these goals depends on their design. Poorly designed or incomplete parental care policies can make hiring women, relative to men, more expensive for employers, leading to discrimination or reduced job opportunities for women, while exacerbating or emphasizing gender divisions of labor in the household. The factors that affect the success of parental leave include who pays for the leave—the employer, the employee, or the government; whether a country has provisions for both paternity leave and maternity leave, or covers only maternity leave; and what fraction of wages is received by beneficiaries.

Parental benefits in the region vary considerably by the number of days given, the percentage of leave that is paid, who pays for the leave, and whether paternity leave is provided for. Maternity leave is the most commonly found component of work-and-family policies internationally and in the region. A study of 13 East Asian and Pacific countries found that provisions for maternity leave have been instituted in all the countries examined in the region (World Bank 2010b, 2011g). In the majority of countries studied, maternity leave is paid for by the employer; only in Mongolia, the Philippines, and Vietnam does the government fund maternity leave. Where paternity leave provisions are not in place, policies that grant only employer-paid maternity leave likely will reduce employers' incentives to hire female workers because of the additional costs of hiring women, particularly among younger age groups.

Within the region, only Cambodia, Indonesia, and the Philippines currently have policies for paternity leave. Besides introducing differential hiring costs by gender, the provision of maternity leave policies without similar paternity leave policies will likely reinforce gender differences in child-rearing responsibilities, because the asymmetries in the ability to take time off from work reinforces gender divisions of time. However, even in the presence of gender-neutral parental leave, evidence from the OECD suggests that progress has been slow in encouraging fathers to take any leave, let alone equal leave (Gornick and Hegewisch 2010).[9] The majority of evidence suggests, however, that providing paternity leave alone is not sufficient to changing the current gender division of child-rearing responsibilities within households; rather, parental leave policy needs to be combined with other approaches to breaking down gender norms regarding household caregiving roles.

Increasing women's access to resources

Access to land and other productive resources in agriculture and other entrepreneurial sectors is important in all contexts. It becomes an even more salient and binding constraint in many East Asian and Pacific countries where women, despite accumulating sufficient human capital, still face barriers in access to productive assets.

Promoting equal access to land
Policies aimed at promoting equal access to land for women and men require careful thought because complex legal, social, and economic factors are at play. Although economic growth and income can help women acquire more land in some contexts, improving gender equality in access to land and land ownership does not automatically result from single interventions targeting economic factors. Access to land requires attention to formal property rights, customary systems, and the informal norms and practices that structure access to resources at household, community, and land agency levels. Access to land affects women's economic opportunities, since land is an important productive asset and form of collateral. As with education, access to land assets and income generated from land contribute to women's voice and influence. To that extent, policies promoting equal access to land also affect agency.

Leveling the legal playing field with respect to land ownership is essential. Gender-progressive legal review and reform should take place in all areas that legally infringe on land rights, including family code, inheritance rules, and civil legislation regarding women's rights to land, both within a marriage or consensual union and in cases of family dissolution or divorce (Giovarelli 2006). Explicit language ensuring equal property rights in the law helps avoid interpretation unfavorable to women. For example, Bolivia's 1996 land reform law specifies equity criteria in all land transactions, independent of the civil status of the party involved. The reference to civil status is important because it does not require that a woman be head of household or married to be eligible for land rights (World Bank 2010a).

In addition, countries are likely to benefit from adopting legal changes that actively promote better gender equality in access to land. For example, in 2003, Vietnam passed the Land Law, which requests certification of land under the name of both spouses if plots are used by both spouses. Qualitative analysis of impacts in Vietnam suggests that joint titling improves procedures and opportunities for women to access loans; empowers women in case of disputes, given the security of land use rights; and leads to higher mutual decision making (World Bank 2008). Using national survey data from 2004 to 2006, Trung (2008) found that joint land certification increased the working time and decreased the domestic labor burden for women belonging to the Kinh ethnic group. The study also found that joint land certification increased education expenses for daughters relative to sons in rural areas.

Experience from projects suggests several guiding principles to ensure that on-the-ground implementation of land titling and land administration does not put women at a disadvantage. These guidelines include gender- and socially inclusive land titling policy, procedures, and service delivery, and provision of complementary inputs to new landholders, particularly women (see World Bank 2010a for details). Important capacity building to increase women's legal literacy in relation to land may include sufficient campaigning and outreach to women and men regarding upcoming titling programs, the benefits and risks of land title, and procedures for redress (Harrington 2008). The adoption of these project guidelines in, for example, the Lao Land Titling Project (LTP1 and LTP2) yielded impressive results in improving women's access to land. The project had a special focus on women's awareness of land certificates and titling to women, including the involvement of the Lao Women's Union and a significant number of female staff in land titling agencies. The projects resulted in 38 percent of titles going to women, with a further 29 percent issued jointly to both spouses (Bell 2011). Project procedures should also ensure that women are represented in local titling and adjudication and community mapping processes, for example, through the use of female paralegals and parasurveyors

(World Bank 2010a). Training women in the use of new ICT resources can give them access to tools such as handheld global positioning system devices or mobile land registries for help in developing land cadastres and registering land transactions. The First Kyrgyz Land and Real Estate Registration Project put in place mobile registries for villages located 25 kilometers or more from the local registry office, and the registries traveled to the village either weekly or at the request of the village chief (Harrington 2008). Sufficient training of land agency and titling agency staff in women's and ethnic or other minorities' rights to land, as well as in local norms and practices, can help remove barriers faced by these groups in accessing titling services (World Bank 2010a).

Apart from securing land titles for women, policy makers planning for formal land allocation and designing interventions must understand and consider the gender implications of customary land tenure and use-rights regimes operating in specific communities.[10] Land titling projects can support the issuance of titles to all rightful heirs to secure women's legal rights, which women can then transfer, gift, trade, or sell as desired (World Bank 2010a). Education regarding the benefits of land ownership is important for potential beneficiaries. Other target audiences for this kind of education include formal organizations—such as state institutions, labor organizations, legal aid organizations, women's organizations, and donors—and informal, customary institutions—such as councils of elders and neighborhood committees (World Bank 2010a).

Public policies should make specific efforts to support women's access to justice in formal and informal systems (including for individual and collective land rights), and to improve women's position in customary, as well as collective, land negotiations. The World Bank's Justice for the Poor program (J4P) aims to improve the delivery of justice services and to support inclusive and fair negotiation processes over customary land dealings, in which women usually are excluded from the decision-making process and receive fewer benefits (World Bank 2011c). In Vanuatu, for example, land leasing for tourism, agriculture, and extractive industries has excluded women during formal lease registration and overlooked women's customary access to garden plots or other productive uses. The J4P program used consultations and outreach, including through community drama, to bring the community together to discuss and debate sensitive land issues (Stefanova, Nixon, and Porter 2010). Evaluations of the impacts of such approaches are under way.

Increasing women's access to other productive resources in agriculture

Improving female farmers' access to agricultural extension and other services that promote rural livelihoods will improve their income. In rural areas, women often play a leading role in agricultural activities when

large numbers of men leave in search of employment in cities. Acquiring knowledge and skills in agriculture and other rural livelihoods can be extremely challenging, particularly when extension services are oriented around traditionally male activities or when training occurs outside of rural villages, limiting female participation. Recent projects have sought to address these issues (box 6.3).

Consultations with female farmers can help to identify gaps in existing service delivery. For example, the Integrated Agriculture Training Project in Papua New Guinea focused especially on gender concerns. The project expanded the number of female agricultural extension agents, incorporated women's needs in the design of the training, and included gender monitoring in the program (box 6.3).[11]

BOX 6.3 Leveling the playing field in access to resources: Lessons from Papua New Guinea and Mongolia

In Papua New Guinea, men and women have had differentiated access to agricultural extension services for multiple reasons. First, extension and training activities have traditionally focused on the formal commercial agriculture sector, dominated by men, rather than the subsistence and informal agricultural sectors that women are concentrated in (Cahn and Liu 2008). Second, contact between unrelated men and women is not culturally acceptable in Papua New Guinea, further reducing women's access to extension services, which are predominantly provided by men. Third, traveling to training courses can be difficult for women because of time limitations, financial constraints, and fears about safety, particularly for women in very remote villages, where limited transport options are available.

The Integrated Agriculture Training Program (IATP) provided targeted training and information services to men and women in rural areas. The program included a focus on women's needs. Participatory workshops aimed to ensure that women's voices were heard, and that the training needs highlighted by women were included in the selection of topics for training modules. The training content included

livelihood, credit and savings, accounting, and management of poultry and commercial vegetables, which particularly reflected women's interests (Cahn and Liu 2008). The program trained both male and female trainers to deliver gender-sensitive modules— by 2005, 30 percent of all trainers were female. This allowed course participants to be split into different groups when discussing topics on which men and women had different perspectives or when discussing more sensitive topics, and it also represented women as being capable of taking important roles in agriculture.

In Mongolia, a component of the Gansu and Inner Mongolia Poverty Reduction Project provided improved agricultural and livestock technology packages and upgraded agricultural and livestock support services (World Bank 2007b). Women were consulted during the development of different farm models to ensure that the gender divisions of labor were taken into account in the choice and development of activities. Approximately half of the total project beneficiaries were women, who benefited from women-specific project training and production activities such as cropping and livestock raising.

ICTs have been successfully used to increase women's access to agricultural technologies. In Indonesia, Nokia Life Tools aims to bridge information gaps for farmers through a text message–based interface. The program works with local agricultural boards to distribute crop and market information and with meteorological departments to get climate and weather information to farmers. The Kenya Farmer Helpline, introduced in 2009 by Kencell, Kenya's largest call center, provides free advice to small-scale farmers. Call center operators provide expert advice in local languages on agricultural practices, from controlling pests, to raising livestock and poultry, to marketing products. Nearly half of the 30,000 farmers reached by the program are women, a substantially higher fraction than are reached through standard agricultural extension services (World Bank 2011f).

Promoting equal access to finance and skills for entrepreneurs

Distinguishing between systemic constraints to agricultural or entrepreneurial development and gender-specific constraints is crucial for identifying policies that can reduce gender disparities in economic opportunities. In chapter 3, the evidence from the formal sector suggested that broader constraints to business development, such as cumbersome registration procedures, affect both female- and male-led enterprises. Evidence from the informal sector is more limited but, again, suggests that gender differences in self-reported constraints are not necessarily as large as the systemic constraints facing enterprises. Because both the types of constraints facing enterprises and the gender differences in constraints vary substantially across countries, policy makers should focus on gaining a better understanding of the environment facing male- and female-run firms in both formal and informal sectors.[12]

Where gender-specific constraints to enterprise development are found, promoting gender equality in the control of productive assets—including land, financial capital, and information and technology—can help to enhance the productivity of female-led enterprises. Microfinance is the most pervasive approach to increasing access to finance among entrepreneurs, and it has been used extensively throughout the region to target female entrepreneurs.

Efforts to improve access to finance among female entrepreneurs should consider how to move their financial portfolios away from informal sources of credit and toward more formal credit institutions. Because women in many parts of the region have more limited assets that could serve as collateral and also often have more limited credit histories than men, they are more likely to be restricted in their sources of borrowing. For example, in the Pacific, family law and inheritance law have been identified as important constraints affecting women's ability to acquire and harness assets that may be used as collateral (Hedditch and Manuel 2010). Customary laws, in particular, introduce discriminatory practices against women with respect to access to land and property rights.

Beyond addressing financial constraints, governments could implement training programs to improve women's business skills and thereby address gender differences in entrepreneurial capital. For example, the results of a survey of 500 female enterprise owners in Vietnam suggest that female entrepreneurs feel the need to improve their business skills through training and education (IFC and MPDF 2006).

Although greater numbers of women are acquiring higher levels of education and business training, converting the skills acquired through these programs into productive outcomes has remained challenging. Therefore, training programs should address the appropriate skill gaps in target populations. For example, evidence shows that improving life skills had a greater impact on economic outcomes than improving vocational training skills in the Dominican Republic (Ibarrarán, Ripani, and Villa-Lora 2011). The evidence also suggests that the effects of training may vary across existing and aspiring entrepreneurs, and among existing entrepreneurs training may be needed in combination with other assistance to improve outcomes. For example, an impact evaluation of business training to female entrepreneurs in Sri Lanka suggests that business training alone may be insufficient to raise outcomes among existing female entrepreneurs, but may be more effective for new entrepreneurs (de Mel, McKenzie, and Woodruff, 2012). The study finds that combining business training with grants has a large impact on short-run business profitability among existing entrepreneurs, but no longer-term impact. Among new entrants, business training was found to increase profitability and improve business practices.

Programs that improve women's access to productive resources have been shown to increase women's self-confidence and empower them to be more actively engaged in society. In Cambodia, the Women's Empowerment Program provides poor women with literacy, leadership, and financial training. Apart from gains in endowments and economic opportunities, the program reports a positive impact on increasing women's voice and influence. For example, several participating women later joined leadership posts at the local level (Rosenbloom 2004). Similarly, an impact assessment of a women's entrepreneurship training program in Aceh, Indonesia, showed that business planning and management training helped to promote greater confidence among women trainees, create or strengthen new social networks, and identify ways to improve the business environment for women (ILO 2008). Other examples of programs that increase women's agency include community-driven development programs and self-help groups. In Indonesia, a community-driven development program named the Kecamantan Development Program contributed to developing the leadership skills of local women through its activities. The program helped develop the capacity of beneficiaries to plan, prioritize, and manage local community investments, as well as developed various livelihood skills to make them self-reliant (World Bank 2009). In India, self-help groups have increased women's assertiveness by offering support mechanisms and resources that develop women's confidence (Suguna 2006).

Establishing an enabling environment for gender equality in the workplace

Even though the formal sector in many East Asian and Pacific countries is small, governments have a role in strengthening the formal sector work environment to promote gender equality. Policy instruments available to the government include labor regulations, active labor market policies, and even affirmative action policies. Governments can work with the private sector using promising approaches that include improving business procedures or establishing firm certification for gender-friendly work environments.

Promoting gender-equitable labor institutions and practices

Labor regulations that result in asymmetries in the employability and costs of hiring male and female workers can be found throughout the region. In countries where gaps in discrimination legislation exist, closing those gaps should be made a priority. For example, although the Tongan constitution guarantees equality, it does not contain a specific prohibition on discrimination on the basis of gender (Hedditch and Manuel 2010). The lack of antidiscrimination legislation raises the possibility for women to be treated unfairly by employers.

Protective legislation, though well intentioned, can restrict the employability of women and should be tackled more directly. For example, restrictions on night work for women in the Philippines can reduce the attractiveness of hiring female employees. In addition, several countries in the region restrict the industries that women can work in as a means of protecting them from hazardous conditions. These protective measures can reduce the labor market opportunities of women.

Governments should give priority to tackling the motivating factors behind protective legislation more directly. Employment conditions and protections in industries continuously evolve, thus giving policy makers the opportunity to regularly revisit limitations on women's work in restricted industries in light of current safety standards and practices, while increasing protections for both male and female employees. Where the original concerns motivating these policies continue to be valid, for example, where transport safety issues restrict women's movement, governments can undertake measures such as providing safe and reliable transport infrastructure to ensure that women are able to overcome these concerns.

Gaps in the coverage of labor protections imply that certain groups may be at risk of exploitation. Addressing the lack of labor protections covering these groups can help to reduce these infringements. Female migrants in the region tend to be segregated into informal occupations such as domestic work, where they have few labor protections and are more at risk for becoming victims of exploitation. Improving the legal and social protections of female migrants working abroad will better protect those women in isolating and informal positions. Approaches to improve the well-being of migrants are discussed in greater detail later in this chapter.

Establishing active labor market policies

Active labor market policies can be used to overcome gender-based differences in access to job opportunities. For example, wage subsidies may allow individuals to signal their abilities to future employers and reduce employers' costs to hire female workers whom they may not otherwise have considered, albeit temporarily. This incentive for hiring gives employers the opportunity to reduce stereotypes by directly observing their female workers' skills, and also gives women valuable labor market experience. Promoting female participation in all sectors and in all jobs, at management level and below, is likely to increase information on the competencies of women as leaders within an organization, and may also be a way to establish environments that are more attuned to the issues faced by women. Skills training programs can be used to reduce occupational segregation by encouraging women and men to move into professions outside of gender stereotypes, particularly when paired

with apprenticeship opportunities. Evidence within the East Asia and Pacific region on these programs remains limited; however, evidence from other regions suggests some promising lessons (box 6.4).

Efforts to promote female participation and gender equality in the private sector—by training women in leadership and giving them a space to be activists for gender equality—can also play an important role in strengthening women's voice. The Adolescent Girls Initiative (AGI) in countries worldwide is a public partnership with a core focus on leadership that helps young women transition from school to productive employment and active economic participation (World Bank 2011d). In Lao PDR, the AGI has led to the creation of career counseling offices where young women can get information on entering the labor force and long-term career planning.

Using affirmative action policies

Affirmative action policies, both voluntary and mandatory, can be used as a mechanism to increase the representation of women at all levels of the hierarchy, from entry-level to managerial positions. Evidence on affirmative action from developed countries suggests that voluntary programs may have limited effects on female employment and that, to be effective, programs need to be mandatory and have a credible enforcement mechanism (Holzer and Neumark 2000, World Bank 2011f). In Korea, affirmative action policies were introduced to expand women's employment and to reduce discriminatory practices in 2006. They were initially implemented for public enterprises and private firms with more than a thousand employees, and was extended after a two year grace period to smaller private firms (Jung and Sung 2012). Firms who failed to meet the specified female employment or female manager ratio criteria were asked to submit an implementation plan, with the goal of raising female representation within the firm. An empirical evaluation of the program has found that it had no significant effect on female participation or on firm performance. The limited impact on female managerial or

BOX 6.4 **The impact of active labor market policies on female employment: Evidence from across the world**

The evidence on the effectiveness of training programs in developing countries is mixed. However, the effects of training programs in developing countries are greater than those in developed countries, where benefits have been modest overall, albeit higher for female workers (Betcherman, Olivas, and Dar 2004; Card et al. 2011). For example, Jóvenes en Acción, a subsidized vocational training program, was found to have had a positive effect on paid employment and earnings, where the effects are principally driven by women (Attanasio, Kugler, and Meghir 2011). In contrast, the first phase of the Juventud y Empleo program in the Dominican Republic, which aimed to increase the employment probability of disadvantaged youth through basic skills training and vocational training, was found to have no effect on employment rates and hours of work (Card et al. 2011).

Broadening skills programs to include softer skills may have positive labor market effects. The Juventud y Empleo program in the Dominican Republic was modified in the second phase to include new life skills. A randomized evaluation of the modified program suggests that the life skills component of the program plays a central role in improving the employment outcomes of young women, who experienced a higher likelihood of having a job, higher wages, and higher job satisfaction approximately 12 to 18 months after the program. The program also had an important effect on reducing pregnancy in young women.

Programs that encourage both men and women to think outside of gender stereotypes are likely to increase the efficiency of allocating talent toward jobs. As discussed earlier in this chapter, just as policies in the education system can help change the traditional gender patterns of what to study and, consequently, what job to do, policies to address occupational segmentation directly in the labor market and in the business environment can also change outcomes for women. For example, a vocational and technical training voucher program in Kenya noted that men almost exclusively choose male-dominated courses while females almost exclusively choose female-dominated courses such as hairdressing. To address misconceptions about the returns to

vocational training, the program provided its female beneficiaries with information highlighting the large discrepancy between expected earnings from graduates in traditionally male-dominated trades (such as electricians) versus traditionally female-dominated trades (such as seamstresses). In addition, the intervention also used more subjective methods, including a presentation showing successful female car mechanics in Kenya, to encourage women to select training for more lucrative male-dominated trades. An impact evaluation of the program suggests that women given this information were almost 9 percentage points more likely to express interest in a male-dominated course (especially younger and more educated women) and 5 percentage points more likely to enroll in one (Hicks et al. 2011).

Reducing information barriers about female youth may also have a positive effect on labor market outcomes. In Jordan, young female graduates have a higher rate of unemployment than men, in part because of perceptions that women have lower levels of interpersonal and decision-making skills, that women are less productive than men, and that their commitment to the labor market is lower, potentially because of marriage and child rearing (Groh et al. 2012). The New Opportunities for Women pilot program puts forward two policies to overcome these perceived constraints: short-term wage subsidies, in the form of job vouchers equivalent to the minimum wage for six months, and employability skills training, which focuses on interpersonal and professional skills. The skills component focused on softer skills that had been identified as a constraint of young female workers, such as communication, team building, presentations, and business writing. An impact evaluation of the program finds that the soft-skills training had no impact on employment, but that the job vouchers had statistically significant and persistent positive effects on employment of graduates outside the capital, where it almost doubled the employment rates of graduates. In the capital, the wage subsidy was found to have a substantial employment effect in the short-term, but was unable to raise long-term employment prospects for unemployment female youth.

female employee participation is partly attributed to the near voluntary nature of the program, which lacked an effective penalty for non-compliance and had a weak incentive system (Jung and Sung 2012).

In many countries, the public sector may take the lead in promoting gender-based quotas; this support can act as a signal to the private sector of women's productivity in underrepresented jobs. For example, in 2004 Malaysia introduced a public sector gender quota of 30 percent female representation across all decision-making levels, including positions such as department heads or secretary general (ASEAN 2008; *Washington Post* 2011). In 2006, 25 percent of women were holding top positions in the public sector; by 2010, this share had risen to 32 percent. The Malaysian government aims for 30 percent of key corporate board positions to be held by women by 2016 (*Washington Post* 2011).

Evidence suggests that affirmative action policies, particularly at a board level, involve trade-offs. Board diversity has been found to increase the attendance of both male and female board members in the United States, and female board members are more likely to participate as tougher monitors than men on corporate boards (Adams and Ferreira 2009). However, tougher monitoring does not necessarily translate into higher firm value. In the United States, board diversity has had a positive effect on firm value among firms with weaker governance, where greater monitoring may be beneficial, but has had a negative effect among firms with stronger governance.

The design of affirmative action can affect performance, particularly in circumstances where gender differences in the characteristics of the existing workforce imply that individuals with the appropriate skill set may not be available (Pande and Ford 2011). For example, in Norway, corporate board quotas of 40 percent were introduced in 2003 with a short implementation horizon. The Norwegian gender quota was found to decrease firm values, but this effect vanishes once board characteristics are accounted for, suggesting that

the reduction in firm value is attributable to the comparatively young age and low job experience of women on Norwegian boards (Pande and Ford 2011). This finding suggests that policy makers may also want to keep the characteristics of the existing workforce in mind when designing affirmative action policies.

Working with the private sector to foster gender equality

Business formalization procedures that are gender blind may result in gender-differentiated outcomes. In these contexts, procedures should be modified to increase women's ability to operate their businesses on a formal basis. A number of factors, such as the need to juggle household and market roles, cultural restrictions on travel, and lower education levels, may imply that women are less able than men to maneuver complex formalization procedures and may therefore restrict their ability to register their businesses (Simavi, Manuel, and Blackden 2010). Simplifying and reducing the cost of business registration procedures, as well as introducing flexibility in application procedures to minimize the effects of time constraints, are likely to increase the ability of both male and female entrepreneurs to comply with business regulations and registration procedures, but such interventions are likely to have a greater impact on female entrepreneurs.

Encouraging companies to promote greater transparency in recruitment and promotion procedures can level the playing field for women and can help to achieve greater female representation, particularly at a managerial level. Proactive and supportive organizational policies can also help women acquire relevant competencies and move to managerial levels. For example, in 1996, Mongolia Telecom adopted a human resource development plan to reduce the crowding of women at lower levels. The two-pronged strategy included measures to help women cope better with their dual responsibilities at home and at work, and measures to enable the professional development and career growth of women. To help ease the work-family conflict on female employees, the company gave assistance to single mothers and financial aid for childbirth and education of children. As a result of these proactive steps, women now constitute over 20 percent of senior managers, compared with just 9 percent when the plan was introduced.

Firm certification has been used to promote gender equality in companies and organizations and to encourage women to reach their labor market potential. One firm certification tool, the Gender Equity Model (GEM), has proved to be successful in promoting equal opportunities for men and women and in overcoming cultural barriers in business practices. This public-private partnership, which was designed and implemented in Mexico, has been replicated in Argentina, Chile, Colombia, the Dominican Republic, and the Arab Republic of Egypt (World Bank 2011b). Firm certification validates and recognizes gender equity actions in private firms. An impartial and independent agency assesses firms' compliance in four areas: recruitment, career advancement, training, and sexual harassment. Findings from the GEM in Mexico show that participating firms have eliminated pregnancy discrimination from recruitment practices, and 90 percent of participating organizations report that workers' performance and productivity have increased (World Bank 2011b). Furthermore, organizations report that promotion of women to managerial positions has increased, although women at the top tier are still rare (World Bank 2011b). Certified firms also are more likely to have processes in place to deal effectively with harassment cases, although workers in certified firms also are more likely to have been victims of some form of harassment. This may also signal that they are more aware that some behaviors are forms of harassment that are inappropriate in the workplace (World Bank 2010c).

Identifying policy priorities to foster equal economic opportunity

This discussion lays out policy recommendations to address the major barriers constraining women and men in the region from fulfilling their potential in the economic sphere. Although all the policies presented here are important—reducing the trade-offs between women's household and market roles, addressing occupational segmentation, increasing women's access to resources and markets, and establishing an enabling environment for gender equality in the workplace—each country needs to consider its own appropriate policy priorities and policy mix. The guiding principles below recommend a way to prioritize policy actions, depending on the country's economic structure and, therefore, the types of major constraints faced by women.

- In predominantly agricultural economies, the suggested policy priorities are to focus on constraints that affect women in rural areas, such as (a) reducing gender disparities in access to productive resources in agriculture and (b) improving infrastructural services to increase women's access to markets.
- In economies with a heavier focus on manufacturing and services, policies should focus on reducing the constraints faced by women in urban areas and in the formal sector, such as the promotion of women-friendly work practices or affordable child care.
- In all economies, policies need to focus on reducing gender disparities in the constraints faced by enterprises, particularly those in the informal sector. A key priority is to identify constraints that are systemic and those that are gender-specific. Continuing to promote women's access to land will be important in all contexts because land serves as a productive resource, as a form of collateral, and as a status-enhancing asset. In addition, identifying and closing discriminatory labor market regulations and practices to make formal institutions more even handed should be a policy priority at all stages of structural transformation.

Taking measures to strengthen women's voice and influence

The evidence presented in chapter 4 suggests that all countries in the East Asia and Pacific region can create opportunities to improve women's agency. In the Pacific Islands, policy priorities include increasing female voice and participation in the public sphere, and addressing pervasive gender-based violence. East Asia must address the increased vulnerabilities specific groups may face—including those who have been trafficked and those who are not protected by the law as a result of plural legal environments. Strengthening women's agency on multiple fronts is likely to have rewards, since increasing women's voice in one domain of agency is likely to have positive repercussions on others.

Measures to increase women's endowments and economic opportunity, such as those described previously, can strengthen their voice within the household and in society. Educated women in good health, with assets and income, are better able to act on their preferences and influence outcomes that affect themselves and others in society. In addition, several other policy approaches can directly promote women's agency and reduce gender-based violence. This section takes a more detailed look at the following areas of policy identified in chapter 4: (a) supporting initiatives to transform gender norms and practices, (b) strengthening the legal and institutional environment, (c) increasing women's access to justice, (d) enabling women's participation in politics and policy making, (e) pursuing a multipronged approach to reducing gender-based violence, and (f) creating space for women's collective agency.

Supporting initiatives to transform gender norms and practices

Socially constructed gender norms that limit women's ability to act and make choices for themselves are pervasive across the East Asia and Pacific region and affect women's lives through diverse channels—from influencing what they study and the occupations they choose to the societal acceptance of domestic violence and the negative perceptions of women leaders. Although socially constructed norms are difficult to transform directly, evidence from around the world suggests that they can be influenced in several ways, including public information campaigns and the promotion of role models. Policy makers in many East Asia and Pacific countries have an opportunity to promote the evolution of social norms where the societal transformations underpinning rapid urbanization and migration are likely to expose men and women to different models of gender relations.

Approaches that harness the social and structural change occurring in many countries in the region are likely to have a greater influence in changing social norms. Returning migrants are a source of new information and can be influential in changing practices and social gender norms. Returning rural migrants are often better placed than rural residents to question practices and norms that constrain people from exercising their voice and influence. Countries wishing to make positive change in rural areas, where gender stereotypes still predominate, can engage the help of returning migrants. For instance, women from Jiangxi province in China who have experienced migration to the city are more aware of their rights as people and are more knowledgeable about social and health practices, including access to services and modern health care. In Cambodia and China, a number of civil society organizations call upon the experiences and lessons of returning migrants to influence women's decision making on reproductive matters in rural areas.

Providing a forum for successful women to inspire other women and girls can help the evolution of gender perceptions. Women in leadership roles can serve as effective role models for other women and girls. Female teachers are among the first professional women that show girls that being active outside the home is socially acceptable. These women serve as agents of change in the communities they work in by educating and socializing children beyond gender stereo-types (Kirk 2006). In Papua New Guinea, for example, pastors are also involved in settling disputes between husbands and wives. They are viewed as role models for youth, while their wives, who typically engage in leadership positions, are viewed as role models for women (World Bank 2011c). Exposure to positive female role models from an early age can help break the cycle of gender inequalities across generations.

Policy makers can also support and encourage mass media outlets such as television, radio, and virtual outlets to promote positive messaging that will help change gender norms. Mass media outlets are used in countries like the Philippines and Vietnam to regularly disseminate messages raising awareness of women's contribution in society. These countries have adopted a more extreme approach of requiring mass media outlets to abide by gender equality principles. In the case of the Philippines, the requirement states that the media must regularly disseminate messages that raise awareness of women's contribution in society and avoid portraying negative female stereotypes. In Vietnam, the government, with the support of donors, is undertaking an information, education, and communication campaign to help curb gender-based violence by redefining male identity with respect to domestic violence. The message is: "Real men don't hit women."

Increasing access to information through new information technology can improve women's agency by increasing their knowledge base and exposing them to broader experiences and social practices. In Malaysia, women used information technology to create self-help cybercommunities to improve their networking opportunities in an environment that was more suitable to their lives and social gender norms. The Malaysian e-Homemakers project is one example of these self-help communities; through the e-Homemakers, women are able to share information on how to start and run a business, and are able to share their experiences in a safe space. In Japan, the government promotes self-help and external help through cybercommunity initiatives;

the information provided to women through technology ranges from dealing with domestic violence to promoting women entrepreneurs in the agricultural and fishing industries (Farrell and Wachholz 2003).

Strengthening the legal and institutional environment

For a country to achieve gender equality in voice and influence, government and civil society must make concerted efforts across multiple fronts. A key element of this approach includes the institution and enforcement of legislation to create an enabling environment for equality of voice and influence regardless of one's gender. Countries should accede to and ratify international conventions that promote gender equality. Although most countries in the East Asia and Pacific region have acceded to and ratified the Convention on the Elimination of all Forms of Discrimination Against Women (CEDAW), Palau and Tonga have not (UNIFEM 2008).

Another important element is to ensure that domestic legislation and the institutions of implementation and enforcement are aligned with countries' commitments. For instance, despite the fact that CEDAW's international mandates are compulsory, many countries have laws that continue to discriminate against women by limiting their human rights. Some countries that have introduced legislation to reflect CEDAW mandates have made the language gender-neutral or exempted some types of violent acts, thus making laws inconsistent with international standards and leaving them open to broad interpretation. For example, legislation introduced in Cambodia and Thailand to address domestic violence contains subjective language that must be interpreted by the courts. In Thailand, a person must show that the perpetrator demonstrated "unethical dominance"; in Cambodia, a man can discipline his wife as long as it is done with compassion.

Another critical component of a multipronged strategy involves strengthening enforcement institutions to create an environment that enables both men and women to have voice and influence in the public sphere. Enforcement can be particularly challenging in plural legal environments, as seen in several East Asian and Pacific countries. Programs that increase women's knowledge of the law with respect to violence and human rights and the capacity of justice services to address women's issues contribute to greater safety and security among women in society. The Cambodian Women's Crisis Center (CWCC) began a community program that increases the society's awareness of violence against women and the rights of women, especially of the law on domestic violence and protection that was implemented in 2005. The program promotes initiatives to enforce the law by providing information and training to local authorities and developing community networks. The program has sponsored 63 community education sessions, serving a total of 1,638 participants; 6 training sessions for the police, which trained 150 officers; and 300 legal advice cases. The evaluation of the project shows that the program's targets were met or exceeded. Participants believed that, as a result of the program, the severity and rate of violence were reduced as a result of the education provided on the law, human rights, and the rights of women, and because information on the expected roles of men and women in their marriage were made clear (Weaner 2008).

Increasing women's access to justice

Programs that increase women's knowledge of the law help them to access the formal justice system and encourage them to exercise agency through formal mechanisms. In Indonesia, a civil society organization supporting women heads of household (PEKKA) helped shed light on the need to provide women with better access to justice services. Implemented by PEKKA and building on existing work by civil society, the Women's Legal Empowerment program encouraged the poor and marginalized (many of them women-headed households) to know and demand fulfillment of their rights while it worked on increasing the capacity of legal and justice institutions to respond. At the

village level, the program was implemented by trained paralegals who educated people on the law, in particular, family law and laws on gender-based violence, and provided assistance and advocacy. The result of this effort has contributed to the opening up of spaces for women to voice their rights—not only within the home, but also in local communities—and to demand better justice services from their government. In the long term, the program is expected to improve women's livelihoods, and that of their families, through their increased agency (World Bank 2011c).

Programs can make the judicial system more accessible to women in a variety of ways. Technology can help women access the justice system. For example, mobile courts in rural areas of China and Indonesia provide a solution to the problem of accessibility and security for women who wish to exercise their rights in the legal system but lack transportation to the courts. Courts can also be viewed as financially inaccessible and, in Indonesia, the waiving of court fees for poor and marginalized groups has increased the ability of women to bring their cases to court (World Bank 2011c). The justice system can also adapt to better address the specific needs of women. Countries can institute gender-sensitive training for officials in the system, as well as increase the representation of women within all institutions charged with formulating, implementing, and enforcing the laws. For instance, in Papua New Guinea, female local magistrates have helped raise awareness among their male counterparts of the need to adopt gender-sensitive approaches to the cases that come before them (World Bank 2011c). An example of a gender-sensitive consideration is that a female victim of gender-based violence may find it easier to approach a female police officer after an assault than a male police officer.

Enabling women's participation in politics and policy making

Implementation of political reservation systems has contributed to increasing women's participation in electoral politics in a number of countries. Gender-based political reservation systems seek to reduce obstacles in the recruitment process so that more women are attracted to participate in politics, thus adding strength to their collective voice. How these systems work varies across countries. Quotas—one form of reservation—take the form of constitutional changes to reserve a certain number of legislative as well as formal or informal political party posts for women. Such measures have been credited, at least in part, for the increased political and administrative participation of women seen across developing regions. For example, India has endorsed a constitutional amendment that women should constitute one-third of the heads of local governments. In East Asia and the Pacific, China, Indonesia, Korea, Papua New Guinea, Thailand, Vietnam, and, recently, the Cook Islands have all adopted formal or informal measures to promote political representation of women at the local level. For example, in Korea, the Grand National Party voluntarily supports quotas of 30 percent women candidates. At the national level, the law states that political parties must have a list with 50 percent women candidates, and at the local level, the party law of 2002 states that city council elections must aim to have gender parity (Quota Project 2010).

Countries can also consider using informal means of promoting women in politics through incentives instead of formal reservation systems. This route may be more effective in countries where formal reservation systems are likely to be contentious or elicit a political backlash. The argument in favor of formal reservation is that they open doors to women in politics and provide a relatively fast track for women to enter politics. In addition to enabling women to gain experience as office holders, such programs can help to change traditional views about women as political leaders. Evidence from India shows that public opinion about female political leaders improved with increased exposure (Beaman et al. 2009). Similarly, evidence from Italy finds that affirmative action in government has been effective in breaking down stereotypes against women (De Paola et al. 2009). However, active measures to promote women's political participation can prove

controversial, as was the case in Timor-Leste, where quotas were not widely accepted. As with affirmative action in the labor market, the public has concerns about the pipeline of qualified candidates, along with possible perceptions about the qualifications of women elected through a reservation system (independent of their actual qualifications).

Pursuing a multipronged approach to reducing gender-based violence

Reducing discrimination against women and gender-based violence requires action on a number of fronts: efforts to increase women's voice within the household; enactment and enforcement of appropriate legislation and strengthening of women's access to justice; provision of adequate support services for victims of violence; and use of the media to provide information on women's rights, to increase social awareness, and to shift social norms with respect to violence.

Countries should enact and enforce laws that eliminate discrimination and violence against women. By taking a strong stance on gender-based violence legislation and enforcement, countries can make positive strides against gender-based violence in a short time. Cambodia saw a significant decrease in the incidence of domestic violence between 2000 and 2005. This decrease was largely attributed to strong efforts by the country's Ministry of Women's Affairs, which introduced the domestic violence legislation draft in 2001. The legislation was adopted by the National Assembly four years later, in October 2005. The new law criminalized acts of domestic violence, provided for the protection of victims, and allowed neighbors or local organizations to intervene if they witnessed domestic violence. As a complement to the law, women's organizations and other nongovernmental organizations carried out information and dissemination campaigns to help disseminate information on people's basic rights and responsibilities under the law.

Providing adequate access to services and support for victims of violence can include a range of services, from police and judiciary to health and social services. In Malaysia, the government established integrated one-stop crisis centers in hospitals that provide easy access to medical care, various social services, and the opportunity to report the crime with specially trained police officers (World Bank 2011f).

Reducing human trafficking also requires a multipronged approach of prevention, protection, and prosecution. Preventing human trafficking can include education, employment, and other services.[13] People who are likely to be front-line responders in identifying victims of human trafficking should be trained to both identify and safely support victims of such abuse, as well as prepare victims to pass evidence on to investigators (UNIFEM 2005). Governments, with the help of relevant actors, should enact and enforce antitrafficking laws.[14] Enforcement includes effective legislation and policies that hold traffickers accountable for their crimes to help constrict the industry. Countries' protection policies include providing immediate protection for potential or identified trafficked victims. Policies should address immediate needs such as shelter, medical, psychological, and psychosocial care; food and clothing; and longer-term care and reintegration of the victim. For example, in Taiwan, an antitrafficking law that went into effect in June 2009 provides trafficking victims with continued residency and temporary work permits. These measures enhanced victim treatment and improved victim cooperation in trafficking prosecutions (U.S. Department of State 2010).

Creating space for women's collective agency

Partnerships with women's business associations can provide a space for women to interact, learn, and advocate for gender equality. Women's business organizations help their members be better equipped to benefit from commercial opening to international markets by providing them with access to a global network of women's business associations, information, and advocacy on their behalf.

Specifically, these organizations can offer access to contacts for sources of credit, training in international trade issues, and mentors, as well as access to the more basic skills of operations management and marketing. As advocates for women's businesses, these organizations help raise awareness among broader international stakeholders about the importance of incorporating women into policy planning (for example, trade pacts). Also, their activism can help to spur laws that provide a level playing field for businesses owned by both genders (Jalbert 2000).

In Cambodia, the Government Private Sector Forum (G-PSF) was established in 1999 to improve the business environment in the country and promote investment. The G-PSF provides a forum for public sector and private sector collaboration in the country. Even though the majority of small businesses in Cambodia are run by women, until recently, women had little voice in the dialogue on small business policy, because female-owned businesses were barely represented in the G-PSF. In this context, the International Finance Corporation, together with Cambodia's Ministry of Women's Affairs, has sponsored an initiative to increase women's participation in the private sector's dialogue with government on how best to enable the growth and productivity of both male- and female-run enterprises. Soon after the gender composition of the G-PSF was changed, new policies addressing the needs of female entrepreneurs were incorporated in the agenda of the forum. As a result of women's participation in the G-PSF, import tariffs and taxes on silk yarn were cut for a period of three years, helping more than 20,000 silk weavers in the country (Simavi 2011).

Fostering new opportunities, managing emerging risks

A new and important challenge for policy makers will be to help foster emerging opportunities and, in particular, to manage emerging risks associated with increases in economic integration, increased access migration, rapid urbanization, population aging, and increased access to ICTs. Many of the emerging opportunities can be fostered through efforts to promote greater gender equality in endowments, economic opportunity, and agency. For example, where gender gaps in access to ICTs are emerging or growing, ensuring that women are able to benefit fully from these new technologies may require active measures similar to those discussed earlier to ensure equality of access to other types of productive resources. Managing emerging risks, however, may require additional policy approaches, such as those discussed in this section.

Economic integration

Greater economic integration will bring with it increased exposure to employment shocks, which will have gender-differentiated impacts. Addressing the risks associated with greater economic integration will require designing social protection programs that adequately account for the different risks faced by female and male workers. Building on the lessons from recent economic crises, several developing countries, including some within the East Asia and Pacific region, have begun to recognize the gender dimensions of risks and shocks when designing their programs. In Indonesia, for example, a conditional cash transfer program, Keluarga Harapan, targets households with members whose outcomes are particularly vulnerable during times of crisis, such as pregnant and lactating women.

Labor market programs and policies enacted to mitigate crises and their effects are likely to have gender-differentiated impacts. In Cambodia, the construction, garment, and tourism sectors were directly hit by the 2008 global financial crisis. In response, the government promoted short-term vocational training courses for 40,500 laid-off garment factory workers, who were almost entirely women. Furthermore, gender-blind policies in response to crises may not be gender-neutral in the longer term. Indonesia's economic crisis in 1997–98 awakened the pro-labor pressures

that led to better enforcement of minimum wages (Narjoko and Hill 2007). Although the minimum wage policy helped narrow gender wage gaps, it decreased female employment, without affecting male employment (Suryahadi et al. 2003).

Migration and trafficking

Just as female migration throughout the region increases economic opportunities for women, it also creates new concerns about the welfare of female migrants. Protecting female migrants from exploitive situations, including from sex work, human trafficking, and inadequate labor protections, will require a gender-aware approach. Greater protection through improved laws, safety nets, and knowledge transfers will better address the vulnerabilities specific to women traveling abroad. Specific areas for action include improving the legal and social protections of female migrants, strengthening the monitoring and credibility of recruitment agencies, and developing and providing welfare and support services to assist female migrants. Well-developed welfare and support services can provide migrants with gender-specific assistance and safety nets. For example, the Philippines provides counseling, legal assistance, and liaison services to migrant workers in need of assistance. Resource centers with labor officers are established in countries with more than 20,000 Filipino migrant workers; they are open 24 hours a day, 365 days a year, and are staffed with a minimum of four officers. Offices in countries with many Filipina migrants are staffed with more female welfare officers and offer gender-specific assistance (Blokhus 2004).[15]

Governments in both labor-sending and labor-receiving countries should actively address the issue of human trafficking through prevention, protection, and prosecution. Female migrants are the most vulnerable group at risk of being trafficked for prostitution and forced labor. Policies aimed at prevention, protection, and prosecution, as discussed earlier in policies to minimize gender-based violence, can assist victims or those at risk of human trafficking.

Rapid urbanization

Growing urbanization in the East Asia and Pacific region has presented women with increased economic opportunities but also with challenges such as limited access to child care and higher security risks than men in urban areas. The types of policies needed in urban areas are similar to those discussed earlier to ensure equal access to economic opportunity and to reduce gender-based violence. Policy makers need to ensure that child care, education, infrastructure, transportation, and water and sanitation policies take into account women's specific social and cultural needs. Rigorous laws and policies to protect women in urban areas from the risk of violence and exploitation are also warranted.

Population aging

Rapid population aging in the region is likely to have important gender-differentiated effects because older women may increasingly find themselves living as widows. Along with risks from urbanization and the breakdown of extended family support networks, elderly women are likely to find themselves at increased economic risk, having accumulated relatively few assets and mostly lacking access to formal social security. Old-age income security programs can protect women from destitution. Joint annuities and survivor pensions have been argued to be effective means to compensate and secure women's incomes in old age (World Bank 2001).[16] Also, equalizing the retirement age for men and women may increase pension coverage among those working in the formal sector. Simulations from Latin America suggest that equalizing retirement ages between men and women can increase pension benefits for women and decrease the male-female pension gap (James, Edwards, and Wong 2003).[17] In addition, policy makers may consider ways to strengthen elder

care to ensure that younger women do not bear an undue burden from caregiving.

Filling knowledge gaps

Although global and regional evidence on gender and development have provided useful lessons, much remains to be understood empirically to help inform effective public action. Continuing to close data and analytical gaps will be important to better understand policy priorities, the effects of specific interventions, and the costs and benefits of different policy options. Additional gender-disaggregated data and empirical analysis, both on long-standing gender issues and on the gender implications of emerging trends in the region, will enable policy makers to promote both gender equality and more effective development. The following discussion highlights a few priority areas.

First, collecting additional gender-disaggregated data, in many domains, is a key first step to fill knowledge gaps. This action can take place during data collection for administrative purposes or as part of projects, so that countries can monitor progress in gender equality at the national and project levels as well as take corrective actions and perform impact evaluations. Many topics warrant data collection by gender. For example, although land and credit have been identified as important factors affecting productivity and female empowerment, few sources of information are available on individual-level land ownership and access to credit in the region. Similarly, countries need more data on gender differences in access to other productive inputs and services in the agricultural sector. Collecting gender-disaggregated data within the household, including information on individual time use and consumption, would enable researchers to assess the impact of policies such as electrification programs or the provision of child-care services on the well-being and economic outcomes of men and women. Other areas for which countries in the region could collect information include why and to what extent performance of male- and female-owned enterprises differ, and how male and female access to technologies varies. Although household surveys provide information on household-level access to the Internet and mobile phones, very limited evidence is available to assess whether males and females have differences in access to ICTs. Further evidence is much needed on the enforcement of antidiscrimination legislation in the workplace and beyond, to monitor whether legislation translates into actions.

Second, additional empirical analysis should help policy makers to understand whether and why specific policy interventions work. The East Asia and Pacific region is behind the curve on conducting impact evaluations, and even further behind on conducting impact evaluations with gender dimensions. For example, rigorous evaluations of curriculum reforms to reduce gender stereotyping as well as evaluations of promising approaches to promote balanced sex ratios at birth would guide countries in such efforts. More rigorous evidence on the impact of extension services—including providing knowledge about new crop varieties and agricultural technologies—for men and women would inform policy makers of the extent to which such services can improve farmers' productivity. Other areas in which impact analysis could contribute to better understanding of policy effectiveness include the following: the long-run impact of gender quotas on firm performance and the impact women have on corporate boards; the effects of youth unemployment programs on girls' and boys' labor market outcomes; gender-differentiated impacts of community-driven development and public works programs in the region; and the effects of various approaches to increasing agency, such as political quotas, role models, and approaches designed to promote communication between spouses to reduce gender-based violence.

Notes

1. For examples of previous and ongoing research, see http://go.worldbank.org/78EK1G87M0.
2. In the mid-1970s, both the United Kingdom and the United States adopted antidiscrimi-

nation legislation pertaining to education (Arnot, David, and Weiner 1999; Madigan 2009; Salisbury and Riddell 2000). Several states passed laws prohibiting the use of gender-stereotyped curricula, and companies producing textbooks developed guidelines to eliminate bias. Although the literature stresses that the review of teaching materials takes a central role in the process of promoting "equal opportunity education," the causal link between curriculum reforms and more gender-equitable education is not straightforward. In particular, curriculum reforms in the United Kingdom and the United States happened within the context of broader social change, in which greater numbers of women participating in the labor market challenged the notion of women's primary role being in the home.

3. For instance, the Finnish project TiNA included special day courses for girls ages 14–16 to provide information about nontraditional female occupations. The project also used female students as role models in the visits to schools (http://tina.tkk.fi/tina_2003/tina_2004_eng/). Another initiative that was launched in the late 1990s in Norway aimed at increasing the number of female students in computer science at the Norwegian University of Science and Technology. The initiative included an information campaign for girls at all high schools in the country. Girls who were interested in a career in computer science were invited to the university for the day, where they met students and professors and received information (http://www.ercim.eu/publication/Ercim_News/enw38/gjestland.html).

4. OERL Under-Represented Populations Projects (http://oerl.sri.com/reports/up/reportUP_es.html).

5. Evidence has shown that programs that raise the value of women's time in the workforce affect gender divisions of labor within the household and increase the amount of time women have available for market-oriented activities. In Ecuador, women's increased employment in the cut flower industry led to men's increased participation in housework (Newman 2002). In India, a project that increased the value of women's time by introducing new economic opportunities in agricultural procurement resulted in a decrease in their domestic workload, an increase in mobility, and enhanced decision-making powers within the household (World Bank 2007a).

6. Quantitative evidence from South Africa suggests that the expansion of electrification raised female employment by nearly 10 percentage points and increased female earnings, but it had no effect on male employment (Dinkelman 2008). Furthemore, the evidence suggests that electrification increased women's market activities by releasing female time from cooking through altered cooking technologies (away from wood-based toward electricity-based cooking).

7. Child-care centers can also play a key role in reducing inequalities in access to nutritional support and mental stimulation among disadvantaged groups, including ethnic minorities and young girls.

8. Although child care services are found throughout the region in a variety of forms—from publicly provided child-care facilities such as the Early Childhood Development Centers in Thailand to child-care centers in factories in the southeast industrial zones in Vietnam (UNICEF 2004; ADB 2010)—there is clearly a greater and unmet need for affordable child care in many countries.

9. For example, in Austria, France, and Germany, only 2 percent of men participate in leave compared to 90 percent of women (De Henau et al. 2007, cited in Gornick and Hegewisch 2010). Evidence has shown that men are more sensitive to levels of wage replacement than women, suggesting that policies that hope to encourage men's use of parental leave should consider higher wage replacement levels.

10. Individuals often "forum shop" in pressing their land cases, alternating between customary and formal systems, depending on which is expected to decide in their favor. For men, particular advantages can often be gained by using customary systems (Giovarelli 2006).

11. While acknowledging the overall positive effect of the program on women, Cahn and Liu (2008) highlighted several constraints that limited the effectiveness of the project's implementation with regard to gender parity. Except for the credit module, in which men and women participated equally, during 2002–05 female participation was below 20 percent in all other modules. This gap can be explained by the selection process of participants, which disadvantaged women. The selection of participants was made by ward councillors and district rural development officers, who had no gender training and discouraged female participation. Courses were not held in the home village and so involved travel and overnight stays, making women's participation particularly difficult. Finally, educational differences between men and women undermined women's motivation in the mixed learning groups of certain modules.

12. Self-reported constraints might not give the full picture, because firms are likely to report constraints only when they have tried to access a service. For example, firms that have not applied for a loan are unlikely to report the cost of finance as a constraint.

13. The Development and Education Programme for Daughters and Communities (DEPCD), a nongovernmental organization in Thailand, aims to prevent the trafficking of women and children into the sex industry and other exploitive labor situations. The organization targets at-risk youth and their families and educates them through seminars, research workshops, and awareness campaigns at community and local government levels. In addition to education, the organization offers employment alternatives through life skills and vocational skills training programs, in addition to sports activities for children (Smarasinghe and Burton 2007).

14. In the Philippines, the Visayan Foundation works in cooperation with the port authorities to set up institutions that provide protective services against trafficking in seaports of the archipelago. The foundation also promotes information in the seaports about trafficking and engages in local networking to encourage actions against trafficking.

15. To better ensure that women are migrating through legal channels, governments should strengthen the monitoring and credibility of recruitment agencies and overseas employment service providers. In Malaysia, the Private Employment Agency Act of 1981 requires employment agencies to abide by several terms in order for the government to better monitor the recruitment process. Such terms include providing the Labor Department with the details of the migrant domestic workers' and employers' employment contract, checking in on the migrant workers' welfare, and providing the migrant worker with the information of necessary contacts (the employment agency and the

Labor Department) should employment conflicts or emergencies arise (UNIFEM 2005).

16. The effectiveness of the contributory pension system to increase the well-being of elderly women may be limited in the East Asia and Pacific region since the formal labor market is small and predominantly found in urban areas in many countries in the region. Therefore, gaps in pension coverage between rural and urban areas are likely to be as great as or even greater than gender gaps in coverage within rural and urban areas. For example, evidence from Zhejiang and Gansu provinces in China suggests that 79 percent and 54 percent of men and women, respectively, above age 60 have access to pensions in urban areas, whereas in rural areas only 5 percent and less than 1 percent of men and women, respectively, are covered by pensions (Giles, Wang, and Cai 2011).

17. Pensions can be crucial to safeguard incomes for the most vulnerable groups among the elderly, but they may cause adverse labor supply incentives. Flat non-contributory, minimum pensions and targeted benefits have been found to be particularly beneficial for women (James, Edwards and Wong 2003; Aguila, Attanasio, and Meghir 2011; World Bank 2001). However, minimum guaranteed pensions may reduce formal sector labor market participation (Aguila, Attanasio, and Meghir, 2010). Evidence from urban Zhejiang and Gansu provinces suggests that pension eligibility of urban men and women decreases their likelihood of working by 15.2 and 18.3 percent, respectively. In Indonesia, pension eligibility decreases the probability of working by 23.8 and 24.6 percent for urban men and women, and by 13.2 and 12.6 percent for rural men and women, respectively (Giles, Wang, and Cai 2011).

References

Adams, Renée, and Daniel Ferreira. 2009. "Women in the Boardroom and Their Impact on Governance and Performance." *Journal of Financial Economics* 94: 291–309.

ADB (Asian Development Bank). 2010. "Community-Based Early Childhood Care and Development in Viet Nam." Social Protection Project Brief, Asian Development Bank, Manila, Philippines.

Aguila, Emma, Orazio Attanasio, and Costas Meghir. 2011. "Changes in Consumption at Retirement: Evidence from Panel Data." *Review of Economics and Statistics* 93 (3): 1094–99.

Arnot, Madeleine, Miriam David, and Gaby Weiner. 1999. *Closing the Gender Gap: Postwar Education and Social Change.* Cambridge, U.K.: Polity Press.

ASEAN (Association of Southeast Asian Nations). 2008. "ASEAN Continues to Empower Women." ASEAN Bulletin. February. http://www.aseansec.org/Bulletin-Feb-08.htm#Article-2.

Attanasio, Orazio, Luis Carlos Gomez, Ana Gomez Rojas, and Marcos Vera-Hernandez. 2004. "Child Health in Rural Colombia: Determinants and Policy Interventions." *Economics and Human Biology* 2 (3): 411–38.

Attanasio, Orazio, Adriana Kugler, and Costas Meghir. 2011. "Subsidizing Vocational Training for Disadvantaged Youth in Colombia: Evidence from a Randomized Trial." *American Economic Journal: Applied Economics* 3 (3): 188–220.

Baird, Sarah, Ephraim Chirwa, Craig McIntosh, and Berk Ozler. 2009. "The Short-Term Impacts of a Schooling Conditional Cash Transfer Program on the Sexual Behavior of Young Women." Policy Research Working Paper 5089, World Bank, Washington, DC.

Beaman, Lori, Raghabendra Chattopadhyay, Esther Duflo, Rohini Pande, and Petia Topalova. 2009. "Powerful Women: Does Exposure Reduce Bias?" *Quarterly Journal of Economics* 124 (4): 1497–1540.

Bell, Keith. 2011. "Integrating Gender Issues into Bank Land Projects: The Experience of South East Asia." Presentation at the Annual World Bank Conference on Land and Poverty, Washington, DC, April 18–20.

Betcherman, Gordon, Karina Olivas, and Amit Dar. 2004. "Impacts of Active Labor Market Programs: New Evidence from Evaluations with Particular Attention to Developing and Transition Countries." Social Protection Discussion Paper 0402, World Bank, Washington, DC.

Björkman, Martina, and Jakob Svensson. 2009. "Power to the People: Evidence from a Randomized Field Experiment on Community-Based Monitoring in Uganda." *Quarterly Journal of Economics* 124 (2): 735–69.

Blau, David, and Janet Currie. 2006. "Preschool, Day Care, and Afterschool Care: Who's Minding the Kids?" In *Handbook of the Economics of Education, Volume 2*, edited by Eric A. Hanushek and Finis Welch. Amsterdam: North-Holland.

Blokhus, Ingrid. 2004. "Social Protection for Women Migrant Workers: A Comparative Study among Sending Countries." Paper prepared for the World Bank, Jakarta.

Bloom, Erik, Indu Bhushan, David Clingingsmith, Rathavuth Hong, Elizabeth King, Michael Kremer, Benjamin Loevinsohn, and J. Brad Schwartz. 2006. "Contracting for Health: Evidence from Cambodia." http://www.brookings.edu/views/papers/kremer/20060720cambodia.pdf.

Boatman, Mette Rohr, and Vilaythong Chanthalinh. 2009. "Rapid Assessment of 'Power to the Poor' Pilot Project, Lao PDR Rural Electrification." Unpublished report prepared for the World Bank.

Braverman, Marc T., and Leif Edvard Aarø. 2004. "Adolescent Smoking and Exposure to Tobacco Marketing Under a Tobacco Advertising Ban: Findings From 2 Norwegian National Samples." *American Journal of Public Health* 94 (7): 1230–38. http://ajph.aphapublications.org/cgi/content/abstract/94/7/1230.

Cahn, Miranda, and Mathias Liu. 2008. "Women and Rural Livelihood Training: A Case Study from Papua New Guinea." *Gender and Development* 16 (1): 133–46.

Card, David, Pablo Ibarrarán, Ferdinando Regalia, David Rosas-Shady, and Yuri Soares. 2011. "The Labor Market Impacts of Youth Training in the Dominican Republic." *Journal of Labor Economics* 29 (2).

Chi, Wei, and Bo Li. 2007. "Glass Ceiling or Sticky Floor? Examining the Gender Pay Gap across the Wage Distribution in Urban China, 1987–2004." MPRA Paper 3544, University Library of Munich, Germany.

Chung, W., and Monica. Das Gupta. 2007. "The Decline of Son Preference in South Korea: The Roles of Development and Public Policy." *Population and Development Review* 33 (4): 757–83.

Cotlear, Daniel. 2006. *A New Social Contract for Peru: An Agenda for Improving Education, Health Care, and the Social Safety Net.* Washington, DC: World Bank.

Dawson, Angela. 2010. "Towards a Comprehensive Approach to Enhancing the Performance of Health Workers In Maternal, Neonatal and Reproductive Health at Community Level: Learning from Experiences in the Asia and Pacific Regions." Discussion Paper 2, Human Resources for Health Knowledge Hub, University of New South Wales, Sydney.

de Mel, Suresh, David McKenzie, and Christopher Woodruff. "Business Training and Female Enterprise Start-up, Growth and Dynamics: Experimental Evidence from Sri Lanka." Policy Research Working Paper No. 6145, World Bank, Washington, DC.

De Paola, Maria, Roseeta Lombardo, and Vincenzo Scoppa 2009. "Can Gender Quotas Break Down Negative Stereotypes? Evidence from Changes in Electoral Rules." Working Paper no. 10-2009. Department of Economics and Statistics, University of Calabria, Cosenza, Italy.

Deutsch, R. 1998. "Does Child Care Pay?: Labor Force Participation and Earnings Effects of Access to Child Care in the Favelas of Rio de Janeiro." IDB Working Paper 384, Inter-American Development Bank, Washington, DC.

Dinkelman, Taryn. 2008. "The Effects of Rural Electrification on Employment: New Evidence from South Africa." PSC Report 08-653, University of Michigan, Population Studies Center, Ann Arbor.

Duflo, Esther. 2000. "Child Health and Household Resources in South Africa: Evidence from the Old Age Pension Program." *American Economic Review* 90(2): 393–98.

Ebenstein, Avraham, and Steven Leung. 2010. "Son Preferences and Access to Social Insurance: Evidence from China's Rural Pension Program." *Population and Development Review* 36 (1): 47–70.

Efroymson, Debra, and Menchi Velasco. 2007. *Tobacco Use in Southeast Asia: Key Evidences for Policy Development.* Bangkok, Thailand: Southeast Asia Tobacco Control Alliance (SEATCA).

Elder, R. W., R. A. Shults, D. A. Sleet, J. L. Nichols, S. Zaza, R. S. Thompson. 2002. "Effectiveness of Sobriety Checkpoints for Reducing Alcohol-Involved Crashes." *Traffic Injury Prevention* 3: 266–74.

Farrell, Glen, and Cédric Wachholz. 2003. *Meta-survey on the Use of Technologies in Education in Asia and the Pacific 2003–2004.* Bangkok, Thailand: UNESCO Asia and Pacific Regional Bureau for Education.

Filmer, Deon, and Norbert Schady. 2008. Getting Girls into School: Evidence from a Scholarship Program in Cambodia, *Economic Development and Cultural Change* 56: 581–617.

———. 2009. "School Enrolment, Selection and Test Scores." Policy Research Working Paper 4998, World Bank, Washington, DC.

Fiszbein, Ariel, and Norbert Schady. 2009. *Conditional Cash Transfers: Reducing Present and Future Poverty.* Policy Research Report Series. Washington, DC: World Bank.

Frankenberg, Elizabeth, and Duncan Thomas. 2001. "Women's Health and Pregnancy Outcomes: Do Services Make a Difference?" *Demography* 38 (2): 253–65.

Giles, John, Dewen Wang, and Wei Cai. 2011. "The Labor Supply and Retirement Behavior of China's Older Workers and Elderly in Comparative Perspective." Policy Research Working Paper 5853, World Bank, Washington, DC.

Giovarelli, Renée. 2006. "Overcoming Gender Biases in Established and Transitional Property Rights Systems." In *Land Law Reform: Achieving Development Policy Objectives*, edited by John Bruce, 67–106. Washington, DC: World Bank.

Gornick, Janet, and Ariane Hegewisch. 2010. "The Impact of 'Family-Friendly Policies' on Women's Employment Outcomes and on the Costs and Benefits of Doing Business." Commissioned report, World Bank, Washington, DC.

Groh, Matthew, Nandini Krishnan, David McKenzie, and Tara Vishwanath "Soft Skills or Hard Cash? The Impact of Training and Wage Subsidy Programs on Female Youth Employment in Jordan." Policy Research Working Paper No. 6141, World Bank, Washington, DC.

Harrington, Andrew. 2008. "Women's Access to Land and Property Rights: A Review of the Literature. Justice for the Poor Kenya." Report 53001, World Bank, Washington, DC.

Hedditch, Sonali, and Clare Manuel. 2010. "Gender and Investment Climate Reform Assessment: Pacific Regional Executive Summary." International Finance Corporation, Washington, DC.

Heymann, Jody. 2004. "How Are Workers with Family Responsibilities Faring in the Workplace?" International Labour Organization, Geneva. http://ilo-mirror.library.cornell.edu/public/english/protection/condtrav/pdf/wf-jh-04.pdf.

Hicks, Joan Hamory, Michael Kremer, Isaac Mbiti, and Edward Miguel. 2011. "Vocational Education Voucher Delivery and Labor Market Returns: A Randomized Evaluation Among Kenyan Youth." Report for the Spanish Impact Evaluation Fund (SIEF) Phase II, World Bank, Washington, DC.

Hillier, Dawn. 2003. *Childbirth in the Global Village: Implications for Midwifery Education and Practice.* London: Routledge.

Holzer, Harry, and David Neumark. 2000. "What Does Affirmative Action Do?" *Industrial and Labor Relations Review* 53 (2): 240–71.

Hu, Teh-wei, Zhengzhong Mao, and Jian Shi. 2010. "Recent Tobacco Tax Rate Adjustment and Its Potential Impact on Tobacco Control in China." *Tobacco Control* 19 (1): 80–82.

Ibarrarán, Pablo, Laura Ripani, and Juan Miguel Villa-Lora. 2011. "Youth Training in the Dominican Republic: New Evidence from a Randomized Evaluation Design." Technical evaluation prepared for the Inter-American Development Bank, Washington, DC. http://www.iza.org/conference_files/worldb2011/ripani_l6206.pdf.

IEG (Independent Evaluation Group). 2011. *Do Conditional Cash Transfers Lead to Medium-Term Impacts? Evidence from a Female School Stipend Program in Pakistan.* Washington, DC: World Bank.

IFC (International Finance Corporation) and MPDF (Mekong Private Sector Development Facility). 2006. "Women Business Owners in Vietnam: A National Survey." IFC Gender Entrepreneurship Markets Initiative and MPDF, Washington, DC.

ILO (International Labour Organization). 2008. "Women's Entrepreneurship Development in Aceh: Impact Assessment." ILO, Jakarta, Indonesia.

IOM (International Organization for Migration). 2000. "Combating Trafficking in South-East Asia: A Review of Policy and Programme Responses." Migration Research Series 2/2000, IOM, Geneva.

Ir, Por, Dirk Horemans, Narin Souk, and Wim Van Damme. 2010. "Using Targeted Vouchers and Health Equity Funds to Improve Access to Skilled Birth Attendants for Poor Women: A Case Study in Three Rural Health Districts in Cambodia." *BMC Pregnancy Childbirth* 10: 1. doi:10.1186/1471-2393-10-1.

Jalbert, Susanne. 2000. "Women Entrepreneurs in the Global Economy." Center for International Private Enterprise, Washington, DC.

James, Estelle, Alejandra Cox Edwards, and Rebeca Wong. 2003. "The Gender Impact of Pension Reform: A Cross-Country Analysis." Policy Research Working Paper 3074, World Bank, Washington, DC.

Jung, Jin Hwa, and Hyo-Yung Sung. 2012. "Affirmative Action in Korea: Its Impact on Women's Employment, Corporate Performance and Economic Growth." Presented at the 2012 Annual Economic Association in Chicago, January 6–8, 2012.

Jusi, S. Asigau, W. and Laatunen N. 2007. "Social Impact Benefits of Road Rehabilitation Projects in Six Provinces in Papua New Guinea, South Pacific." Photocopy. Finnroad Ltd, in association with SMEC, Helsinki, Finland.

Kirk, Jackie. 2006. *The Impact of Women Teachers on Girls' Education—Advocacy Brief.* Bangkok: UNESCO Bangkok.

Lefebvre, Pierre, Philip Merrigan, and Matthieu Verstraete. 2009. "Dynamic Labour Supply Effects of Childcare Subsidies: Evidence from a Canadian Natural Experiment on Low-Fee Universal Child Care." *Labour Economics* 16 (5): 490–502.

Li, Shuzhuo. 2007. "Imbalanced Sex Ratio at Birth and Comprehensive Intervention in China." Paper presented at the "4th Asia Pacific Conference on Reproductive and Sexual Health and Rights," Hyderabad, India, October 29–31.

Madigan, Jennifer. 2009. "The Education of Girls and Women in the United States: A Historical Perspective." *Advances in Gender and Education* 1: 11–13.

Makowiecka, Krystyna, Endang Achadi, Yulia Izati, and Carine Ronsmans. 2008. "Midwifery Provision in Two Districts in Indonesia: How Well Are Rural Areas Served?" *Health Policy and Planning* 23: 67–75.

Narjoko, Dionisius, and Hal Hill. 2007. "Winners and Losers during a Deep Economic Crisis: Firm-Level Evidence from Indonesian Manufacturing." *Asian Economic Journal* 21 (4): 343–68.

Newman, Constance. 2002. "Gender, Time Use and Change: The Impact of the Cut Flower Industry in Ecuador." *World Bank Economic Review* 16 (3): 375–95.

OECD (Organisation for Economic Co-operation and Development). 2009. "Contracting Out Government Services and Functions." OECD Partnership for Democratic Governance, OECD, Paris.

Paes de Barros, R., P. Olinto, M. de Carvalho, T. Lunde, Norbert Schady, S. Santos, and A. Rosalem. 2010. "Impact of Free Childcare on Women's Labor Market Behavior: Evidence from Low-Income Neighborhoods in Rio de Janeiro." Presented at the GAP Workshop on "World Bank Regional Study on Gender Issues in Latin America and the Caribbean," Washington, DC, June 14.

Pande, Rohini, and D. Ford. 2011. "Gender Quotas and Female Leadership: A Review." Background paper for the *World Development Report 2012*, World Bank, Washington, DC.

Peek-Asa, Corinne. 1999. "The Effect of Random Alcohol Screening in Reducing Motor Vehicle Crash Injuries." *American Journal of Preventive Medicine* 1999 16 (1 Suppl): 57–67. http://www.ncbi.nlm.nih.gov/pubmed/9921387.

Quota Project. 2010. "Republic of Korea Country Overview." Global Database of Quotas for Women. Institute for Democracy and Electoral Assistance (IDEA) and Stockholm University. http://www.quotaproject.org/uid/countryview.cfm?ul=en&country=122.

Rosenbloom, Jackie. 2004. "Adult Literacy in Cambodia." Pact Cambodia, Phnom Penh. http://www.pactcambodia.org/Publications/WORTH_Education/Adult_Literacy_in_Cambodia.pdf.

Salisbury, Jane, and Sheila Riddell, eds. 2000. *Gender, Policy and Educational Change: Shifting Agendas in the UK and Europe.* London: Routledge.

Shankar, Anuraj, Susy Sebayang, Laura Guarenti, Budi Utomo, Monir Islam, Vincent Fauveau, and Fasli Jalal. 2008. "The Village-Based Midwife Programme in Indonesia." *Lancet* 371: 1226–28.

Shrestha, Ranjan. 2007. "The Village Midwife Program and the Reduction in Infant Mortality in Indonesia." Paper prepared for the 2007 Meeting of the Population Association of America, New York, March 29–31.

Simavi, Sevi. 2011. "Gender Dimensions of Investment Climate Reform: Operational Perspective." Presentation at the World Bank conference on "Female Entrepreneurship: What Do We Know? What Is Next?" World Bank, Washington, DC, April 6.

Simavi, Sevi, Clare Manuel, and Mark Blackden. 2010. *Gender Dimensions of Investment Climate Reform: A Guide for Policy Makers and Practitioners.* Washington, DC: World Bank.

Smarasinghe, Vidyamali, and Barbara Burton. 2007. "Strategising Prevention: A Critical Review of Local Initiatives to Prevent Female Sex Trafficking." *Development and Practice* 17 (1): 51–64.

Stefanova, Milena, Rod Nixon, and Raewyn Porter. 2010. "Leasing in Vanuatu: Findings and Community Dissemination on Epi Island." Justice for the Poor Briefing Note 5 (4), World Bank, Washington, DC.

Suguna, B. 2006. *Empowerent of Rural Women Through Self-Help Groups.* Grand Rapids, MI: Discovery.

Suryahadi, Asep, Wenefrida Widyanti, Daniel Perwira, and Sudarno Sumarto. 2003. "Minimum Wage Policy and Its Impact on Employment in the Urban Formal Sector." *Bulletin of Indonesian Economic Studies* 39 (1): 29–50.

Trung, Le Dang. 2008. "Two-Name Land Use Certificates and Gender Inequality: An Empirical Investigation for Vietnam." DEPOCEN Working Paper 19, Development and Policies Research Center, Hanoi.

UNFPA (United Nations Population Fund). 2005. "Cultural Programming, Reproductive Health Challenges and Strategies in East and South East Asia." UNFPA Country Technical Services Team for East and South-East Asia, Bangkok, Thailand.

UNICEF (United Nations Children's Fund). 2004. "Regional Experience on Integrated Approach to Early Childhood: Six Case Studies in East Asia." UNICEF East Asia and Pacific Regional Office: Bangkok.

———. 2009. "Gender Equality in Education, East Asia and the Pacific, Progress Note." UNICEF East Asia and Pacific Regional Office, Bangkok.

UNIFEM (United Nations Development Fund for Women). 2005. "Good Practices to Protect Women Migrant Workers." Presented at the High-Level Government Meeting of Countries of Employment, co-hosted by the Ministry of Labor, Royal Thai Government, and UNIFEM East and South-East Asia, Bangkok.

———. 2008. "A Life Free of Violence: Unleashing the Power of Women's Empowerment and Gender Equality Strategy 2008–2013." United Nations Development Fund for Women, New York, NY. http://www.unifem.org/attachments/products/EVAWkit_03_UNIFEMstrategy_en.pdf.

U.S. Department of State. 2010. *Trafficking in Persons Report 2010.* Washington, DC: U.S. Department of State.

Utomo, Iwu, Peter McDonald, Terence Hull, Ida Rosyidah, Tati Hattimah, Nurul Idrus, Saparinah Sadli, and Jamhari Makruj. 2009. "Gender Depiction in Indonesian School Text Books: Progress or Deterioration." Australian Demographic and Social Research Institute, Australian National University, Canberra.

Valdivia, Martin. 2010. "Contracting the Road to Development: Early Impacts of a Rural Roads Program." PMMA Working Paper 18. http://ideas.repec.org/p/lvl/pmmacr/2010-18.html.

Washington Post. 2011. "Malaysia Says Women Must Hold 30 Percent of Top Corporate Posts by 2016 for Gender Equality." June 27.

Weaner, J. 2008. "End of Violence against Women Project: Evaluation Report." UNIFEM, Cambodia Women's Crisis Center, Phnom Penh.

WHO (World Health Organization). 2005. "Gender Health and Alcohol Use." Gender and health information sheet, Department of Gender, Women and Health, WHO, Geneva.

———. 2007. "Gender and Tobacco Control: A Policy Brief." Department of Gender, Women and Health, WHO, Geneva.

———. 2008. "Strategies to Reduce the Harmful Use of Alcohol, Report by the Secretariat." Sixty-first World Health Assembly, WHO, Geneva.

———. 2011. "Global Status Report on Alcohol and Health." http://www.who.int/substance_abuse/publications/global_alcohol_report/en/.

World Bank. 2001. *Engendering Development: Through Gender Equality in Rights, Resources, and Voice.* Washington, DC: World Bank.

———. 2003. *World Development Report 2004: Making Services Work for the Poor.* Washington DC: World Bank.

———. 2007a. *World Development Report 2008: Agriculture for Development.* Washington, DC: World Bank.

———. 2007b. "Gansu and Inner Mongolia Poverty Reduction Project, Implementation Completion and Results Report." World Bank, Washington, DC.

———. 2008. "Analysis of the Impact of Land Tenure Certificates with Both the Names of

Wife and Husband in Vietnam: Final Report." World Bank, Washington, DC.

———. 2009. "Gender Equality in East Asia: Progress and the Challenges of Economic Growth and Political Change, East Asia Update." World Bank, Washington, DC.

———. 2010a. "A Practical Guide for Addressing Gender Concerns in Land Titling Projects." http://siteresources.worldbank.org/INTARD/Resources/genderinlandguide.pdf .

———. 2010b. *Women, Business and the Law 2010: Measuring Legal Gender Parity for Entrepreneurs and Workers in 128 Economies*. Washington, DC: World Bank.

———. 2010c. *Women's Economic Opportunities in the Formal Private Sector in Latin America and the Caribbean: A Focus on Entrepreneurship*. Washington, DC: World Bank.

———. 2011a. "Gender Disparities in Mongolian Labor Markets and Policy Suggestions." World Bank, Washington, DC.

———. 2011b. "Gender Equality as Smart Economics: A Work in Progress." GAP Booklet, World Bank, Washington, DC.

———. 2011c. Justice for the Poor. World Bank, Washington, DC. http://web.worldbank.org/WBSITE/EXTERNAL/TOPICS/EXTLAWJUSTICE/EXTJUSFORPOOR/0,,contentMDK:21172652~menuPK:3282951~pagePK:210058~piPK:210062~theSitePK:3282787,00.html.

———. 2011d. "Lao PDR Country Gender Assessment." World Bank, Washington, DC.

———. 2011e. Toward a Healthy and Harmonious Life in China: Stemming the Rising Tide of Non-Communicable Diseases. World Bank *News & Views* July 26. http://www.worldbank.org/en/news/2011/07/26/toward-health-harmonious-life-china-stemming-rising-tide-of-non-communicable-diseases.

———. 2011f. *World Development Report 2012: Gender Equality and Development*. Washington, DC: World Bank.

———. 2011g. "Women, Business and the Law in East Asia and Pacific Region." A background paper commissioned for *Toward Gender Equality in East Asia and the Pacific: A Companion to the* World Development Report. Washington, DC: World Bank.

Yadamsuren, Buyanjargal, Mario Merialdi, Ishnyam Davaadorj, Jennifer Harris Requejo, Ana Pilar Betrán, Asima Ahmad, Pagvajav Nymadawa, Tudevdorj Erkhembaatar, Delia Barcelona, Katherine Ba-thike, Robert J Hagan, Richard Prado, Wolf Wagner, Seded Khishgee, Tserendorj Sodnompil, Baatar Tsedmaa, Baldan Jav, Salik R. Govind, Genden Purevsuren, Baldan Tsevelmaa, Bayaraa Soyoltuya, Brooke R. Johnson, Peter Fajans, Paul F. A. Van Look, and Altankhuyag Otgonbold. 2010. "Tracking Maternal Mortality Declines in Mongolia between 1992 and 2007: The Importance of Collaboration." *Bulletin of the World Health Organization* 88: 192–98.

Yun-Suk, Lee, and Eun Ki-Soo. 2005. "Attitudes towards Married Women's Employment in Korea and Japan: Implications from Latent Class Analyses." *Development and Society* 34 (1): 125–45.